598.8

D1632503

THE
WHEATEAR

For
Emma, William, Hannah and Lucy

THE
WHEATEAR

Peter Conder

CHRISTOPHER HELM
London

© 1989 Peter Conder

Illustrations by Peter Conder and John Busby

Christopher Helm (Publishers) Ltd, Imperial House,
21-25 North Street, Bromley, Kent BR1 1SD

ISBN 0-7470-0406-4

A CIP catalogue record for this book
is available from the British Library

Typeset by Leaper & Gard Ltd, Bristol
Printed and bound in Great Britain by
Biddles Ltd, Guildford and Kings Lynn

CONTENTS

List of Figures

TABLES

PREFACE

I came to Skokholm, an island off the Pembrokeshire coast, to take up an appointment as honorary warden of the Bird Observatory in 1947 while the Bureau of Animal Population, Oxford, awaited a grant from the Agricultural Research Council to employ me as an assistant to H. N. Southern in his Tawny Owl work. I enjoyed that long hot summer of 1947 on Skokholm. It was, and remains, a beautiful island, a mile long and half a mile wide (1.6 × 0.8 km) of Old Red Sandstone, lying about three miles (4.8 km) off the Pembrokeshire (Dyfed) coast just north of the entrance to Milford Haven and south of Skomer. In spring the top of the island was covered with vast spreads of bluebells, or in summer with acres and acres of pink thrift; and along the cliffs there were thousands of Puffins, Razorbills and Guillemots by day and tens of thousands of Manx Shearwaters and Storm Petrels by night.

The grant from the Agricultural Research Council was slow in materialising and, in 1948, I was offered the full-time post of Warden of Skokholm Bird Observatory by the Council for the Promotion of Field Studies (now Field Studies Council). Although the salary was low, this was the start of a further six, wonderful years of island living, of studying bird migration by ringing and by diurnal observation, of learning about all kinds of plants and animals; and I came to know a large number of people who stayed on the island, many of whom helped with my Wheatear study and many of whom remain my friends to this day.

When it finally became clear that I was staying on Skokholm as Warden in 1948, I had wondered what species I should study in those moments when the ringing of migrant and resident birds, looking after our visitors and other duties left time to spare. Knowing from 1947 how common were Meadow Pipits, how relatively tame they were, and how little was known about them at that time, I rather fancied the idea of studying them, although distinguishing between sexes was a problem. After discussing the point with John Buxton, who had taught me a lot about bird study when we were prisoners of war in Germany, and who had introduced me to his brother-in-law, Ronald Lockley, writer, naturalist and former tenant of Skokholm, I changed my mind and decided to look at Wheatears.

During July and August 1947, I had begun to look at the way in which Wheatears foraged, counting the number of times that they hopped and pecked down for food and working out how successful they were in capturing something. But other than trapping resident and migrant Wheatears, finding Wheatear nests and colour-ringing their young as part of my duties as a Warden, I had not paid much attention to them. I began studying them in earnest as soon as I returned to the island in March 1948, and I continued watching them every season until I left to take up an appointment with the Royal Society for the Protection of Birds in May 1954. For the next few years I had little time to look at them: my job as Assistant Secretary of the RSPB in charge of all the outside activities of the Society was very demanding, taking me all over the British Isles, and what little time was left was taken up by a growing family. As the young people became more independent, however, it left me more time for birds and I returned to Wheatears.

In the late 1960s we had begun to take family holidays in Alderney, the northernmost of the Channel Islands, and it was here that I was able to pick up the Wheatear threads. Longis Common, which became my second study area, was good for watching migration behaviour since Wheatears did not breed on the main island and I could be certain that the behaviour I was seeing was that of a migrant rather than that of a local breeder. I continued

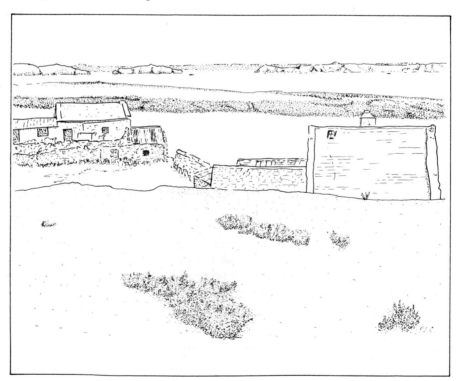

watching them there until the time came to complete this book.

In the early spring of 1978 I returned to Skokholm, and from then until 1983 spent short periods on the island at different stages of the breeding season to refresh my memory and to put life into the notes that I had written some 25 years earlier; on at least one occasion I understood the significance of some behaviour that I had not understood earlier.

Cambridgeshire does not have many breeding Wheatears, so after I had retired from the RSPB I spent some time annually on Weeting Heath National Nature Reserve, owned by the Norfolk Naturalists' Trust, where hides give good visibility over the breeding grounds.

During the course of the study I also visited East Africa to watch Wheatears in their winter quarters.

One of the advantages of the Wheatear as an object of study is that distinguishing adult males from females is fairly easy. On the other hand, one of the disadvantages is that they are shy, take alarm at a fairly long distance and fly away. They are particularly shy near their nests — which are chiefly in rabbit burrows on Skokholm — and deserted their eggs easily if I put my hand in the burrow and the female was still in it. One female deserted her newly-hatched chicks. By contrast, one female let me put my finger between her belly and her eggs — but I did that only once. As soon as I realised the extent of their shyness, I never went to a Wheatear's nest without checking first that neither parent was in the burrow, even though this extended the time that it took to weigh and measure eggs, to check whether the eggs were hatching and so on. Sometimes, because the female might be anywhere in a large territory, I had to wait for her to go back on the eggs, do her 20-minute incubating stint and emerge again before I dared go to the nest.

Because of this shyness, binoculars and telescopes were an essential aid to the study of Wheatears. When I wanted to watch the way they behaved, I would sit with my back against a rock and try to make myself as inconspicuous as possible. For observation of nests at close quarters I used a so-called portable hide, built by one of the visitors, which needed four strong men to carry it short distances at a time. To see what went on in the nests, however, I constructed a 'coffin' hide: this had a low rectangular frame, about 0.5 m high, 1 m wide and 2 m long, which was covered with heavy canvas which made the interior of the hide very dark; it had no observation slits through the side. This was slowly brought into position over a nest at the end of a longer burrow than usual. A small hole was excavated down to the nest and a metal-polish tin with its bottom removed inserted. I covered the hole at the top of the tin with a microscope slide to stop the draught blowing in my eye. It was so dark in the hide that neither parent noticed me about 25 cm from them.

For many years the Wheatear has been known as simply 'the Wheatear' with the scientific name of *Oenanthe oenanthe* (Linnaeus) 1758, and there have been few problems in understanding which bird was meant by that English name. As birdwatchers travelled more widely and encountered several different wheatear species, however, it was thought important to add a second name, or forename as it were, to make clear which species was being referred to. To distinguish it from other members of the genus *Oenanthe* our Wheatear was usually called the Common, European or Eurasian Wheatear. The Common Wheatear, however, breeds in the northern parts of the American continent so that it is more than European. Recently American authors, and now a British author, Peter Clement (1987), have called it 'Northern' Wheatear, a name very descriptive of its almost circumpolar distribution which is well to the north of the range of any other species of wheatear.

Having called this species the Wheatear or the Common Wheatear for the last 50 years and more, I shall change only a little now and am glad that the Editors of *The Birds of Western Palearctic* (Cramp 1988) continue the traditional name and relegate the name 'Northern Wheatear' to the synonymy (of English names). In the pages that follow, when I refer to Wheatear with a capital letter I am still referring to the Wheatear *Oenanthe oenanthe*; when I use the word wheatear or wheatears with a small initial letter then I am referring to any member of the genus *Oenanthe*. Periodically, however, I have, regretfully, found it necessary for clarification to use the term Northern Wheatear, particularly when discussing this species in relation to other wheatear species. Also for simplicity's sake, when referring to that vast compendium of knowledge *The Birds of the Western Palearctic*, I shall use the abbreviation *BWP*.

Skokholm already had a tradition for serious bird study. Before the war Ronald Lockley, when not shepherding, studied Manx Shearwaters, a study which resulted in his classic book *Shearwaters* (1942). In 1932 he established the first bird observatory in Britain on the island, and by using a number of different types of traps he caught and ringed large numbers of migrant birds and studied their migration patterns, which, in its turn, generated a number of publications in the scientific literature.

I ringed and marked as many Wheatears as I could, catching them either in the large, funnel-shaped Heligoland traps into which one normally drives migrants which may be feeding in front of the trap, or else in a variety of smaller traps which Wheatears often seemed to enter out of curiosity. Occasionally, in order to catch a specific bird, I set a clap net in its territory baited with a mealworm, although I discovered that, if I set the trap on flat ground and tilted the bait tray upwards so that it was fairly conspicuous and then painted the top edge white to look like a bird dropping, Wheatears would perch on it apparently taking it for a much-used and safe perch; mealworms were not really necessary.

For reasons which I mentioned earlier, I never caught adults on the nest. When they were away from the burrow, however, I put my hand in to check what was happening to the eggs and, once they had hatched, I weighed the nestlings daily and eventually colour-ringed each with a unique combination of four rings (two on each leg), of which one was a metal ring on which were printed a number and an address and the other three plastic coloured rings. Colour-ringing caused no desertions by adults. As a result of this ringing, in the last two or three years of the study over 90% of Wheatears were individually recognisable by the combination of their colour rings.

I appreciate that some people are worried by the apparent threat to an individual bird posed by rings and colour-rings. My Wheatears carried a metal ring and three plastic rings, making two rings on each leg; the combined weight of these four rings was about 0.15 g, or much less than the weight gained daily or lost nightly by a Wheatear. The real answer to those worries comes, I think, from the population figures. In 1947 the Wheatear population was 14 pairs, whereas at the end of the study in 1952 it was 38 pairs. At no time since Lockley began his census in 1928, nor since my study ended, has the Wheatear population been higher (see Appendix 5).

The ringing scheme is run in Britain by the British Trust for Ornithology, but after the war rings were in such short supply that E. John M. Buxton had some made for use on Skokholm with a New College, Oxford, address. Most of the Wheatears I studied carried these rings.

One of the great benefits of colour-ringing was that without having to retrap an individual I was able to identify it and to look up its life history, which I kept on Filofax sheets. Half way through this study I would know who were the parents of an individual bird, whether any of its siblings were still on the island, and whether it was occupying the same territory that it had in a previous year; indeed, any statement made in this book about the age of an individual bird or which asserts that the same bird was involved in some incident at some time or place is based on a colour-ringed individual.

Every year, each pair of Wheatears on Skokholm was given a number, such as W22, and its nest, its clutch and its brood were known by that number. Should a pair produce a replacement clutch for a preyed-on first clutch, or a second clutch, it would be known as $W22^R$ or $W22^2$ respectively.

ACKNOWLEDGEMENTS

Over the years a considerable number of people have been a great help in a number of ways, in the field, in libraries, with advice and so on, and to them I am enormously grateful. John Buxton encouraged my interest in birds when we were prisoners of war together and was responsible with his brother-in-law, Ronald Luckley, Chairman of the West Wales Field Society, for my appearance on the island in the first place. John Barrett was another ornithological POW, who later, as Warden of Dale Fort Field Centre (Field Studies Council), helped with advice and in many practical ways: his staff kept us furnished with essential supplies. On the island I am grateful to a number of staff (some honorary) of the period 1947-54: J. H. R. Boswall, D. Bradley (née Sims), E. M. Cordiner, Joan Jenkins (née Keighley), R. M. Nedderman, C. J. Pennycuick, S. Y. Townend and R. Vernon; and later, in the years 1976-83, the Wardens G. and E. Gynn, R. and J. Lawman, and S. and C. Warman. All these people helped in a number of different ways.

Many visitors sat for hours, sometimes in the cold and wet but also on warm and sunny days, recording Wheatears or various aspects of their ecology for me: N. Bell, C. Benson, D. Boddington, F. M. Boston, B. Brown (née Harvey), O. Brown, R. Culverwell, J. Cohen, Dr P. Collett (née McMorran), P. Ellicott, H. Ennion, V. Field, H. Gale, D. Gibling, M. E. Gillham, G. T. Goodman, D. Hall, R. J. Harrison, H. Horder, M. Jellicoe, T. E. Jenkins (Skokholm Lighthouse), R. Jenkinson, J. Jeremy, M. Jervis, M. Hope Jones, R. Lovegrove, D. Low, A. Maunder, P. E. Naylor, A. E. A. Pearson, M. W. Reade, R. A. Richardson, J. C. S. Robinson, D. Rook (who was later helpful as RSPB Librarian), V. Ruffles, A. Scott-Lewis, A. Shaw, F. Shaxson, B. Snow (née Whittaker). M. Sutherland, F. Tetley, E. M. Thomas, T. Walsh, P. Walters, Mr and Mrs S. Westcott, M. White, K. Williamson, and M. H. Williamson.

Since those days, many more have become involved. In particular I wish to mention H. E. Axell, also interested in Wheatears, especially when Warden of the RSPB's Reserve at Dungeness, Kent; J. Andrews of the RSPB staff, who translated Russian papers for me; D. Brooks, Executive Editor of *The Birds of the Western Palearctic*, for permission to use sonagrams; many members of the BTO staff, particularly J. J. D. Greenwood for permission to use maps from *Ringing and Migration*, and R. Spencer and C. Mead of the Ringing section and H. Mayer-Gross of nest Records for help and advice over the years; S. Cowdy for information from Bardsey; E. H. Chater for help on plant ecology; S. Cramp for much general help; I. Dawson, L. Gydding, C. Harbard and A. Scott of the RSPB Library; E. Dunn for help with my Wheatear submission for *The Birds of the Western Palearctic*; R. G. Frankum for information on heights; J. Hall-Craggs for help with voice and sonagrams; on Alderney, V. and M. Mendham; for data from Scotland, J. R. Mitchell (Nature Conservancy Council); for data from Saltee Island, R. F. Routtlege; M. G. Wilson for help with Russian papers.

I owe an enormous debt to Alan Tye, who spent part of his leave from the BOU Expedition to Colombia reading the entire manuscript and gave generously of his unpublished information and advice, which I believe has greatly increased the interest of this book. Chris Mead kindly read and commented on Chapters 3, 4, 15 and 17, which deal largely with migration or involved applications of ringing; his comments and suggestions have helped me to avoid many problems. C. J. Bibby kindly commented on an early version of Chapter 5.

Finally, to my wife, Pat, I am immeasurably grateful; since 1952, when she joined me on Skokholm, she has encouraged my pursuit of Wheatears, accompanying me to many parts of its range. She read, typed and improved the grammar of many drafts, and urged me on to the completion of this book.

I am also grateful to Jo Hemmings, Carolyn Burch, David Christie, Linda McVinnie and their colleagues at Christopher Helm for their work in producing the published book.

Without the help of these people, the book would have had far less value. Where faults remain they are entirely mine.

Chapter 1

INTRODUCING THE WHEATEAR

Almost every year that I lived in Pembrokeshire, I saw my first Wheatear of the spring on the Old Deer Park at Marloes, that snout of hard, dark grey, Ordivician rock that thrusts out towards Middleholm and Skomer. In the early mornings of the first days of March I searched for Wheatears in the short grass along the cliffs, and it was usually about the 11th that I discovered the first. Three miles (4.8 km) to the southwest across Broad Sound I could see Skokholm, and a week or so later, with others of the Observatory staff, I crossed to the island to scrub, clean and repair the cottage and dormitories in readiness for the coming season and to repair the gale-torn wire netting of the Heligoland traps. After working for several hours, I would walk out over the island's sere and wintery sward to discover which of the summer migrants had arrived and which of the birds that had wintered here were still waiting for the most opportune moment to begin the journey to their breeding grounds to the north and east.

Mid March could be bitterly cold and windy on the open plateau of Skokholm, which slopes gently from the 50 m high cliffs at the west end to about 15 m in the east. Rocky ridges and dry-stone walls gave some shelter, but when the wind blew from the northwest — across the seas from the Greenland icecap — at 30 to 35 mph (48-56 kph) they did not seem to be all that effective and you really had to squeeze yourself into the cracks in the rocks to avoid it. Looking for Wheatears in those early days could be a chilling business: my eyes watered, my nose dripped, fingers were frozen and the wind shook my binoculars so that, through my tears, the first sight of a Wheatear was often blurred.

Not all March days were like that. Some were bright and sunny, balmy even, and, although the wind hardly ever dropped, the roughness of the winter was dying away and we could anticipate rushes of Chiffchaffs, Willow Warblers, Whitethroats and Swallows, and that, soon, the island's cliffs would be thronged with Puffins, Razorbills and Guillemots and the wind- and rain-flattened grasses would show the first shoots of spring squill, dog violets, wild pansy, tormentil and the pink flowers of thrift lining the cliffs.

I had to force myself out in the wind to look for those early Wheatears.

1

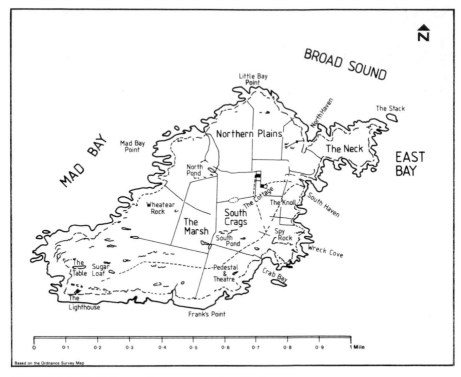

FIGURE 1.1 *Skokholm — place names mentioned in the text*

Finally, among the thrift tussocks on top of the western cliffs, where I usually found my first island Wheatears, I would see the pulsating whiteness of a male Wheatear's rump and tail feathers as he flew away from me in a deeply undulating flight with his tail flashing open to show aggression as I entered the territory which he had occupied the previous year. When he alighted, he showed the characteristic behaviour of a newly-arrived resident, standing upright, flicking up his wings and fanning his tail, momentarily revealing the white rump, showing both alertness and anxiety in his new surroundings, and, while looking at me, presenting the black and white head pattern and the black and grey pattern of the wings and mantle.

I sat in the lee of a rock to watch him. After a minute or so he ignored me and, still erect and alert, began to forage by half-hopping and half-running over the rabbit-grazed red fescue and thrift. Close by was a Black Redstart, another early migrant which passed through Skokholm in small numbers. It showed many of the same characteristics as the Wheatear, including its upright stance and its method of hopping. It, too, moved, almost vibrated, its tail and rump up and down; the tail was rust-coloured rather than black and white.

The Wheatear flew at and chased the Black Redstart, which moved away a few feet. Again the Wheatear flew at the redstart and once more it withdrew, flying a short distance, but it resumed foraging. The Black Redstart, when resting on the island, occupied a very similar habitat to that of the Wheatear — short grass and thrift in the southwestern corner — and their methods of feeding were similar. Consequently, the two frequently came into contact. Like some Wheatears which only passed through the island, Black Redstarts remained only a few days during which they established individual territories and hunted, building up their weight until ultimately they were stimulated to re-orient themselves and fly towards their breeding areas. .

After the Wheatear's aggressive tendency had been temporarily satisfied and the Black Redstart had been driven from its immediate vicinity, the Wheatear perched on a grass tussock and warbled a very quiet subsong, so soft that at times I could hardly hear it: sometimes hard and scratchy and sometimes soft musical piping. The warbling subsong generally lacked the volume and carrying power of the Wheatear's loud territorial song and was used before the male was mated and, later, between pairs in sexual circumstances. The territorial song was often delivered in song flight, chiefly after the male had been joined by a mate, but sometimes in intense territorial battles before she arrived.

For a minute the Wheatear stood on the tussock warbling, then he began hunting again, looking around, hopping five or six paces, poking down vigorously among the grasses, then resting again. Suddenly he flew up about 6-7 m as if in the song flight and, carried by the westerly breeze, flew eastwards for about 100 m and for a further six minutes fed, except for short spells when he remained still and was perhaps warbling — I was too far away to hear.

This behaviour was typical of a male Skokholm Wheatear in the day or two immediately after its arrival and before the arrival of its mate, or before the

FIGURE 1.2 *Male Wheatear showing aggression to Black Redstart*

arrival of other male Wheatears which might be either neighbours or competitors for territory, or of other passage birds which, in their search for places to
rest and regain weight lost on the last stage of their flight before they flew on,
often disrupted the daily routine of established residents.

The male would move around the area in which it had settled, learning
where it could find food, which were the easiest places to hunt over, where it
could find shelter from strong winds or gales which blew up fairly commonly
through March and early April, learning the best song posts, learning where it
could roost, occasionally entering burrows (although the final choice of nest
site was the female's), at times picking up nest material in its bill and fiddling
with it and even taking it into holes. Indeed, it was learning much about its
territory that could help it survive.

Although the earliest residents which arrived in the last half of March were
usually the older males which had bred on the island the previous year and
which tended to return to the same territory, they did not necessarily centre
their activities on a previous year's nest burrow or within the boundaries of a
previous territory which had been contested with neighbours but wandered
over a much larger area, which might have a diameter as large as 200-300 m.
They became localised in their territories within a few hours of their arrival.

THE SPRING PLUMAGE OF THE WHEATEAR

Wheatears are about 14.5-15.5 cm long, with a wingspan of 26-32 cm, and
weigh 24-25 g, or less than an ounce. They are long-winged, ten-primaried
songbirds of the thrush family, Turdidae, and within that family are related to
a group of genera jointly known as the chats; these include such birds as
Whinchat, stonechats, redstarts, other wheatears and other species which
occur in Africa and the Middle East, most of which have harsh 'chacking' call
notes which give this group their name.

When Wheatears first arrive in late March or early April, some still have
traces of their winter plumage, particularly the year-old individuals which are
returning to the island to breed for the first time. Some young males are even
difficult to separate from females, because the brown tips to the feathers which
they acquired in their moult from juvenile to first-winter plumage have not yet
worn off, but occasionally only a trace of the brown tips overlies the grey of the
breeding plumage even of the older males. Similarly, the breast feathers may
well be a warm peach colour at first but, as the season progresses, the tips of
the feathers abrade, so that the breast becomes a pale sandy colour and what
had been grey-brown on the upperparts is now plain grey.

Once the last traces of the brown tips of winter plumage have worn off, the
crown of the male's head, the nape of his neck and the mantle are grey. His
white forehead and supercilium vary in conspicuousness with age, origin and
season, older birds and Wheatears of the Greenland race *leucorrhoa* usually
having much whiter foreheads and supercilia. The bill, lores and ear-coverts
are black, giving a masked appearance, as are the wing-coverts and flight

feathers. On year-old males the ear-coverts are slightly browner than on the older birds. The throat and breast are sandy-brown, while the chin, belly and undertail-coverts are usually paler. The feathers of the underwing-coverts and axillaries are dark grey, edged with white, but they are difficult to see except when they are exposed by the Wheatear in some of the displays.

The most conspicuous features of the Wheatear are the rump, uppertail-coverts and tail, which are white, although the central tail feathers and a broad terminal band across all the tail feathers are black; thus, when the tail is spread, the black markings look like ⅃. It is this white rump, prominent on adults and young, which gave the Wheatear the name we now use; over the years it has collected a mass of local names (some of which are listed in Appendix 7), but I think it is clear that, while 'Wheat' is a corruption of the Anglo-Saxon *hwit*, meaning white, 'ears' is a corruption, or bowdlerisation, of *aers*, meaning rump or backside — hence an old English name of 'white arse'. It has a shortish square tail; its bill, which has nasal and rictal bristles, and its gape are black; and it has longish black legs with a short hind claw.

While on the subject of names, the Wheatear's scientific name *Oenanthe*

FIGURE 1.3 *Male and female Wheatear*

oenanthe has nothing to do with its appearance, but is said to be related to the time of its arrival in Greece when the vine is in blossom (*oinanthe* means wine-blossom). This name was applied to a bird, thought to be the Wheatear, by Aristotle: he said that it arrived at the time of the setting of the Dog Star.

Throughout the year, plumage colour changes almost imperceptibly as the tips of feathers wear away. By the time Wheatears start growing new feathers at the end of the breeding season their plumage has become very ragged; the change is most noticeable when you return to a population of known individuals after an absence of three or four weeks.

Females, which generally arrive later than the males, are more sombrely coloured, even though they have the white rump, white uppertail-coverts and the black and white tail pattern. On females two or more years old (which tend to arrive before the year-old females) the crown and mantle are a greyer brown, similar to a male, whereas on first-summer females the mantle is buff-brown. Confusion between elderly females with their greyer mantle and first-year males which still have the brown tips to their grey feathers is quite possible. The superciliary stripe, chin, belly and undertail-coverts are pale buff, but the shade and the extent of brown on the throat and breast are very variable, some having a pale creamy-brown breast whereas others are quite dark; in addition, the extent of the darker brown on the breast varies with individuals, on some terminating quite abruptly in a marked pectoral band but on others shading gradually onto the belly. The ear-coverts are darker brown on older females and lighter on first-summer females.

I found the variability of the shade, and the size, of the various patches of colour useful because it allowed me to identify and keep track, for a few days, of individuals which I had been unable to trap and ring but which were localised for the short time they rested on the island. However, I could not rely on that method of identification for longer periods because, as I mentioned above, feather wear gradually changed the individual's appearance.

In late July and August both sexes undergo a complete moult, losing all their old feathers and growing new ones over a matter of six weeks. The male usually shows new feathers before the female. Although I describe the progress of the moult in greater detail in a later chapter, I must point out here that, except for the head and tail patterns, the remainder of the male's plumage at this time is dun-coloured like the female's, though usually slightly darker.

There has been much discussion in the past as to whether the striking black and white colour pattern of a male Wheatear in the breeding season is cryptic and hides the bird or whether it is aposematic and warns possible predators that its flesh is distasteful. Cott (1947) considered that Wheatears were conspicuous and, rather surprisingly, he claimed that their flesh was distasteful to predators — he chose hornets to represent predators — and that the colour patterns, rather than affording protection to Wheatears, warned possible predators of their unpalatability.

Cott's conclusion that Wheatear flesh was unpalatable to predators was surprising in view of the fact that many writers from Yarrell (1871) onwards quote Pennant (1766), writing about a hundred years before Yarrell, who said that the flesh of the Wheatear was delicate and tasty and that large numbers were caught annually on the Sussex Downs. The catching season opened every year about St James's day, 25 July, and ended in the third week of September. One shepherd and his lad could look after 500 to 700 traps. Pennant wrote that numbers snared above Eastbourne amounted annually to about 1,840 dozens, which were usually sold for sixpence a dozen; one shepherd said he had caught 84 dozens in one day. Sir William Wilson, who lived near Eastbourne, used to supply Wheatears to Charles II, 'because of which, no doubt, *Oenanthe* figures conspicuously in the former family's coat of arms' (Walpole-Bond 1938).

In later years, Dresser (1871) wrote that comparatively few Wheatears were caught owing probably to the fact that large tracts of open common and waste land had been broken up and brought under tillage so that the birds could no longer breed there. The price by then had been inflated to three or four shillings a dozen. Although there is no accounting for human tastes, I think one must have some doubt about Cott's conclusion that the Wheatear's plumage warned potential predators of distasteful flesh. It is also surprising that Cott did not apparently know that both juveniles and adults were regularly taken and eaten by predators of various kinds (see Chapter 5).

The other interesting point that arises from Pennant's account is the apparently vast numbers that appeared on the Sussex Downs at that time, many of which must have been migrants (although Dresser seemed to think that they were locally bred). It is also interesting to see that in the mid-nineteenth century authors were describing how Wheatear breeding grounds were ploughed up and that, as a result, numbers were declining.

In the lowland areas of Britain, Wheatears 'positively assert themselves' (Venables 1937) against the general greenness of sheepwalks and heavily grazed pastures: it is a most conspicuous bird on Skokholm and easy to see at a range of 100 m or more. Mayr and Streseman (1950), in describing a Wheatear's plumage, stressed the conspicuousness of males of the genus *Oenanthe*, saying that, 'whilst most desert birds have a cryptic colouring, it is remarkable that so many species of *Oenanthe* have a plumage pattern composed of contrasting colours black and white'.

On the rocky spoil slopes of the Presely quarries in Pembrokshire, however, the Wheatear's grey mantle merged with the background of grey rocks and stones and the black patches merged with shadows so that I had difficulty in spotting it even when I knew from its calls that the male was bobbing and flirting its wings and black and white tail continuously and vigorously. Even on Skokholm, a Wheatear would behave in a way that ensured that part of its plumage helped to hide it. When disturbed by a human intruder, a Wheatear often flew to the top of a rock and landed on the far side of it and just below the top; there it would sit with its eye just peering over, most difficult to see because the facial mask merged with the rock behind which it was hiding and the superciliary stripe and the grey crown merged with the sky.

It might be thought that the white rump-and-tail pattern of the female would be a hazard since it could attract attention to her. Like the male, she makes it conspicuous by spreading her tail when involved in territorial skirmishes with other females. When flying toward the nest, however, she closes up her tail so that the extent of the white is diminished and the tail then appears black.

The Wheatear can make itself conspicuous by using these black and white patterns in other ways: the facial mask, the black gape, the black wings, the black and white tail, the grey underwing pattern and the peach breast are all used in one or other of the territorial or sexual displays which are described in Chapter 6. The black and white facial mask, which is very reminiscent of that of a Turnstone, or that of a Great Grey Shrike which it meets in Africa, is quite fearsome if, through binoculars, you watch a male flying directly towards you.

It becomes clear in time that all these patterns have a dual role: when used in a certain way and moved about vigorously they have value in sexual or aggressive displays, but if the bird is in its primary habitat the patterns are similar to shadows and highlights of broken ground and help to hide it. The question as to what is the Wheatear's primary habitat I hope to answer in the following chapter and there show that, during the breeding season, it is at high altitudes or at high latitudes.

Cornwallis (1975) suggested that the males of eleven species of wheatears that he studied in southwest Iran were black and white mainly for a reason other than the fact that they afforded the bird some protection, and I would suggest that the reason lies in their use in self-advertisement. He said that his eleven species, which included the Northern Wheatear, were largely confined to rocky areas which afforded them some protection from predators in the form of escape holes (used when attacked by hawks or falcons) and variegated backgrounds (against which they were much less conspicuous than they were against those which were dun-coloured).

In discussing the significance of the Wheatear's plumage it is sometimes forgotten that, after the adults' complete moult at the end of the breeding season and as they begin to disperse from their breeding territories, the males' plumage is basically dun-coloured and not unlike that of the females and first-winter birds: a plumage which is generally judged to have a protective value. We tend to think that birds which breed in Britain are British or European, which, in a chauvinistic sort of way, is true. Nevertheless, the Wheatear spends seven months of the year away from its breeding habitat and in its ancestral home, where, on the African plains and sandy savannas, it wears a dun-coloured plumage. When it is arriving in its primary habitat for breeding purposes, it puts on a striking plumage not only to help it advertise itself to a mate and competitors but also to hide it when anxious in the presence of potential predators. In fact, for, say, eleven months of the year the general plumage patterning helps to conceal it except when it flashes the bolder patterns. Altogether a very utilitarian dress.

THE CHARACTERISTICS OF WHEATEARS

The male Wheatear has relatively long legs, a short hind claw and generally adopts an upright stance. It is a ground-living bird which prefers short vegetation, bare soils or rocky surfaces. It requires extensive vistas — to be able to see over wide areas — and it has a longer alarm distance than other species, such as Meadow Pipits or Skylarks, which nest in the same type of habitat. In this respect Wheatears have a similar requirement to Lapwings, which also nested on Skokholm until gulls became too numerous, or to Stone-curlews, which nest in the same habitat as Wheatears in Breckland, Norfolk. All these birds prefer short vegetation of a type which used to be produced in southeast England by very heavy rabbit grazing prior to myxomatosis in the mid 1950s.

Although the Wheatear may have an arctic-alpine distribution over part of its range at present, it would seem that the genus *Oenanthe* originated within the area of the Mediterranean basin in arid areas or steppe or savanna-like grasslands around which the other wheatear species are presently more or less grouped. It would seem likely that the Stone-curlew also originated in the same kind of habitat, since it, too, is a bird preferring long vistas, and, being a walking bird, it also prefers the same short vegetation as the Wheatear.

It is sometimes said that the Greenland race *leucorrhoa* of the Wheatear can be distinguished in the field from our breeding Wheatear of the race *oenanthe* because it stands more erect. I never felt this to be a reliable characteristic: Wheatears of both races stand very erect when they are alarmed or uneasy, such as when they arrive on migration in a new locality, so as to get a good view of a potential enemy; and both can adopt very hunched-up postures when relaxed. By the time that *leucorrhoa* arrived on Skokholm on their way northwards and showed their uneasiness in their new surroundings by standing very erect, the local *oenanthe* had become more accustomed to theirs and were generally more relaxed, although they would stand as erect as *leucorrhoa* — if occasion demanded.

The Wheatear has various other ways of getting a better view. On sunny days before or after the more hectic parts of the breeding season, individuals often 'belly-doze', or sun-bask, on an anthill or on the side of a sheltered rock with the belly actually on the ground. At the very first sign of something unusual, the bird stretches up its neck without moving the remainder of its body. Similarly, when standing in a relaxed position, rather hunched, and something in the distance catches its attention, it will stretch up its head and neck but without lifting its body, so that the eye reaches a slightly higher level. When the Wheatear becomes slightly more alarmed, it lifts the fore part of its body and then stands erect as it watches. On each occasion the view is improved.

Wheatears freely made use of perches for observation, including grass tussocks, anthills, rocks and rock outcrops, walls, fences, telephone or power wires and herbaceous plants, bushes and trees. It was said by MacGillivray (1839), but contested by Saxby in his book on the birds of Shetland (1874), that one of the field characteristics that distinguishes Greenland Wheatears, race *leucorrhoa*, from the race that breeds in Britain, *oenanthe*, is that the former perch more often on plants. On the other hand Nicholson, writing in 1930 about the birds he had observed in Greenland, commented that 'apparently the habit for which the Greenland Wheatear is chiefly celebrated — that of perching on trees — is more or less confined to the migration season: during the breeding season very little of perching in bushes 3 to 4 feet high was seen; they almost always alighted on rocks and stones'. In my experience on Skokholm and Alderney, migrant *oenanthe* perched just as freely as *leucorrhoa* on plants and bushes if these were available, including the upper stems of bramble bushes, the corymbs of ragwort, the upper stems of wood sage, bare twigs of elder, the umbels of hogweed, the stiff, branched stems of sea radish, and so on; in fact, both races would perch on anything which was strong enough to hold them and which did not obstruct the deployment of their wings when they had to take flight. This they did particularly when the ground layer was wet. Alan Tye tells me that in Senegal both races perch on bushes much more often than in Britain, and from these they hunt by the 'perch-and-pounce' method.

On most of the windswept islands where I have studied them, Wheatears usually perched on more substantial objects such as anthills, grass tussocks, horse droppings, stones, rocks, walls, piles of earth, and other prominent objects usually less than 2 m high. In the breeding season, males in particular sometimes perched on buildings or rocks as high as 10 m: obviously, in areas where rocks, or the house, restricted the vista at ground level, the Wheatear had to climb higher to see over or around the obstacle. Later in the year, migrant Wheatears of all ages and both sexes perched on electric-light or power wires.

I have stressed the point that Wheatears used these high points to keep a lookout. Moreau (1972), however, quoted G. Morel as saying that, because the surface of the unshaded sand is so hot (between 40°C and 80°C) in some parts of its winter quarters in Africa, it is not surprising that wheatears (*Oenanthe* species), when not actually feeding, perch on even as slight an elevation as a small stone. The implication here is that the habit has evolved so that the bird escapes some of the heat radiating from sand. Yet the same habit occurs in Greenland, where it exposes the Wheatear to bitter winds. I think that good vistas are the more important function of high perches, although the latter may serve other ends.

When a Wheatear was watching a potential enemy from such a position, it often stood just below and behind the top of a rock or stone or even a tuft of grass and peered over. Only the crown, supercilium and eye were visible to the potential enemy; and even they were difficult to see since the grey of the crown and the superciliary stripe appeared to be part of the rock. To obtain a better view, the Wheatear would sometimes hop up to the top of the rock for a few seconds before hopping into cover again. Of course the Wheatear was perfectly obvious to an observer on the other side of the rock.

I know of a few passerines which deliberately creep behind cover to peer at the observer. Some warblers of the genera *Sylvia* and *Acrocephalus* are very adept at creeping about in cover, and you can occasionally catch sight of an eye looking at you through a gap in the leaves and you see the tail through another gap; so you realise that, while the individual is trying to hide, it is insufficiently self-conscious to be aware that its tail may be visible. The Great Spotted Woodpecker is another species that, clinging on to the far side of a tree trunk, where its body cannot be seen by you, pokes its head around every few seconds to see what you are doing.

Wheatears began to take shelter when winds reached force 4 on the Beaufort scale (13-18 mph/21-29 kph); they tended to stop in the lee of or among a group of plants, behind a rock or a wall, in the entrance to a rabbit burrow — but not down it — or down the cliff face. During a gale, those Wheatears which had not started laying, and whose territories were on the exposed side of the island would leave them and find shelter downwind. In heavy rain Wheatears did occasionally take shelter under stones, and juveniles three or four weeks old went down burrows but usually emerged after about a minute. The effect of rain on incubation and feeding of the nestlings is described in later chapters.

On the ground a Wheatear hops, runs or uses an odd combination of both for a few paces in an asymmetric hop, which Tye (1982) called galloping, and then stops with its tail beating slowly up and down, or it pecks down after prey (its gait when foraging I discuss in Chapter 8).

It flies fast and low and, except during its song flight, when flying across the territory of another pair or on migration, it seldom flies higher than 2 m and more usually at a height of 25 cm or less. Quite often it flies so low along the rabbit-grazed lanes between the heather shrubs that it is hidden for several metres at a time.

Both male and female fly up 2-3 m to observe potential intruders, enemies or predators. A female watching a weasel on Weeting Heath, Norfolk, continually fluttered up to about 1 m and then dropped back again almost immediately, although it occasionally hovered; Wheatears watching snakes in Africa behaved in the same way.

A very similar behaviour pattern was what I called the 'bouncy' flight, in which a territory-holding male or female flew towards the intruder with a deeply undulating, or 'bouncy', flight, seldom rising above 2-3 m and during which it was apparently trying to locate an intruder. Unmated males also used this flight to approach females entering their territories, but, instead of attacking as they did males, they continued the flight and alighted within 1-2 m of the females and behaved as though soliciting pair formation. I return to this behaviour when describing pair formation and territory defence.

Hovering is a common method of watching all kinds of possible dangers. At one time it was thought to be a sexual display, like the song flight, but a detailed study of the circumstances showed that it was used to observe potential enemies. As a rule, a Wheatear hovers with rapidly and continually

FIGURE 1.4 *Hovering*

beating wings about 3-5 m above ground (although I have seen them hovering as high as 10 m), without changing its position relative to the ground, for anything up to ten seconds before descending to the ground or changing its position: an action very similar to the hovering of a Kestrel. The legs frequently dangle below the body. By contrast, the wings beat intermittently in the song flight so that, at its apex, a Wheatear tends to dance up and down.

Machell Cox (1921) first realised the significance of hovering when he watched a male and female Wheatear hovering in a curious manner. Through his fieldglasses he was able to detect a grass snake. The birds were fluttering 2-3 feet (60-90 cm) up and down over the snake, almost touching it. Thomas (1922), who had originally described it as a nuptial display, changed his mind when he watched a pair hovering near a weasel; he also records that another pair of Wheatears 'came to help drive the weasel out of the territory of the first pair'. Cornish (1947) recorded a male hovering near a Kestrel perched in a bush when fledged young Wheatears were about.

A second interpretation of hovering was that it was used for hunting. Seton Gordon (1942) records hovering as a means of procuring food. He wrote that a Wheatear was hovering at 10 m and, seeing an insect, it dived to the ground, but was at once in the air again. Pound (1942) also saw a Wheatear hovering and each time it dropped it picked something off the grass. Williamson (1949), while discussing hovering and reviewing other records, considered that in many cases it was obvious that hovering was practised as an aid to hunting, the birds swooping quickly to the ground to pick up food. David Low told me that, on 9 May 1951, he saw a female Greenland Wheatear hovering and then dropping into the grass, where it would pick up something and then hop to the nearest stone or clod of earth, from which it would fly up and hover again.

Several of my own records of Wheatears hovering refer to birds dropping to the ground to pick up food after they had been hovering. A final group of interpretations maintained that there was no apparent reason for hovering except 'sheer exuberance'.

It is interesting to note two points about records of hovering: first, there are some, but very few, records of first-winter Wheatears hovering; and, second, most of the records refer to May and June and I have only two in July. This suggests to me that hovering is not used primarily for procuring food and, although, at first sight, it lends support to the interpretation that hovering is some form of, or has some connection with, courtship, I think that it is the connection with territory that is important. In Africa, Wheatears commonly hover over predators in the months November to March when holding a territory (Tye, pers. comm.).

Looking at my own records, in one or two cases it was obvious that I was the cause of alarm: both male and female parents hovered over me as I went to the nest to weigh the nestlings. One nest was in the edge of the bracken and it was difficult for the adults to see what I was doing unless they were above me: they would hover about 3 m above me, calling the 'weet' alarm note and displaying the dark grey, almost black part of the underwing at me. At other times when I was in a hide watching Wheatear nests, people used to pass through the territory and very often the adults would fly up and hover, presumably watching them pass through. Once, when walking across the island, I turned to look back the way I had walked and a Wheatear was hovering over some flat ground beyond a strip of bracken which would have hidden me from the bird if the latter had remained on the ground. As far away as Alaska, Murie recorded a Wheatear hovering: he thought it was watching him (Gabrielson and Lincoln 1959).

While a Wheatear hovers chiefly to get a better observation point, it will break off and catch an insect if one should present itself. Furthermore, it will break off and hover in the course of sexual displays in the same way that they will interrupt displays and pick up an insect or two. Hovering for food is largely opportunistic.

I feel certain that where the observer has seen no obvious reason for hovering he has failed to realise that he is not possessed of Perseus's cap of invisibility and that a Wheatear becomes alarmed and watches a birdwatcher for, at times, distances of 100 m and more.

I should point out that, in the food section of *BWP* (Cramp 1988), the authors mention only those records which suggest that hovering is a hunting technique.

FEATHER MAINTENANCE AND COMFORT MOVEMENTS

Wheatears are host to a variety of parasites — fleas, feather lice, mites and hippoboscid flies — and the daily preening to which they subject their feathers presumably helps to control their numbers. In my study I found that Wheatears might preen their feathers at any time of the day, but they would often have a session in the morning after their first burst of energetic feeding, after midday, and then again in the late afternoon. These were periods when they were less active anyway and when they might otherwise be loafing about or sunning. Of course, much depended upon the stage of the breeding season since they rarely had time for much preening when feeding the nestlings. Their preening sessions were hardly ever undisturbed: another Wheatear might approach closely and they would move away; or they broke off preening to catch some tempting prey.

In the manner of most birds, Wheatears preen by running their contour and flight feathers through their beaks and scratching their head with legs and claws over their wings. Rarely, I saw one wiping its head or beak against some hard object such as the strand of wire on which it was perched.

Open water was in short supply in most Wheatear habitats, but when any was present they bathed occasionally; one whose territory included a fresh-water well bathed in the outflow, and others occasionally bathed in shallow puddles formed after rain. One female which had just finished a spell of incubating first drank and then bathed when she came to a shallow stream formed by heavy rain that morning. At first her efforts were very tentative, her legs hardly bending as she went through the motions of bathing, but later she got down to it, poking her head into the water and shuffling it over her back with beating wings. I do not think that this female sought out the stream to drink and bathe in and she reacted rather slowly when she came upon it by chance. Probably few Wheatears ever bathe in pools or streams during the breeding season (or even during their lives), but their plumage is often soaked by rain and they preen after that.

Most of the sunning, or sun-bathing, behaviour that I observed by Northern Wheatears was of a rather subdued nature. Simmons (1986) has suggested five categories for the different intensities of sunning. They are (1) simple sunning behaviour, (2) the 'wings-down' postures, (3) 'lateral' postures, (4) 'raised-wing' postures, and (5) 'spread-wing' postures. The commonest form of sunning adopted by Wheatears on Skokholm was for an adult or a bird of the year to sit belly to the ground in some form of shelter, on the side of an anthill or a thrift tussock, at the entrance to a rabbit scrape, in the lee of a rock or a wall, or even among a heap of stones. There it sat for as little as five seconds or for several minutes. This was the posture that I originally called 'belly-dozing': it then seemed to me that the individual was loafing and enjoying the warmth of the sun in a comfortably sheltered position. This was, according to Simmons, the lowest-intensity, simple sunning behaviour, or sun-basking, and was a means of absorbing heat and hence a form of thermoregulation.

A second form of sunning which I saw only infrequently involved movement of limbs and feathers: it seemed to be related to Simmon's 'raised-wing' posture. It usually began with the individual belly-dozing, or sunning, in a sheltered position. Then the feathers around the head were ruffled, and the wings raised and slightly outstretched, which exposed to the sun the white of the rump (and perhaps the preen gland) and the flank under the wing. Finally, the wing-coverts were raised.

On other occasions I saw a third posture: the wings were spread out fully with mantle feathers raised, which again exposed the white rump, in Simmons's 'spread-wing' posture. Once a Wheatear had got to this stage, it usually remained like this for two or three minutes.

The two last postures indicated an increase in the intensity of sunning. They were more related to what Simmons called 'sun exposure', which probably functions in feather maintenance and related ways, and hence is a form of comfort behaviour.

It is apparently thought that the functions and the internal motivation of sun-basking and sun exposure differ, even though, from what I have seen of these behaviour patterns in Skokholm Wheatears, they each seemed to develop from the simple sunning position, and the longer Wheatears remained in a sun-basking position the more likely it was for them to develop one of the sun-exposure positions. Wheatears sunned only in a sheltered position on this island and never in the sun in an exposed position.

There is controversy as to whether it is the warmth or the light that stimulates sun exposure. Simmons thinks warmth, and Wheatears certainly demonstrate that point. Experience with sunning Blackbirds in a corner of my garden in Cambridgeshire also supports the idea that warmth starts them off: for about 25 years, male Blackbirds, obviously different territory-holders over the years, have sunned — sun-basking and spread-wing sun exposure — in the same small, sunny, very sheltered corner. Yet, although the same individuals are present in winter, they do not sun.

THE GEOGRAPHICAL RACES OF THE WHEATEAR

Since the Wheatear *Oenanthe oenanthe* has an almost circumpolar distribution, with only about 1,000 miles (1,600 km) of the Canadian Arctic separating the two wings of its breeding range (see Figure 3.1), one might expect that this species was polytypic and that a number of geographical races would have developed in different parts of its range. There are various opinions among systematists as to which of those that have been described are 'good' races. The most recent review of the races, or subspecies, of the Wheatear is that of Tye in *The Birds of Africa* (Keith *et al.* in press), from which source it would appear that four races of the Wheatear are now recognised within the boundaries of the Palearctic. *Oenanthe phillipsi*, or the Somali Wheatear, which used to be regarded as a race of the Wheatear, is now regarded as a separate species (Tye 1986).

Of the four Western Palearctic races, the typical race *Oenanthe oenanthe oenanthe* (Linnaeus 1758) has a breeding distribution from northwestern and central Europe, including Britain and Ireland, as far north as Novaya Zemlya, Siberia, Pribilof Islands, Alaska, and northwest Canada as far east as 137° in the north of Yukon and in northwestern Mackenzie (Godfrey 1986). In the southern part of its range, it breeds from Spain and the Balearic Islands eastwards north of the Black and Caspian Seas and Lake Aral to about latitude 66°30'N on the Yenisei River and to about latitude 65°N on the Koyara River and Anadyr Bay.

In the non-breeding season the typical race travels to Arabia and to Africa as far south as Senegal, northern Nigeria, northern Malawi, Tanzania and Mozambique to Zumbo and the Zambesi River.

Another race which passes through Britain and Ireland quite commonly in spring and autumn is the Greenland Wheatear *O. o. leucorrhoa*, which, as its name suggests, breeds from northeast Canada as far west as 93°W and 70°N on the Boothia Peninsula to the northwest of Hudson Bay, the Rankin Inlet (District of Keewatin), through Baffin Island and Ellesmere Island (Godfrey 1986), through Greenland, probably as far north as 82°46'N in Peary Land (Håkansson *et al.* 1981), Iceland and Jan Mayen. This race, like *oenanthe*, winters in the Senegambia area of West Africa. The Wheatears that breed in Iceland are sometimes treated as a separate subspecies *O. o. schioleri*, but the general opinion is that the case for separation is not very good.

In addition to the races of Wheatear that breed in Eurasia and Greenland, there are two further races: *O. o. libanotica* covers southern Europe and southwest and central Asia, and *O. o. seebohmi* Morocco and Algeria. Two races which have recently been lumped in with *libanotica* were *O. o. virago* of the eastern Mediterranean islands through to Israel and Egypt (the race was named *virago* by Meinertzhagen in 1920 because of the females' 'presumptuous' plumage!) and *O. o. nivea*, which occupied southern Spain and the Balearics. Voous (1977), in his survey of recent Holarctic species, includes *seebohmi* as a race of *O. oenanthe* but points out that it is sometimes treated as specifically distinct. It is an 'incipient species' (Hall and Moreau 1970). Some East European authors call the population of Wheatears which inhabits the arctic regions of Europe *oenanthoides* to distinguish it from the typical race which is found further south in Russia, but this name is not generally accepted.

The most westerly races of Wheatear demonstrate Bergmann's rule rather neatly: the rule states that, among the forms of polytypic species, body size tends to be larger in the cooler parts of the total range and smaller in the warmer parts (see Table 1.1). Because of their small size, birds lose heat very rapidly. The increase in body size shown by the population occupying the coldest parts of the range of the species can be explained as an adaptation to climate as a result of the demands of thermoregulation. In a way, a bird's weight is a more reliable measure of the effect of Bergmann's rule, since populations of birds which migrate long distances tend to have longer wings than those which travel only short distances.

Most Wheatear races are difficult to identify in the field and they are normally separated with certainty only when they have been examined in the hand or against skins in a museum collection. Svensson (1975), for instance,

TABLE 1.1: *Wing lengths in millimetres of three races of the Northern Wheatear*

| | MALE | | FEMALE | |
	Range	Mean	Range	Mean
Greenland *leucorrhoa*	101-109	= 105.2	99-105	= 101.6
Britain *oenanthe*	95-102	= 98.1	93-97	= 94.7
North Africa *seebohmi*	93-100	= 96.4	91-96	= 93.8

Source: Cramp (1988)

suggesting criteria for separating *leucorrhoa* from *oenanthe*, states that Northern Wheatears trapped in spring in northern Europe which have a wing length greater than 101 mm and which have a prominent rufous tinge should belong to *leucorrhoa*. In autumn this distinction is not so clear, since some northern specimens of *oenanthe* have a distinctly rusty-buff plumage. Furthermore, several members of one brood of *oenanthe* ringed as nestlings on Skokholm and retrapped later in first-winter plumage had wings which measured over 101 mm.

Nevertheless, an experienced observer can separate, in spring, some *leucorrhoa* from *oenanthe* in the field because the former are clearly larger than the typical race which breeds in Britain, they have a heavier bill (which I have never been certain about in the field) and they are more strikingly coloured. They appear in Britain when the local birds have just started to build or to lay and have already begun to lose the richer colours they had when they first arrived. As a ringer handling several hundred Wheatears on Skokholm, I concluded that after 15 April, and particularly after 1 May, most Wheatears seen on spring migration were likely to be birds on passage for Greenland or northeast Canada or Iceland.

Chapter 2

WHERE THE WHEATEAR LIVES

In the first part of this chapter, I describe some of the environmental factors that influenced the distribution of Wheatears on Skokholm from 1947 to 1954 and between 1978 and 1983 in both the breeding and the post-breeding seasons, and in other parts of the species' range at a later date.

A Wheatear's breeding habitat must provide a sufficient and available food supply, shelter from the climate and protection from predators not only for the eggs and the nests but also for the pair and their recently fledged offspring. Even if a sufficient food supply existed in the different habitats on Skokholm, a Wheatear was able to take its arthropod prey only where the physiognomy of the vegetation allowed it to move freely and easily when foraging (Conder 1952). As I have mentioned earlier, Wheatears are long-legged, have a short hind claw, and they stand upright. On the ground they move by hopping or a kind of running hop. They are fast, low-flying birds which, except during the song flight or when flying across the territory of another pair or across unsuitable habitats, seldom fly higher than 2 m and usually lower than 25-50 cm above the vegetation or rocks. I have also argued that their plumage is cryptically coloured. These are points to which I shall return later in this chapter.

Being primarily a hopping or running-hopping bird, a Wheatear needs a fairly firm surface with sparse vegetation (preferably about 1-3 cm and less than 6 cm), bare earth or a rocky terrain of one kind or another. If the surface is soft a Wheatear will not be able to hop efficiently, since much of the energy put into a hop would be absorbed by the resilience of the grass blades; hopping would be difficult or impossible in longer and thicker grasses. Meadow Pipits and Skylarks, both of which forage among the longer grasses, have long hind claws which act like snow shoes and prevent them from sinking deeply into the grasses.

When foraging in short grass, a Wheatear on Skokholm usually hopped about four to six paces and then looked down to pick up its prey or search for the next item. Sometimes it might see prey 1-2 m away and would hop and almost run towards it, sometimes even using its wings. It might hop for as much as 3 m or as little as one or two hops.

A second method of hunting was to fly up after a slow-flying insect: on the more windswept Skokholm a Wheatear might reach 2-3 m, but on calm days on Alderney I have seen them flying up to catch insects at 10-20 m. A third method of hunting where grasses were taller (because grazing was light) was to perch on plants such as ragwort and from there drop down on prey, or to hover like a Whinchat over the grass blades and again drop down on prey. Wheatears foraged like this, too, when they were hunting in the open bracken communities. I shall discuss these and other foraging methods in Chapter 8 when writing about food and feeding.

At this stage I am going to assume that there was an adequate food supply for a Wheatear population on Skokholm which more than doubled itself during the years of my study. Except in 1980, I had no evidence that starvation or shortage of food affected the development of the young. In 1980, the Wardens, Graham and Liz Gynn, told me that an early drought had reduced fledging success and, when I arrived on the island a week or two later, it was clear that the number of independent juveniles was markedly lower than in 1978 and 1979. Presumably these early droughts affected the abundance of lepidopterous caterpillars and tipulid larvae which feed on the grass roots and which were the chief food items brought to the nestlings. Droughts later in the season when the young are independent and free-flying are perhaps not so serious for Wheatears since they can disperse to another area.

In addition to an available food supply Wheatears require holes for nesting, and for these they were largely dependent upon rabbits, whose numbers on the island varied from about 400, as in spring 1978 (Ray Lawman, pers. comm.), to an estimated 15,000 (Conder 1952). For a more detailed description of nest sites see Chapter 9.

Wheatears did not use the wide-mouthed burrows that were found in warrens and which showed constant use, but those whose entrances had a diameter of about 10 cm or less. Some had been excavated by rabbits and used as nurseries, but others had been taken over by Manx Shearwaters or Puffins. A few Wheatears nested in holes in the eighteenth-century earth and

FIGURE 2.1 *Thrift near Lighthouse*

stone walls from which the earth had been washed out. They never dug their own burrows, although they removed a certain amount of debris. Nor did they build a second nest in the same burrow on Skokholm unless I had first removed the old material (usually for examination for parasites) or after two years had elapsed which allowed the old material to decay and the infestation by fleas or other nest-dwelling parasites to be eliminated. In other parts of their range, however, Wheatears have re-used nest cups for second clutches. At Dungeness, Kent, where there was a shortage of natural holes, they repeatedly used artificial sites, even though material from the first nest was still present (Axell 1954). Wynne-Edwards (1952), working on Baffin Island, discovered one Wheatear nest, again in an area where sites were scarce, which appeared to have the remains of eight previous nests packed down one on top of another.

Water did not seem to be necessary for drinking and apparently Skokholm Wheatears obtained all the moisture they needed from their food, but on the other hand they bathed when streams and rainwater puddles became available. Indeed, water could be dangerous to a Wheatear: if a bird is to nest successfully in an underground burrow, then the soil must be well drained and safe from flooding during normal summer rains. Many burrows used by Wheatears on Skokholm flooded in autumn and winter, but in 1952 two burrows that had been used as nest sites in 1951 were partially flooded on 6 May; rainfall during the first ten days of May that year was 32.3 mm (1.27 in) more than in the same period in 1951, and since the patch of bare earth in the meteorological enclosure had been recorded as dry 36.6 mm of rain had fallen. In 1981, a nest with eggs was flooded and deserted by the adults after a night of very heavy rain. In many parts of the island there was little depth of soil over the Old Red Sandstone and in heavy rain the soil and burrows in that shallow soil soon become waterlogged. On the same day in 1981, I saw two Manx Shearwaters leave their burrows in daylight — a most unusual occurrence — and fly out to sea; I assumed that rainwater had flooded them out too.

Wheatears did not need song posts since they usually advertised their presence in a song flight, although when outcrops of rocks, walls, or slight rises in the ground and so on were present Wheatears used them.

SKOKHOLM'S VEGETATION

Having discussed some of the Wheatear's habitat requirements, I shall now attempt to relate how Skokholm's vegetation, described in detail by Goodman and Gillham (1954), was able to satisfy these needs. In this section I shall be looking at the vegetation from the Wheatear's viewpoint.

The vegetation of the island could be divided into three main types: the cliff face, the cliff top, and the submaritime heathland. Wheatears were found only rarely on the cliff face, usually when they were sheltering from gales, and only once, in 1982, did I find a nest down the cliffs.

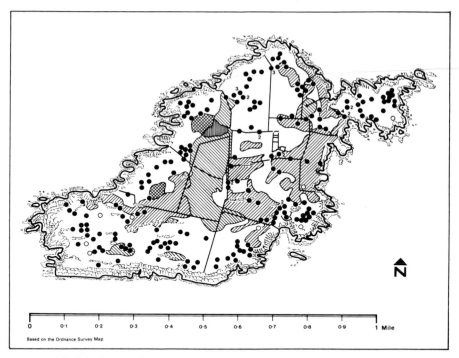

FIGURE 2.2 *Distribution of Wheatear nest burrows in relation to vegetation types. Closed circles indicate known burrows and a numeral the number of times the burrow re-used; open circle shows area where young fledged but exact burrow unknown. Vertical hatching = open water; hatching sloping to right = marsh; hatching to left = bracken; blank = grazed grassland or thrift; cross-hatched = sand sedge.*

The vegetation types preferred by breeding Wheatears during the period of my study was the cliff-top zone (see Figure 2.2). A considerable quantity of spray fell on the top of the island in the gales of autumn, winter and early spring. In the strongest gales (force 10-12 on the Beaufort scale) the waves strike the bottom of the cliffs and their white tops, or even green seas, fall on top of the cliffs. Only salt-tolerant plants such as thrift, sea campion and red fescue can survive there. The width of this zone varied according to the formation of the cliffs and exposure to the prevailing winds. On the southwest, west and northwest coasts the zone was as much as 100 m: here the cliffs were steepest, and the prevailing winds, blowing from the southwest and west (from which directions were the longest distances the wind could blow over open sea without interruption), raised huge waves which crashed against the cliffs and the spray was lifted by the updraught on to the plateau of the island — and even across the island in strong gales.

On the eastern side of the island the width of the cliff-top zone decreased until, in the most sheltered parts of South Haven, it was only about 3 m wide. Along the south coast the cliffs were less steep, and the largest waves that struck the island directly, without being deflected in any way by the island

21

itself, were those that came from the south or southwest, having a 'fetch' of only 100 miles (160 km), so that less spray was carried up the cliffs there than on the western and northwestern coasts. Those cliffs that faced the east and northeast were relatively close to the mainland and were usually spray-covered only in strong westerly gales when spray was blown across the island from the western or northwestern cliffs.

The third plant association was submaritime grass heath. On this vegetation some salt spray fell but only in the heaviest gales, so that some salt-tolerant plants were present and many plants normally found on inland heaths were absent. The main communities were dominated by bracken or heather, and by a grassland area in which common bent-grass, Yorkshire fog and red fescue were the commonest grasses in various combinations. In the centre of the island was a marsh lying on the thickest deposits of boulder clay in which various species of rush or purple moor-grass were present. Sand sedge, which although taxonomically not a grass has the physiognomy of a grass, also formed dense communities in the drier parts of the island.

In the last 30 or 40 years there have been some marked changes in the distribution of thrift and sea campion. In 1947, it was apparent from colour photographs of The Neck taken by E. J. M. Buxton in 1939 that the area covered by thrift had decreased. From a comparison of vegetation maps drawn by Mary Gillham, Gordon Goodman and Ron Williams in 1952 and, rather roughly, by me in 1981, it can be seen that the decline of thrift had continued and that sea campion had spread markedly. Described by Goodman and Gillham (1954) as being a minor community found only in sheltered spots, sea campion was a major community in 1983 even in very exposed parts of the island, forming extensive and uniform cushions 10-15 cm thick. This was not a good surface on which some bird species could forage: whereas Wheatears could hop easily over thrift, they had difficulty with sea campion since their legs would tend to sink into the cushion.

Rabbit Pressure

In addition to the effects of salt spray and wind, practically all the plant communities on top of the island were modified, even dominated, by a rabbit population whose numbers have been estimated to vary between 5,000 and 10,000 individuals under normal circumstances, but dropping as low as 400 in 1939 after Lockley had tried hard to eliminate them altogether since they were taking all the food for his sheep, and again in 1978, when Ray Lawman told me that the hard winter of 1977/78 had resulted in another decrease to about 400 individuals. Myxomatosis was unknown on the island: Lockley had attempted to introduce it before the war, but Skokholm rabbits are unusual in that they have no fleas to transmit the disease. The natural factors which reduced the population from time to time were coccidiosis, a disease caused by a parasitic protozoan, which killed large numbers of young rabbits in late summer, and hard or wet winters. Owen Jones, of Trinity House, told me that in the hard winter of 1947 the island was covered with snow which drifted to a height of 2 m; later, hundreds of burrows were filled to overflowing with meltwater, and the island was then covered with the bodies of birds of many species that had succumbed to the conditions. When a hard winter such as this followed a severe outbreak of coccidiosis the previous autumn, grass grew thickly in spring, which did not directly benefit Wheatears, although it helped the rabbit population to recover.

Under the more usual heavy rabbit pressure the grass was short and tufted. In most places common bent-grass was not more than 1-3 cm high, and Yorkshire fog with its less palatable, softly hairy leaves was able to withstand grazing and become somewhat tufted; but even that was grazed when the rabbit population had grown particularly large. Then the sward was so heavily grazed that the commonest plants were Buck's-horn plantain, sea storksbill, and various mosses and lichens, with much bare earth. This was the *Plantago* sward of Tansley (1939) and Goodman and Gillham (1954). Heather was dominant in some of the central and less exposed parts of the island, and the low shrubs were heavily wind-stunted and rabbit-grazed. Rabbits foraged between the shrubs and left 'green lanes' along which Wheatears could hop without hindrance in search of food.

In the cliff-top zone, too, rabbits also had an important effect. Gillham (1956) showed by means of rabbit-proof enclosures that if rabbits were removed from the island thrift would be replaced by red fescue. In this way the vegetation would resemble that of Grassholm and Cardigan Islands, on which there were no rabbits and no Wheatears bred. On both these islands red fescue excluded all but the strongest competing plants and grew to a height of 20 cm or more, so that it was entirely unsuitable for Wheatears.

FIGURE 2.3 *Clifftop above Crab Bay*

Indeed, the thickness of the fescue could soon hide the entrances to smaller burrows used by Wheatears for nesting.

While the actual number of rabbit-burrow entrances might, as potential nest sites, make the cliff-top zone attractive to Wheatears, the platform of earth outside them also attracted Wheatears. These platforms consisted of the spoil (earth or sand) which had been dug out of the burrows and pulled back like a doormat for as much as 2 m. They were associated with most rabbit-burrow entrances except 'nurseries' in which Wheatears preferred to nest.

Longis Common, Alderney, lies on an extensive area of windblown sand, and these platforms were fresh sand-coloured and contrasted markedly with the green of the sward. When a resting migrant, which had been foraging, was disturbed by passers-by, it almost invariably flew to and alighted on one of the sandy platforms within its individual territory. There it stood alert and erect, often bobbing and wing-flicking, until whatever had caused the alarm had moved on and it could relax and resume hunting. Other Wheatears regularly alighted on these sandy platforms when loafing and preening and sometimes remained on them for two or three minutes.

In Chapter 6 I have also mentioned that holes or depressions in the ground seemed important because sexual displays often occurred in or over them. Sometimes a pair actually flew to a depression and displayed, as though the place in which they had been when the drive to display came upon them was unsuitable for that purpose. So far as I could see, the pair always copulated in such places: the dancing display was almost always performed over a bird crouched in some form of depression. Even a juvenile, which had left the nest two or three days previously, performed the dancing display, perhaps precociously, over a sibling at the entrance to a burrow into which the two had been peering.

Although this habit of alighting on sandy platforms was very common on Alderney, I could find no records of Wheatears doing it on Skokholm, where the platform is the typical red-brown of soils derived from Old Red Sandstone and therefore not very conspicuous.

In spite of the fact that Wheatears often alighted on these sandy platforms, and that outside the breeding season they rarely entered burrows, I think that the close proximity of a burrow to a sandy platform attracted them and, when they had flown there in times of anxiety, apparently gave them confidence. I have shown that on spring migration Wheatears are attracted to areas with plenty of holes which are potential nest sites, whereas in autumn they tend to choose good feeding areas. Sometimes these burrows near sandy platforms have a practical value. On Longis Common, a hunting Sparrowhawk, flying so low that its wings almost touched the ground, surprised a Wheatear as it rested on a platform: the Wheatear dived into the burrow 1 m or less ahead of the Sparrowhawk; while the Sparrowhawk remained inside the burrow for about 15 seconds, three minutes elapsed before the Wheatear emerged. In winter quarters the Red-necked Falcon, also hunting very close to the ground, caused Wheatears in the open to fly to cover (Tye 1984).

The Effect of the Wind

In previous sections I have dealt with problems for ground-nesting species caused by heavy rain. In this section I shall be looking at the influence of wind on Wheatears, not only on their day-to-day movements and the choice of nest sites within a territory but also on the selection of a breeding habitat and territory by pairs breeding on the island for the first time.

It is inevitable that, when living in exposed places, strong winds have to be taken into account. I got the impression that force 4 on the Beaufort scale (13-18 mph/21-29 kph) was the point at which Wheatears, and other passerines, became reluctant to fly and began to take advantage of any local shelter such as grass tussocks and even anthills.

I shall be pointing out in Chapter 3 that, when Wheatears arrived on Longis Common, Alderney, on both spring and autumn migration and wind was blowing at force 4 or stronger, they localised themselves in the lee of the low hills which surrounded the common on three sides. If another group of Wheatears was already established there, most of the new arrivals had to settle in more exposed areas. I shall also show that in gales on both Alderney and Skokholm individuals in exposed territories, whether an individual territory or a breeding territory before the eggs were laid, left it and moved to shelter if none was available in their own.

In the opening paragraphs of this Chapter 1 reiterated that Wheatears tended to fly low, where I assume they escaped the full force of the wind. Goodman and Gillham (1954) provided some measurements of wind speeds at different heights above ground. Using a portable Short and Mason anemometer in a west-southwesterly wind, they showed that at 66 cm (26 in) above ground the most sheltered area of the island in South Haven was only one-tenth as exposed as a position on a rock at the southwest corner of the island. The northwest coast was only seven-tenths as exposed. At 6 cm (2 in) above ground, the northwest coast was only three-tenths as exposed as the southwest corner. In general it can be said that at 6 cm, the height of a hopping Wheat-

FIGURE 2.4 *Wheatear sheltering from strong winds*

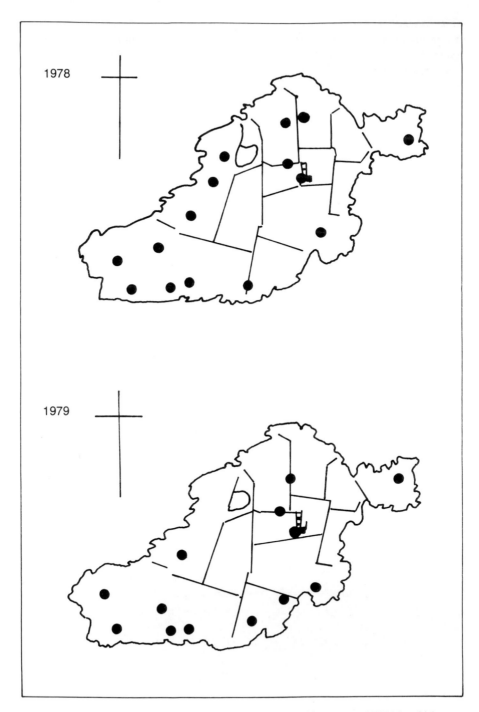

FIGURE 2.5 *Nest distribution in an average season (1978) and in a season (1979) in which persistent cold northwesterly winds predominated in April and early May.*

ear, exposure was half that found in the open places. As one would expect, exposure was greatest on the highest ground or where the wind had blown over wide areas of smooth ground. Thus, by flying within a few centimetres of the ground, Wheatears avoided the full exposure to the wind, even on the relatively smooth ground of the Northern Plains.

In 1951, persistent and strong northwesterly winds temporarily held up migrants and delayed the return of residents to Skokholm (see Table 4.1). In early May 1979, according to the Meteorological Office the coldest early May since records began, northwesterlies again blew cold and persistently. The mean wind speed at 09.00 GMT between 1 April and 5 May was 12.3 mph/ 19.8 kph (range 2-30 mph/3-48 kph; S.D. = ± 7.3 mph/11.7 kph), and during the same period winds were recorded from the north-to-west quadrant on 21 days (59%). On 2 May, when normally many pairs of Wheatears would have been egg-laying, northwesterly winds were blowing at about 30 mph (48 kph) and spume was being carried across the island. One pair I was watching did nothing for 27 minutes except forage among the thrift tussocks, where they were partly sheltered, but even in that relatively sheltered spot they held their bodies horizontal to the ground and occasionally flicked out their tails for balance. Another pair was standing in a slight hole in the ground, occasionally poking their heads above the rim of the hole. A third pair seemed to spend the day down the cliff.

The persistent northwesterly winds of 1979 apparently affected the distribution of nest sites, and a comparison of Figures 2.5a and b shows that more nests than usual were established in the southern part of the island. An examination of Figure 2.2, which shows all nest sites between 1948 and 1953, indicates that the northwest coastal area between The Head and Mad Bay Point was generally a favoured area. It was also in 1979 that the female which had begun building in a hole in the west side of a wall into which these winds are blowing deserted the half-built nest and selected a more sheltered site close by.

While there is some evidence that persistent strong winds can affect the distribution of territories particularly of birds arriving to breed for the first time, the physiognomy of the ground together with wind direction seemed, at least in a general way, to influence the selection of nest sites. Certain ridges of high ground and marshy depressions could act as natural barriers to territories, which remained the same year after year although occupied by different pairs (Conder 1956). In this case these outcrops, although they provided some shelter from strong winds, really lessened visual contact, and thus aggression, between neighbouring pairs.

Even so, the actual slope of the ground might be important in providing a more congenial microclimate. Figure 2.2, giving the distribution of all Wheatear nests between 1948 and 1953, also shows that there is a tendency for some nests to be grouped. While some of these groupings were due to pairs which returned for several years to the same territory and the same area for nesting, other nests were built by different pairs in later years in areas of local shelter.

In Chapter 9 I show that some burrows were apparently 'ideal' and were used for more than one nest by one or more pairs.

Table 2.1a gives the percentage frequency of winds from each direction on Skokholm in April, May and June between 1948 and 1952 and indicates a preponderance of winds from southwest to north. Table 2.1b shows that 62% of the burrow entrances were sheltered from south to northwest winds either by rock outcrops or by other abrupt alterations in gradient, suggesting that at some stage in nest-site selection, probably fairly early on, Wheatears chose a site sheltered from a wind direction that was incommoding them. I have already mentioned the 1979 female that changed her very exposed site in which she was building for one that was more sheltered.

What I have shown so far is that because of its hopping gait the Wheatear has a very restricted habitat type, choosing open country with wide vistas and short or sparse vegetation and with holes in which to nest. On Skokholm, the Wheatear is dependent on rabbits and the climate to produce these conditions. The climate and the soil produce a vegetation which can tolerate salt, and the rabbits grazed it until it was short enough for the Wheatear's foraging methods. Rabbit 'stops', or nurseries, were used as nest burrows. Puffins and Manx Shearwaters also provided some suitable burrows, but in all probability if there had been no rabbits on the island there would have been very few Wheatears.

The majority of Wheatear nests was close to the cliff edge — 52% were within 100 m of the cliff edge — where burrows occupied by rabbits and seabirds were most numerous and where most of the early territories, occupied by older birds, were usually established. Most nest sites were to some extent sheltered from the prevailing winds by rock outcrops or by sudden alterations in gradient.

TABLE 2.1 a: *Percentage frequency of winds from each direction on Skokholm in April, May and June during the years 1948-52*

	N	NE	E	SE	S	SW	W	NW	Calm
	15	5	7	13	6	16	17	20	1

TABLE 2.1 b: *Directions from which individual nest burrows were sheltered*

	N	NE	E	SE	S	SW	W	NW	Exposed
Total	71	60	31	31	92	93	84	83	14
%	12	10	5	9	16	16	15	15	2

Based on Trinity House records in 1948 by courtesy of Trinity House, and on Skokholm Climatological Station records by courtesy of the Meteorological Office.

THE HABITAT OF THE WHEATEAR ELSEWHERE IN BRITAIN

Without rabbits, Skokholm might well have been an island covered by scrubby woodland with little or no habitat suitable for Wheatears. Remembering that rabbits had been introduced into Britain just before AD 1200 from their native Iberia and North Africa (Matthews 1982), I wondered about the extent to which Wheatear habitat was modified by rabbits in other parts of Britain and whether the birds were always so dependent on them. I decided in the 1950s, before the spread of myxomatosis, to check as many Wheatear habitats as I could, either by visiting them personally or by searching through the literature for Wheatear habitat descriptions in Britain and elsewhere within the species' breeding range.

In Pembrokeshire I searched 13 miles (21 km) of coastline and found only seven pairs; and I traversed about 10 miles (16 km) of the Presely Hills, whose highest point was about 530 m, where I located 25 pairs. It became clear that in neither of these areas was the vegetation a climax community but was heavily modified either by rabbit grazing along the cliff tops and below about 200 m on the hills or, above that height, by grazing sheep.

Over most of the Presely Hills the vegetation was fairly dense and about 15 cm high. The most frequent plants were heather, bilberry, purple moorgrass, fescues and mat-grass. Much of it had been burnt during the previous winter. The sheep had not, however, grazed the sward sufficiently hard to produce a habitat which suited the Wheatear's foraging habits. The climate did not keep the vegetation short, but bad weather drove the sheep down the hills to semicircular shelters cut into the hillsides by shepherds or to piles of boulders in various localities along the summit. Consequently, close to the rocks or the semicircular shelters the turf was as short as it was on Skokholm, and here Wheatears nested either under boulders or in the scree.

Since those days in Pembrokeshire, my field experience of Wheatear habitats and my searches in the literature have confirmed that in most of lowland Britain Wheatears nest either in the coastal regions or on heaths and moors, on grassy downlands and so on and are dependent upon rabbits or sheep for short grass and on rabbits for burrows. When myxomatosis swept the country in 1954 and 1955 the rabbit population was severely reduced, which, in turn, allowed the vegetation of previously good Wheatear habitats to grow so tall and dense that the birds could no longer forage over them efficiently or find burrows in which to nest. As a result, they deserted many habitats in which they had formerly nested.

In some parts of Britain, however, some rabbits survived or developed resistance to the disease, and their numbers increased again locally. Tye (1980) has shown that many of the Breckland heaths in Norfolk and Suffolk still hold substantial numbers of breeding Wheatears. Breckland is an area of dry sandy soil, basically a glacial deposit, overlying chalk of the Cretaceous period. Much of the area is now cultivated or forested, and one large section is a military training area — and very good for wildlife, too. So fragments of the old heaths remain. Today they are usually sheep-grazed, but in some parts rabbits have persisted; on Weeting Heath National Nature Reserve, the Norfolk Naturalists' Trust has fenced rabbits in so that, uncontrolled by man, they can maintain the pre-myxomatosis Breckland habitats without damaging the interests of the neighbouring farmers. The vegetation of these heaths ranges from bare sand, through lichens and very short grass to sand sedge and ling; and, most importantly, rabbit burrows are numerous.

Fire over grass and heather can also produce habitats for nesting Wheatears, but these lose their suitability as soon as the heather recovers; the use by Wheatears of these burnt heaths in Surrey and Dorset for breeding was described by Lack and Venables (1937), Venables (1937) and Darling (1947). In some parts of the species' African winter range, too, the burning of grassland to improve cattle grazing, or before ploughing and planting, is a common practice and one which attracts wintering Wheatears (Bannerman 1936).

Dungeness, Kent, is another interesting nesting habitat. The southernmost tip of the ness is almost pure shingle where vegetation is patchily distributed and there are few natural holes. Here some Wheatears have nested under jetsam and other debris (Axell 1954). Further inland, still on the RSPB Reserve, some soil overlies the shingle — but, although it is not deep enough for cultivation, rabbits have burrowed here. Herbert Axell, and later Peter Makepeace, greatly increased the Wheatear population on the bare shingle by setting out nesting boxes specifically adapted for them. The breeding success of Wheatears at Dungeness, however, was lower than that of Wheatears on Skokholm (see Chapter 10), which suggests that the scanty vegetation did not produce a food supply sufficient to feed large broods.

In upland Britain, particularly in north England and Scotland, most nests are in scree, rock outcrops or dry-stone walls. Most grazing here is by sheep because rabbits are less common and their burrows scarcer (Tye, pers. comm.).

Above the tree limit, where vegetation is kept short or sparse by the climate and a relatively small amount of grazing, we also see Wheatears feeding and nesting among rocks and stones and on stony scree slopes. In Scotland generally, Wheatears breed up to 1,200 m or even higher (Thom 1986): indeed, Seton Gordon found them on Braeriach, Cairngorm, at about 1,300 m. In the Welsh mountains, Campbell (1949) found them nesting on screes at about 1,085 m and feeding on the grassland some distance from the spot. W. M. Condry told me of Wheatears breeding at the highest top of Cader Idris, Powys, where the vegetation was sparse and where open, bare rock made up much of the terrain.

HABITATS OUTSIDE BRITAIN

My own records of the breeding distribution of Wheatears and my searches through the literature confirmed that in temperate lowlands, of the species' outside Britain, grazing by mammals of one kind or another was, with few exceptions, a vital factor in reducing vegetation to a height and density over which it was possible for a Wheatear to forage: rabbits being the most effective at grazing the turf closer and in digging burrows. Rather as at Dungeness, the majority of the exceptions occurred in coastal areas of the Netherlands, Belgium or France, where Wheatears nested in stony slopes, piles of stones and earth walls on sand dunes, sandy heaths or barren stretches of ground (van Ijzendoorn 1950; Lippens and Wille 1972; Yeatman 1976).

By far the greatest number of references to the Wheatear's breeding habitat outside Britain refer to its nesting at high altitudes or high latitudes, in fact showing an arctic-alpine distribution.

High Altitudes

Looking first at the altitudes — not all of which are within the alpine zone, or even subalpine zones — at which Wheatears are found nesting, E. J. M. Buxton told me that he found Wheatears nesting at 4,280 ft (1,305 m) in the Dovrefjell, Norway. In Germany, Wheatears nest from sea level up to 2,500 m in the vicinity of glaciers (Niethammer 1937); and in Switzerland from 1,100 m upwards, but their greatest density is between 2,000 m and 2,500 m (Glutz von Blotzheim 1964).

Alexander (1917) found Wheatears inhabiting the subalpine and alpine zones of Latium, Italy, from about 1,700 m, where the vegetation of the sub-alpine zone consisted almost entirely of grassland, more or less rocky with some scrub of juniper (ssp. *nana*) and bearberry, up to exposed peaks of about 2,150 m, where mountain-top detritus and rock exposures had a purely alpine vegetation. He stated that the Wheatear did not breed below 1,000 m, yet about the middle of August Wheatears appear in numbers on the parched Campagna on their way southwards. In Corsica they live chiefly in the sub-alpine zone, but they are also found higher, at 2,000 m or more, in the same zones as Alpine Accentors and Rock Thrushes (Thibault 1979).

Cornwallis (1975) studied the habits and ecology of eleven species of wheatears in southwest Iran and, of particular interest to me, 13 breeding territories of Northern Wheatears. He considered that six of these territories had two main elements: (a) areas of flat turf closely grazed by sheep or goats, where they fed; and (b) adjacent rocky areas, where he thought the nests were probably sited. The remaining seven territories were in stony mountainous areas on sloping ground with patchy vegetation of shrublets and rock outcrops which were used as song posts and for cover. In southwest Iran, Northern Wheatears bred at a higher altitude (2,500-3,500 m or even higher) than other wheatear species. Cornwallis also emphasises the importance of grazing by domestic stock and states that without these there would be much less habitat for Wheatears in southwest Iran. Trott (1947) also collected a Wheatear at 2,380 m in the Lar Valley, north of Tehran, Iran, where the species was very common.

Further east, Carruthers (1910) found Wheatears nesting at 9,000 ft (2,743 m) in the Hirsar Mountains of Russian Turkestan. Panov (1974) studied ten species of wheatear of the northern Palearctic that inhabit the USSR. In the arctic tundra and on the bleak coasts of the Arctic Ocean, the Wheatear preferred short-grass areas on more or less level ground with screes or boulders. In central Asia and Kazakhskaya SSR, it sometimes bred in very arid areas, dry steppe and semi-desert in the mountains, but occasionally in steep river banks, in the ballast of railway lines, deserted buildings and so on; in forested areas it also nested in shingle banks of rivers. In the south of its range, the Wheatear is a high-mountain bird, a typical inhabitant of alpine meadows. In the Elburz Mountains, Iran, it is frequently found over 4,500 m, in the Pamir-Altai range between 1,800 m and 4,000 m, and in the Tien-shan up to 3,500 m. In Kazakhskaya SSR and in the Altai, it breeds from about 3,300 m down to the foot of the mountains.

In northwest Africa, Seebohm's Wheatear, *seebohmi*, breeds in rocky terrain with low rock and sparse scrub, usually above 1,500 m; at lower altitudes it adapts to degraded bush country (Tye, in press).

All in all, the Wheatear is able to breed from sea level to about 4,500 m.

High Latitudes

The tundra of the Arctic is at the other end of the arctic-alpine spectrum. Typically, tundra is an open, forestless stretch of country and it covers the huge tract of land lying north of the Arctic Circle, frozen much of the time though it thaws in summer (Cloudesley-Thompson, 1975). Although some parts of the tundra are swampy, others are quite dry, with vast tracts of higher ground whose only vegetation is lichen interspersed with coarse grass. The surface of the land is rough and broken. Even so, in damper areas dwarf birches and willows are characteristic of the arctic tundra. Summers are short, cool and very windy.

Twenty thousand years ago, a flora which bore striking resemblances to that occurring today in the arctic regions or high on the European Alps covered much of northern Europe as far south as southern England (Raven and Walters 1956). It was possibly in this period that Wheatears extended their range northwards to roughly the limit it reaches today. In the Holocene, as the climate became warmer and more humid, this flora was displaced northwards by forests on the plains and above the forest limits in the mountains (Kishchinskii 1974).

In the tundra regions of the USSR, Wheatears occupy dry open sites with heaps of stones, screes, steep banks or animal burrows for nest sites: in fact similar to the species' requirements in other parts of its range. They feed mainly on quite large surface invertebrates such as craneflies, bumble bees, carabids and house flies and their larvae. Nesting places coincide with areas having the greatest number of big insects (e.g. bumble bees).

In northeast Siberia, Wheatears make considerable use of souslik and marmot burrows, where they nest and even eat the larvae of flies developing in the excrement of these rodents (Portenko 1939). Apparently the presence of the souslik and marmot (which probably appeared in the lower Quaternary period) helped Wheatears to occupy the montane tundra landscapes. One is reminded of the introduction of rabbits into lowland Britain which enabled Wheatears to descend from the mountains to the lowlands.

The same Wheatear population which occupies the Russian tundra also inhabits the arctic-alpine zone of Alaska and the Yukon in the northwest of the American continent, where it is at the eastern end of the Wheatear's circumpolar range. By all accounts it is distributed here rather erratically from the coastal plains to mountains and barren hilltops above the tree line, which lies at approximately 750 m on the Brooks Range. It frequents screes and heaps of broken rocks, particularly where rock fields are juxtaposed with tundra. Annual numbers fluctuate markedly (Gabrielson and Lincoln 1959; Kessel and Gibson 1978).

Switching our attention 1,000 miles (1,600 km) east across the Canadian tundra to northeastern Canada, we find the western limit of the range of the Northern Wheatear. At this point it is represented by the Greenland race *leucorrhoa*. Its range extends as far west as 92°W on the west side of Hudson Bay. In this part of Canada, it inhabits the same sort of stony tundra and arctic wastelands as it does in other parts of the far north (Godfrey 1986).

Wheatears in Greenland have a distinctive habitat, preferring the drier and warmer interior of Greenland to the cool coastal belt. They frequent hillsides with scree slopes, moraines, ravines, dry river beds, low rolling heathland with scattered boulders or large rocks, and even dense shrub growth with stony clearings, sometimes perching on the twigs of birch and willow (Salomonsen 1950).

SUMMARY

Because of its hopping gait the Northern Wheatear has a restricted habitat type, choosing open country with sparse vegetation over which it can hop easily and which provides a fairly wide vista. The short vegetation of Skokholm results from the effects of heavy rabbit pressure in a vegetation already modified by climate, the sea and the rock structure of the island. The Wheatear on Skokholm is dependent directly and indirectly on rabbits for nest burrows. Some of the burrows are rabbit nurseries; other burrows have been used and modified by Manx Shearwaters or Puffins. A map of the distribution of nest sites shows that 56% of the nests were situated within 100 m of the cliff face in areas where there was the greatest number of burrows. It seems, too, that the slope of the ground as well as the rocks and rock outcrops are important in distributing Wheatear nests by providing more congenial microclimates. Further surveys in Pembrokeshire and other lowland parts of Britain again showed the dependence of the Wheatear chiefly on rabbits in the coastal zone, or on sheep and cattle in the uplands of Scotland, Wales and northern England.

In other parts of its range, the Wheatear's primary habitat is in the arctic-alpine zone in Europe or at high altitudes or high latitudes. Only comparatively recently has it begun to inhabit the lowland zones of temperate areas after the arrival of sheep, rabbits and other close-grazing animals. I have suggested that one function of the Wheatear's plumage colour is procryptic, affording it some protection against predators, particularly in its primary habitat.

Chapter 3

FROM AFRICA TO ALASKA

At this point I shall leave the story of the Wheatears on Skokholm and trace whence they have come to take up their breeding territories in Britain and elsewhere within their breeding range. It seems extraordinary that they should have travelled so many thousands of miles from a warm land just to spend five months or less here to raise a family and, when they have done that, face the same hazards again to return to their winter habitat.

The Wheatear is one of the world's smallest long-distance migrants: it winters in Africa south of the Sahara in an area of about 8 million km^2 (Moreau 1972), and summers from a point just west of Hudson Bay, northeast Canada, eastwards around the globe through Alaska and into northwest Canada — which Moreau reckons comprises a breeding area of about 25 million km^2. Only about 1,000 miles (1,600 km) of the Canadian Arctic separate the two edges of a breeding range that would otherwise be completely circumpolar (Figure 3.1). Wheatears of the Greenland race that migrate from West Africa to Greenland and the Canadian Arctic fly at least 7,000 miles (11,265 km), crossing a vast desert and an ocean, but, in spring, probably even further than that when they follow a more overland route. The Wheatears that migrate from East Africa to Alaska travel about 13,000 miles (21,000 km) each way (Moreau 1972). In doing this, they probably follow something like the ancient route by which Wheatears, which originated as a species in the Mediterranean region, gradually expanded their range and fanned northwards from Africa (Stejneger 1901).

These are the bare facts and now I shall try to piece together the whole story. Wheatears winter chiefly in the semi-arid belt from west Mauritania, the Gambia, Senegal (including the Sahel zone), south of Mali, Niger and Chad, to Sudan in the north; the southern limits of the wintering range are in Nigeria, northern Cameroon, north of the Central African Republic, Sudan, southern Ethiopia, Kenya, Uganda and south to Zimbabwe (Tye, in press). Apparently they may also winter in the Tigris and Euphrates Valleys of Iraq (Cramp 1988). In southern Africa they are occasional vagrants, penetrating further south in some years than in others (Borrett and Jackson 1970). Wheatears, probably of the Greenland race, appear irregularly in various parts of

35

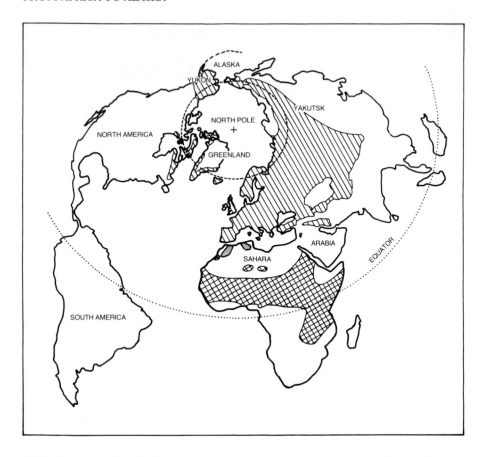

FIGURE 3.1 *Breeding distribution (hatched) and wintering areas (cross-hatched) of the Northern Wheatear; stippled areas show breeding distribution of Seebohm's Wheatear.*

America and the West Indies, but there is no suggestion yet that a permanent wintering area has been established. The world population of the Northern Wheatear, which winters almost entirely within the African savanna, was estimated by the late R. E. Moreau as being in the neighbourhood of 125 million individuals.

In winter quarters, Wheatears, like other Palearctic migrants, tend to inhabit those areas whose structure has a broad resemblance to their summer habitats; they winter either in arid country or, if in more humid places, in overgrazed areas or where the soil is blackened by bush fires. In West Africa, where the vegetation forms distinct latitudinal belts south of the Sahara, they avoid the northern Sahel belt just south of the desert edge where the Desert or the Isabelline Wheatears are more at home (Moreau 1972). When the migrants arrive from the north in the autumn, the vegetation is already beginning to dry up and grain crops are being gathered. Thereafter conditions steadily deteriorate: the ground gets barer and most trees lose their leaves. In these circumstances, Moreau comments, the carrying capacity for birds of this semi-arid belt is amazing.

In Senegal, Wheatears inhabit savanna with bushes and some trees, where they perch on low branches and termite mounds. They like burnt areas and recently harvested fields equally, and they perch on some commanding mound or a bundle of dried guinea-corn stalks. Both sexes establish individual territories and manifest great intolerance towards other species of wheatear (Morel and Roux 1966; Tye 1984 and in press). These open habitats with short vegetation, bare ground and scattered bushes and trees are similar to those in which Wheatears are often found in the British Isles.

In the north crater highlands of northern Tanzania, our Wheatears were found chiefly in the treeless and closely grown grassland areas between 2,255 m (7,400 ft) and 2,560 m (8,400 ft) in the Rift Valley, although they may occasionally descend to lower levels (Elliott and Fuggles-Couchman 1948).

In November 1970, my wife and I spent a month looking for Wheatears in Kenya from the coast up to about 8,000 ft (2,500 m) above sea level. I recorded them in any numbers only between 9 and 16 November when, quite clearly, many migrants from eastern Europe were passing through the area around Lake Naivasha: Northern, Pied and Isabelline Wheatears, Richard's Pipits and Ashy-headed Wagtails were all seen frequently. The habitat which Wheatears particularly favoured was the newly-sown fields with a light soil. They were also seen on grassy plains which had patches of bare ground between grass tussocks, and post and wire fences provided the perches that all the chats used freely. They also congregated where Africans were burning the bush to create new fields. Where they occurred in the thorn country of Samburu, it was the more open patches that attracted them.

BEHAVIOUR IN WINTER QUARTERS

Wheatears are generally considered to be silent, or nearly so, in their winter quarters, although M. F. M. Meiklejohn (in Witherby et al. 1938) heard one singing but did not say whether the song was the loud territorial song usually performed by a mated male in its breeding territory or a quiet subsong which is used by a Wheatear holding an individual territory; I would have suspected the latter. Borrett and Jackson (1970) recorded a Wheatear singing the quiet subsong in South Africa. Tye (in press) has described the vocalisations and territorial behaviour of the Wheatear and the Isabelline Wheatear in winter quarters. He first established that individual Wheatears of both sexes defended territories for long periods, mostly continuously from November to January, from which they attempted to exclude members of their own and other species; some individuals spent up to 10% of their time singing. Tye concludes that the context of winter song together with a comparison with the behaviour in the breeding areas suggest that the main function of song is that of territorial defence. I shall discuss this further in Chapter 5.

LEAVING AFRICA

At the end of January the first Wheatears begin their passage northwards: a journey that is to take some of them more than half way around the world, across deserts and oceans to lands where snow may still be lying. They travel from an area where the average temperature in winter is about 23°C to breed in a land where the summer mean temperature may be as low as 3°C, although the upper limit in another part of their breeding range may be as high as 32°C (Moreau 1972). By March the majority of Wheatears are on the move, and by the end of April practically all have gone.

The most up-to-date account of the spring departure of Wheatears for their breeding grounds is given by Tye in volume 4 of *The Birds of Africa* (Keith *et al.* in press). He shows that Wheatears begin leaving the western Sahara at the end of January. The main passage through Morocco starts at the end of February, reaches a peak from the end of March to early May, and is finished by late May. This is, as one would expect from a broad-front movement which passes quickly across the Sahara and ultimately crosses the Mediterranean, similar to the pattern we see in Algeria and Tunisia and indeed, with small modifications, throughout northern Africa. Tye's records also show a tendency for later departures from the east side of Africa: Mauritania, Gambia, Senegal, etc. from January and February, compared with March to mid April departures from Rwanda, Uganda, Kenya and Tanzania.

The first hazard Wheatears have to face in their flight north is the Sahara desert, which is virtually waterless except for widely scattered oases for about 1,500 km (1,200 miles) across on a direct south-to-north route. Moreau judged that this journey could take as much as 40 hours non-stop if the bird was travelling from south to north and there was no wind. In spring, however, there is a greater chance of head winds, and it is then possible that the flight — at altitudes of over 2,000 m — would take some 50-60 hours. Moreau points out that because of the cooler air at such heights the bird's loss of water would be much reduced. On the other hand, Wheatears found dead by Haas and Beck (1979) had exhausted fat supplies, not water. (Tye: in Keith *et al.* in press.)

In 1968, with three friends, I had the opportunity to stay from 31 March to 5 April in the Kufra Oasis in the Libyan desert 550 miles (885 km) south of the Gulf of Syrte in Cyrenaica and about the same distance north of Lake Chad (Cramp and Conder 1970). The oasis is an extensive wadi about 7 km across and about 21 km long, bounded by a steep escarpment of sandstone hills rising to 30-60 m in the north. The hills are broken in the south by small wadis, jebels covered with broken stones. The oasis encompasses several villages in which by far the commonest trees were date palms. Planted in the well-irrigated gardens were banana plants, grape vines, apple and various citrus trees beneath which barley, cotton and vegetables of different kinds were often grown. The four saline lagoons and one which had relatively fresh water were the focus of the oasis.

FIGURE 3.2 *Wheatear sheltering from sun in Kufra Oasis, Africa*

Four species of wheatear were recorded. The commonest was the Northern Wheatear. The others were the Black-eared Wheatear, which had been recorded from Kufra before, and Isabelline Wheatear and Mourning Wheatear, which were new.

In the oasis I was surprised to see Wheatears standing in the shade of the palms. In more temperate climates they shun cover, but here the heat was so intense that few living things would survive for many hours if exposed directly to this sun and to radiation from the sand. Indeed, at Kufra Wheatears were often as much garden birds as are Robins in England: in the small, shady and irrigated gardens they foraged with Hoopoes, Nightingales, Whinchats and Red-spotted Bluethroats, as well as with many species of warbler which were also sheltering here from the heat of the sun on their way north. Jackson (1938) mentions that, in Kenya, Wheatears may be seen with open beak and palpitating throat while standing in the shelter of a bush. In the Gambia, Andrews (1969) saw Wheatears behaving in the same way when temperatures were 40°C, and they do it in Senegal, too (Tye, pers. comm.).

It is now speculated (Bairlein 1988) that some migrants, such as Wheatears, actually come down to forage and rest at oases like Kufra as a part of their regular migratory strategy. Certainly, warblers which are caught at such sites and which are weighed on arrival and again three or four days later have been able to put on enough weight for further movement.

Even so, Wheatears, more than any other passerine, foraged on the salt-encrusted mudflats of Kufra, where they became the too-easy prey of small Arab boys who were very skilful at catching them: before the boys had walked more than 200 m from their home-made trap baited with a mole-cricket, the trap had been sprung by a Wheatear and, as quick as a flash, that bird's neck was wrung!

Once or twice we drove out into the desert and found Wheatears on the sides of the rocky jebels making the best of the shade afforded by the stones and boulders from a sun which was almost directly overhead. Birds which for one reason or another alighted in the desert far from shade were in a particularly hazardous situation: throughout the day in the open desert there would be no food and certainly no moisture; no insects or other invertebrates venture out in that sun. Death was almost as inevitable as it would be from dropping into the Atlantic, which Greenland wheatears had to cross a little later.

The earliest Wheatears reach Egypt and the northern coasts of Africa at the beginning of March, and sometimes with scarcely a pause, start crossing the Mediterranean. Bundy (1976), reviewing all records of the birds of Libya, gives the last half of March as the peak period for migrant Wheatears. J.K. Stanford (1953), who watched spring migration in Cyrenaica between 14 March and 20 May, came to the conclusion that a large mass of migratory birds reached the coast on a front of 170 miles (274 km) between Benghazi and the Gulf of Bomba. He found the greatest number visible between 21 March and 10 April, which is about a week before the main passage of Wheatears, which presumably wintered in western Africa, passes through Skokholm. Strong northwest winds across the Mediterranean could hold up migration long enough for quite large numbers of Wheatears to have accumulated along the North African shoreline.

There is less detailed evidence about their passage across the western half of the Sahara desert or about their arrival on the coasts of Morocco, Algeria or Tunisia, although more ringed birds are being recovered there. Previously, the ringing evidence (Hempel 1957; Zink 1973) indicated that in autumn at least Wheatears that had bred in western Europe travelled south or south-south-west across the desert; but more birds have now been recovered, and so far these recoveries indicate that in spring British-ringed Wheatears pass through both Morocco and Algeria (Mead and Hudson 1984), whereas none has yet been recorded in Algeria in autumn. Overall, it would appear that most European and Asiatic Wheatears move west in autumn in varying degrees in order to reach their African winter quarters, and there is no evidence to show that they do not, generally, follow the reverse route in spring.

To the North and East

At this point in their journey northwards Wheatears are beginning to fan out, with the two flanks following the great circle routes to the west and the east (Mead 1983). The western flank of this movement, after crossing the Sahara, flies up the western seaboard of Europe to Scotland and from there (Snow 1953) finally crosses the North Atlantic to Greenland and into northeast Canada as far west as Hudson Bay. A central portion flies northwards, and the eastern flank, having started northwards, changes direction to the east, following the great circle route — thus avoiding the Himalayas — and heads through Russia towards Alaska and northwest Canada.

In central and eastern Europe, the published dates of the first arrivals of migrant Wheatears are later than in the west. On the northern plains of Switzerland the first passage of Wheatears occurs between 20 and 31 March, but the most abundant passage occurs between 4 April and 5 May (Glutz von Blotzheim 1964), which, on average, is four or five days later than on Skokholm. The alpine habitats are occupied at the end of April and the beginning of May.

In the Kurskiy Zaliv (Kurische Haff) area of the Baltic Sea, arrivals recorded by Tischler (1941) are also slightly late compared with arrival dates on Skokholm, which is a little further south: Tischler gives 5-12 April as the peak period. At Ottenby, Sweden, during 1958-62, the dates of the first arrivals varied between 30 March and 12 April.

It is interesting to compare these figures with some given by Pleske (1928) as arrival dates in Russian Lapland, about 1,600 km (1,000 miles) to the northeast. He reports that Wheatears were arriving in fairly large numbers on 31 May 1879, 30 May 1883, 31 May 1884, 24 May 1900 and 15 May 1901. Thus there seems to be an interval of one to one-and-a-half months between the earliest arrivals in the southwest of Russia and those in the northwest. In the Soviet Union, Dementiev and Gladkov (1968) state that in the south Wheatears arrive chiefly in April, although there are rare March occurrences, and that in the northern tundra they occasionally arrive in late May but mostly in early June. It appears that the rate at which Wheatears colonise the northern parts of their breeding range each spring tends to slow down as they head into country where the weather may be dangerously inclement.

Before returning to the better-documented passage of Wheatears up the western seaboard of Europe and their arrival in the British Isles, it would be worth trying to trace what happens to those which travel right across Asia on their way to breed in Alaska. As one might expect, the movement from Africa to the northeast is not well documented. There are records from the Red Sea of Wheatears either coasting up the western shores or actually crossing it. Marchant (1941) watched migration at Hughada in the Gulf of Suez: Wheatears began passing on 28 February and the movement reached its peak on 15 April, when 50 or more birds were seen, with another peak on 9-10 May. In Jordan, between 14 March and 25 April, Wheatears were spread over the whole country from the coast to the Jordan Valley, which is an extension of the Rift Valley of East Africa where many of these Wheatears will have wintered (Meinertzhagen 1920).

Moreau (1972) has traced the passage to Alaska and the Yukon of northern Palearctic migrants which have wintered south of the Sahara. They have a complicated journey even after they have crossed the Sahara and face the crossing of some very arid, if not desert, areas such as are found in Arabia, Anatolia, Iran and Transcaspia. During most of the year these are inhospitable areas, although in spring parts receive a rather scanty rainfall which encourages a temporary growth of vegetation, probably resulting in a flush of insects.

To the north of Arabia is the Black Sea, which is about 1,100 km wide, and its greatest north-to-south measurement is about 650 km. It is clear from the limited reports from this area that migrants which follow this route northwards do not shirk the direct crossing.

In Kuwait, Wheatears are abundant passage migrants, present until April. They establish individual territories on sparsely vegetated or stony deserts, often mixed with similarly territorial Isabelline or other wheatears (Pilcher and Tye, in prep.).

To the northeast of the Black Sea, Moreau reminds us that two formidable complications lie in wait. First comes the Caspian Sea, which is 1,300 km from north to south and at least 300 km from west to east, and since some migrants may well cross it on the diagonal they will fly over water a distance much greater than 300 km. Immediately following that, Wheatears have to cross a barren and waterless area to the east of the Mangyshlak Peninsula. Gladkov (1957) reports that, while some Wheatears, which are quite numerous migrants through this part of the world, follow the east coast of the Caspian Sea, others cross the barren areas but their success in so doing depends upon the weather, and many birds perish in the desert crossing. Incidentally, the total distance that Wheatears have to cross from the north coast of Africa and Sinai to the north end of the Caspian or Black Seas is about 1,400 km, and this immediately after the Libyan desert crossing.

It would seem that the pivot and southern edge of the great circle route is the Himalayas: Ali and Ripley (1973) record the Wheatear as a straggling passage migrant through Baluchistan, North West Frontier Province (Chitral and Gilgit) and Kashmir, occasionally in March but chiefly in April. In fact, at this point Wheatears which breed well to the east in Russia or in Alaska have to change direction, almost pivoting on the Himalayas, and head more to the east-northeast (Baker 1978).

Wheatears arrive in their breeding quarters in northern Mongolia on about 12 April (Kozlova 1933). I have no dates for their arrival in northern China, where they nest in the Chinese Tien-shan range, Dzungaria, Outer Mongolia (except the Gobi desert), and northwest Manchuria to Hsingan. The species' southern limit, both for breeding and as a migrant, is approximately the 40°N parallel (Etchécopar and Hüe 1983).

From the Caspian, Wheatears have had another 800 km of barren country to cross before reaching the southern edge of the open steppe, hostile to many species but not so much to the Wheatear. Those breeding in Alaska have still further to fly and another sea crossing to make. Monk (in Moreau 1972) gave some approximate distances between wintering grounds and the breeding grounds in the northeast: he calculated that from Nairobi to Yakutsk (130°E, 62°N) is about 10,000 km, following the great circle route, and that Alaska is a further 3,000 km. When the Wheatears arrive there from mid May to early June (at which time, incidentally, the first broods of Skokholm Wheatears are just leaving their nest burrows), they have about three months in which to raise five or six young before they have to return the same way. Since they have been doing this for generations, I am sure that this enormous expenditure of energy in travelling some 13,000 km each way in order to raise five young does balance out so that this population of the species is able to survive and prosper.

ON TO GREENLAND

Let us now see how Wheatears on the western flank of their spring movement out of Africa arrive in or pass through the British Isles on their way to breeding grounds in Iceland, Greenland, Labrador and the northeast Canadian Arctic.

Several co-operative efforts have been made to study this movement through Britain and Ireland. One enquiry was organised by a committee of the British Ornithologists' Club between 1905 and 1913. It had the assistance of lighthouse-keepers and ornithologists in various parts of the country who recorded daily whether the numbers of birds increased or decreased. Unfortunately a complete summary of these reports was never published, although Eagle Clark made extensive use of them in his book *Studies in Bird Migration* (1912). Another co-operative attempt was made by the British bird observatories to study the movements of selected species — of which the Wheatear was one — through Britain between the years 1950 and 1952 (Conder 1951, 1952, 1953b). More recently, the Young Ornithologists' Club of the RSPB has been enlisting the help of its members to plot the arrival of spring migrants. The interpretations of the results of these surveys differ slightly, partly because the later surveys were made with a better background knowledge of bird migration and partly because improved methods of weather-recording have given a better understanding of the weather patterns that affect migration.

One of the first conclusions drawn by the BOC committee was that the earliest Wheatears tended to arrive on the western half of the south coast. This was substantiated by later work by the bird observatories. There was also an indication of an easterly movement through the country, with birds flying out over the North Sea from the Norfolk coast. Migrant Wheatears arrived on the inland Breckland heaths earlier than on the north Norfolk coast (Seago 1967), which again suggests a northeasterly movement across the southern part of Britain. The BOC was also sent a very few records of Wheatears arriving on the east coast from the east in spring and flying west inland, but these would

FIGURE 3.3 *Great Grey Shrike chasing Wheatear in Senegal*

seem to be birds re-orienting themselves after they had, perhaps, drifted too far east. I return to the subject of drift later in this chapter.

Of course, the information about the directions in which Wheatears were migrating is necessarily rather slight, since Wheatears usually — but not always — migrate at night. Nevertheless, the conclusions of the BOC have been broadly corroborated both by the bird observatory studies between 1949 and 1953 and by later work.

Although Kenneth Williamson's work on drift migration (1952) would have affected the BOC committee's interpretation of the facts it had at its disposal, the arrival dates it collected are still of interest. Between 1905 and 1913, migration was recorded regularly between 12 March and 16 May. On either side of these dates there were occasional very small movements; indeed, there are isolated records of Wheatears seen in every month of the year. The committee also asked for the wings of Wheatears which had killed themselves against the glazing of the lighthouses. Although few wings were received in some years, the committee reported that no Wheatears of the typical race were recorded after the end of April and that, although individuals of the Greenland race were recorded as early as 1 April, not until towards the end of the second week of April in most years did the main movement of the latter race begin and this reached a peak between 2 and 6 May in the southwestern part of the British Isles. This conclusion was again corroborated by work undertaken on Skokholm.

The basis of Williamson's drift theory is that migrants travelling either north in spring or south in autumn leave on a stage of their migratory flight in anticyclonic conditions, when barometric pressure is high, skies usually clear and winds light or moderate. During the flight the birds may fly into cyclonic conditions, overcast skies and strong winds. Losing sight of the stars, they become dis-oriented and are carried away, or drifted, downwind. In spring, ideal conditions for drift exist when pressure is high over Scandinavia and low over the Atlantic. Now the wind circulates around an anticyclone in a clockwise direction (and anti-clockwise around a cyclone or low), so that under ideal anticyclonic conditions there is often a tendency for a south or southeast airstream, sometimes of good strength, over the British Isles; and it is then that many migrants, including some rare visitors, arrive in or pass through the British Isles. So drift may help returning migrants, but be unhelpful to 'rare' visitors because they should be somewhere else.

Figure 3.4 shows that, in 1952, Lundy recorded an increase of Wheatears on 3 April; Skokholm and Saltee (off the southeast corner of Ireland) recorded an increase on 4 April; Cley (Norfolk), Spurn Point (Yorkshire) and Monk's House (Northumberland) on 5 April; the Isle of May (Firth of Forth) on 6 April; and Fair Isle recorded its first Wheatears of the season on 7 April. It would seem that this 'wave' travelled over 600 miles (965 km) in five days through the British Isles.

The *Daily Weather Reports* published by the Meteorological Office showed that during the previous day or two fairly strong northerly winds had swept over most of western Europe. On 2 April, the winds were beginning to moderate and over western France they were mostly light. On 3 April, conditions for the departure of migrants were appearing to the south of Britain, particularly in Spain and western France: high pressure, light winds and, late on 3 April, rising temperatures. If these Wheatears had left western France and northern Spain, they would have encountered, late on 3 and early on 4 April, strong southwesterly winds circulating around a depression off Iceland with associated occluded and warm fronts moving towards the British Isles from the Atlantic, which probably drifted them to the southwestern bird observatories. Southeast England was comparatively unaffected, but these winds reached Yorkshire on 5 and 6 April and peaks of Wheatear migration were recorded at Spurn Point, Monk's House and the Isle of May. A second depression was moving towards Fair Isle, and further strong southerly winds may have carried this same wave of Wheatears to Fair Isle on 7 April.

The next great wave of Wheatears moved across Britain and Ireland between 9 and 11 April. Increases were recorded on Lundy, Skokholm, Saltee and Cley on 9 April; Gibraltar Point (Lincolnshire), also received some Wheatears that day; and Monk's House reported 'many' on 9 but 'very many' on 10 April, when the Isle of May also recorded an increase. Fair Isle recorded the first birds of a very big influx on 11 April.

On 9 April conditions were ideal for drift: over Scandinavia pressure was high, and a depression was situated about 500 miles (800 km) west of Scotland with an associated trough of low pressure extending far south. This combination produced moderate to fresh southerly winds from northern France to the Faroes. Over east Biscay winds were light and from the south. To lend support to the idea that this wave of Wheatears was drifted up from the south, a Short-

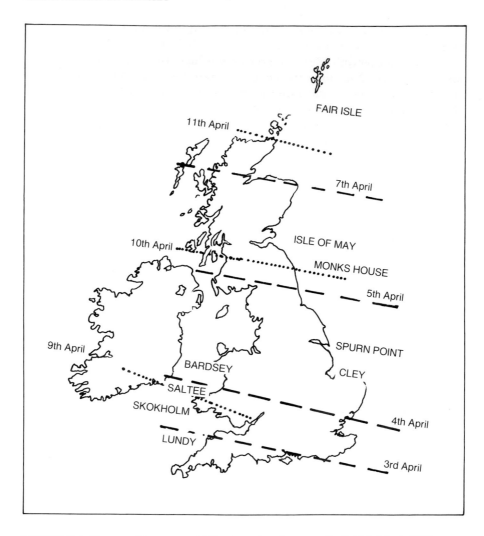

FIGURE 3.4 *Progress of two waves of Wheatear migration across the British Isles in 1952*

toed Lark was recorded on Skokholm (a first record) with many other migrants.

Drift may assist birds, but equally it may also take migrants, such as the Short-toed Lark, a long way off course, and it is certain too that, drift or not, Wheatears reach their breeding grounds in much the same way year after year. In the first place, it became obvious from a study of Skokholm's daily records that few of that island's breeding Wheatears, which were often colour-ringed, arrived on a peak day of migration, and I am now inclined to think that when this did happen it was often a coincidence. In spring 1949, with persistent northwesterly winds, there were few signs of migration and no major peaks; yet, in spite of the apparent lack of migration, the local population built up slowly and steadily and, in the end, it was larger in 1949 than in 1948.

The numbers of Wheatears recorded at the observatories on the southern part of the east coast such as at Cley and Gibraltar Point tended to be very low compared with those at the northeast observatories such as Spurn Point and Monk's House.

The BOC distinguished some movements of Wheatears in an easterly direction, but I have been unable to trace many observations to suggest that a significant proportion of migrant Wheatears cross central Europe and enter Britain from the east. The only records of which I am aware are from light-ships off the East Anglian coast (Rivière 1933, 1934, 1935). Seago (1967) writes that there are few recent records of birds moving into Norfolk from the east, so what Rivière and others recorded might well have been Wheatears which had been drifted off course and which were re-orienting themselves.

It is now time to retrace our steps in order to follow the passage of the Greenland Wheatear *leucorrhoa*. As might be expected, their winter quarters are restricted to the extreme west of Africa, mainly in the Gambia and Senegal but straggling down the coast to Sierra Leone. In these areas it mixes with the typical Wheatear race *oenanthe* and with the Atlas Mountains race, Seebohm's Wheatear *seebohmi*, which winters mainly from south Morocco to Mauritania.

The Birds of Africa (Keith *et al.*, in press) shows that Greenland Wheatears leave Senegal in late January, all having departed by early May, and that they leave the Gambia between February and March. In Britain, the main movement first appears on Skokholm from mid April and birds continue moving through the island until mid May. I was becoming much more involved in finding the nests of the local Wheatears, whose peak of laying also tended to be in the first week of May, and thus had less time to record migration, but my records suggest that the peak dates of Greenland Wheatear migration on Skokholm were 18-25 April and the first week of May. In 1957, Williamson on Fair Isle recorded the first waves of Greenland Wheatears on 4 and 5 May, so it seems possible that some Greenland Wheatears arriving earliest in the southwest, including on Skokholm, did not reach Fair Isle at all but took off northwest from Scotland across the Atlantic to Greenland at a point further south than Fair Isle.

Northwest Scotland is the point on their journey to Iceland, Greenland, Labrador and the northeast Canadian Arctic where, once again, they are faced with a long sea crossing. According to Snow (1953), Greenland Wheatears travel almost to the north of Scotland (approximately on the same latitude as south Greenland) before making the crossing, quite probably, first to Iceland and then on to Greenland. The majority of Greenland Wheatears arrive on their breeding grounds in northwest Greenland in the second half of May.

The Gulf Stream, sweeping up from the Gulf of Mexico, gives Scotland a far higher temperature at this time of year than a place with a similar latitude on the west side of the Atlantic. If Wheatears were about to take off from

Africa or Spain, they might leave in temperatures of 10°C to 20°C and arrive in Greenland where temperatures were only −5°C to 0°C. After a long sea crossing the effects, both direct and indirect, of these low temperatures might well prove disastrous. It is improbable that Wheatears would be in any condition to retrace their steps to avoid extreme cold or storms that they might run into. Bird and Bird (1941) say that great numbers perished in 1937 in northeast Greenland when Wheatears arrived as early as 5 May and died among the snow and ice of a late-winter storm. Freuchen and Salomonsen (1960) also tell of Wheatears arriving early and, caught out by bad weather, hunting around houses for scraps.

Interesting confirmation of the dates of passage and of routes is provided by Snow (1953), who shows that Greenland Wheatears were recorded at sea by weather ships on the following dates: 6 to 15 April, two; 16 April to 5 May, none; 6 to 15 May, nine; 16 to 25 May, eight; 26 May to 4 June, two. These figures show quite clearly that the major movement of Greenland Wheatears at sea was during May, and it is also clear from Snow's maps that most had travelled to north Scotland before crossing to Iceland and Greenland.

Williamson (1958), who studied migration of Wheatears at Fair Isle, commented on the races of Wheatears in relation to Bergmann's rule, which states that warm-blooded animals tend to be larger in colder climates than their relatives in warmer climates. Williamson was aiming to show that natural selection operates most powerfully during the actual period of migration, since survival favours those individuals with the greatest reserves of strength. He suggested that in the case of migratory birds nesting in Greenland or Iceland, which have to make sea crossings of 1,500-2,000 miles (2,400-3,200 km) — and longer on their return south — selection has been in favour of longer wings and greater body weight compared with the more southerly continental wheatears which do not have the long overseas flight. It should not be forgotten, however, that the Wheatear flying from Africa to Alaska covers almost double the distance of the Greenland Wheatear and much of the journey is over very inhospitable desert.

Chapter 4

SPRING ARRIVALS

One we had repaired the worst damage caused by winter gales and repainted some of the rooms, I was able to spend more time looking at Wheatears. In each year from 1949 to 1952, from 20 March until mid April, I mapped the positions of each individual Wheatear daily on a 6-inch base map. By 15 April a large proportion of the local breeding pairs had established themselves in their territories, so I turned my attention to them because some females were already searching for a suitable nest burrow.

Although I had usually seen the first Wheatears on the mainland on about 11 March, I landed on Skokholm before the main movement began. During 1948-52, the peak dates of each spring migration were 25 March 1948, 25 March 1949, 28 March 1950, 30 March 1951 and 27 March 1952, of which the mean date is 27 March. Later, the Young Ornithologists' Club of the RSPB organised a phone-in on migration in 1978 and 1979. In 1978 they received 207 reports and in 1979 266 reports of first sightings of Wheatears from various parts of the British Isles. The pattern of migration revealed by these reports was similar to that recorded in earlier years. In 1978, the first Wheatears were recorded on 25 February in Devon and on 26 February in Kent; numbers built up to a major peak between 28 March and 3 April. In 1979, the first sighting was rather later, on 4 March in Gwynedd, with the bulk of the birds being recorded later too, in the first half of April. In both years the numbers between 25 April and 1 May began to increase again, and from my experience on Skokholm and later on Alderney I would assume that these were Wheat-ears of the Greenland race.

The pattern of arrivals and departures of passage migrants on Skokholm varied from year to year; sometimes this variation was caused by local weather conditions. Figure 4.1 shows the totals of Wheatears present on the island and the number of established pairs. In 1949, for example, the weather had been generally favourable for migration and the highest numbers were recorded on 29 March, after six days of light easterly or southeasterly winds. After this, although in the same quarter, the wind strengthened and the number of migrants declined. Eventually, after the wind had increased to gale force from the northwest, only 13 Wheatears remained on the island on 8 April, of which

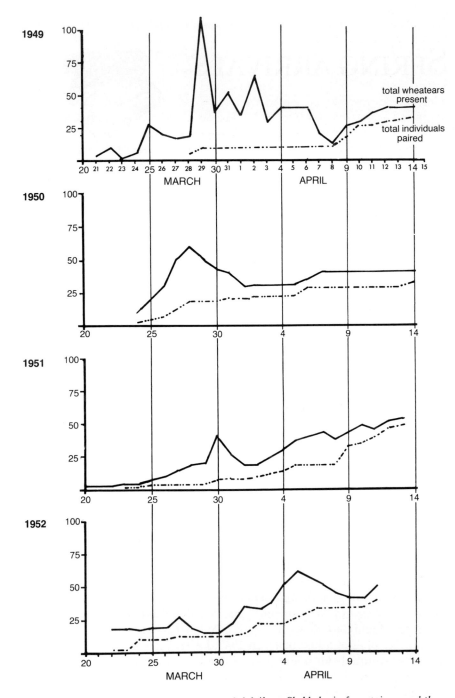

FIGURE 4.1 *Total numbers of Wheatears recorded daily on Skokholm in four springs, and the number of individuals already paired; unbroken line represents the total and broken line the number of individuals already paired.*

12 were known to be local breeders. Light winds followed the gales and passage resumed.

The same period in 1951 provided an interesting contrast: winds blew largely from the west and northwest (to some extent these were contrary winds) and at strengths 4-7 on the Beaufort scale (13-38 mph/21-61 kph), which were normally strong enough to restrict bird activity of one kind or another. Under these conditions the number of passage migrants was low and the build-up of residents on the island was slower than usual; in fact, throughout the first ten days of April the daily total of passage birds never exceeded ten, which was about half the average numbers. Indeed, as time passed I began to wonder if the breeding population might not be reduced that year. In spite of low numbers on passage, however, the local population continued to build up and, ultimately, it was slightly larger than it had been in the previous three years. This indicates the strength of the Wheatears' drive to return to the place of their birth, or where they had nested before, despite adverse weather conditions which had deflected passage migrants.

Figure 4.2 summarises the daily counts of spring migrants between 1949 and 1952. The broken line shows that the average number of breeders already paired (expressed as individuals) increased gradually and unspectacularly. The unbroken line shows the average number of passage migrants on the island, which reached their main peak at the end of March. Thereafter their numbers decline, except for another marked peak on 5 April.

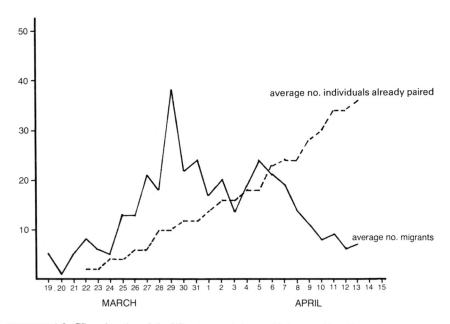

FIGURE 4.2 *The migration of the Wheatear and the establishment of breeding pairs on Skokholm, 1949-52; unbroken line shows average number of migrants and broken line average number of individuals already paired.*

GREENLAND WHEATEARS

In mid April I usually stopped counting migrant Wheatears and turned my attention to the breeding behaviour of residents; this was a pity, since the Greenland Wheatears were just starting to pass through Skokholm on their way up the western seaboard of Europe to breed in Iceland, Greenland and the northeastern parts of the Canadian Arctic.

Because of the difficulty of separating some individuals of the two races with certainty, information on their movements through Skokholm remains incomplete. Between 1948 and 1951, the first dates on which I recorded Greenland Wheatears were 21 April 1948, 19 April 1949, 18 April 1950 and 19 April 1951. Some probably passed through before that, and the late T. A. W. Davies recorded three on the Pembrokeshire mainland on 6 April. The majority passed between 2 and 12 May and the latest was seen on 21 May. Numbers were never as high as those of the typical race; about 25 was the maximum.

Greenland Wheatears behaved in much the same way as migrants of the typical race. They set up individual territories which they defended against other migrants or against attacks by local breeders within whose breeding territory they occasionally established themselves. I discuss the establishment and maintenance of these individual migration territories later in this chapter.

SEX RATIO

Female Wheatears tended to arrive on Skokholm later than males. Figure 4.3 records the average numbers of migrant males and females on the island between 1949 and 1951, as well as the average number of pairs which had already established themselves (this time expressed as pairs). In the last ten days of March and the first 13 days of April males were nearly always more numerous than females and the latter equalled males only on three days in mid April (two in 1949 and one in 1950), but there was some indication that, towards the end of April, numbers of females equalled and were sometimes greater than those of males. By this time, however, Greenland Wheatears had begun to arrive, which confused the situation. Alan Tye tells me that, in Breckland, females also arrive later than males and that in Kuwait, where Northern Wheatears are only passage migrants, males pass through first.

There was also a marked tendency for older females (two years of age or more) to arrive in their breeding territories earlier than year-old females making their first return.

TIME OF ARRIVAL

On Skokholm and Alderney, in both spring and autumn, Wheatears usually arrived overnight. In both places, however, their numbers sometimes increased during the day and, on Alderney, I sometimes saw Wheatears at a height of about 20 m migrating, with a slightly undulating flight, across the

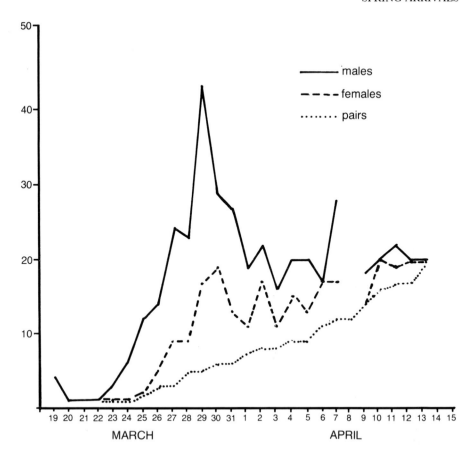

FIGURE 4.3 *Average number of migrant males and females on Skokholm, 1949-52; solid line shows average number of males, broken line average number of females, and dots the average number of local pairs already established.*

island or coasting along the shore. On 25 March 1949, for instance, a big fall of Wheatears occurred on Skokholm after midday, a fall which also included Black Redstarts and Robins. On 26 March 1948, between 10.00 and 15.00 GMT, Wheatears were migrating across Skokholm and by 16.30 GMT the migrants had almost all left. On 2 April a movement was recorded in the morning, and by about 17.00 GMT the passage had finished and most of the migrants had left. On the morning of 22 April 1983 I had found a few Wheat-ears on Longis Common, Alderney, but by 15.00 GMT numbers had increased and the new arrivals were feeding fast. On 10 May 1986, also on Longis Common, there had been an overnight fall of Greenland Wheatears, but by 15.00 GMT these had already gone, leaving behind only those which had arrived a day or so earlier and which had become localised and were hold-ing temporary individual territories: one group of migrant Wheatears had 'leap-frogged' the earlier arrivals.

53

These examples, sometimes involving relatively large numbers, show that, although some Wheatears actually alight on these islands and forage for a time, they still have a tendency to relocate and they leave the island after a few hours: nevertheless, the biggest arrivals and departures were usually overnight. Only once did I see Wheatears leaving Alderney in autumn, when four flew from the northeast corner of the island eastwards towards the coast of France, which was visible 7 miles (11 km) away. They could, of course, have changed direction and flown south when away from land.

The birds' time of arrival at any stopping place may well depend on the type of terrain or ocean over which they had been flying. It is possible that some Wheatears had just crossed over to Skokholm from the Pembrokeshire mainland, a matter of 2-3 miles (3-5 km), but others might have flown the 75 miles (120 km) or more direct from the Cornish peninsula— or, being such strong fliers, they might even have overflown it: the coast of France is only 120 miles (190 km) away. It is by no means impossible for them to have flown direct from Spain or, apparently, from the north coast of Africa some 1,200-1,300 miles (1,900-2,100 km) further south, for they have to overfly the Sahara, a distance of some 1,500 miles (2,400 km). There is also evidence to suggest that in autumn they fly direct from Greenland to the Iberian peninsula, 3,000 miles (4,800 km) or so over the sea. Clearly, if they were undertaking long journeys starting at dusk, the distance and their speed might dictate an arrival time on Skokholm during the hours of daylight. I shall describe this evidence more fully in Chapter 17.

Lee (1963), using radar, observed migrating birds in the Hebrides in the autumn and showed that Wheatears migrated at an air speed of 20 knots, which is about 23 mph (37 kph). This means that they could complete the journey from Spain to Skokholm in 24 hours, which is only half the time that Moreau (1972) suggested a crossing of the Sahara might take. Quite clearly, when travelling such distances over the sea Wheatears could hardly determine when they were going to stop and, therefore, they might arrive on islands such as Skokholm and Alderney at any time of the day or night.

DISTRIBUTION OF WHEATEARS WHEN RESTING ON MIGRATION

By plotting daily the positions of all migrants on Skokholm, I was able to identify their preferred habitats. Later arrivals were forced to use less favoured areas because local pairs had already established and were aggressively defending their breeding territories. In August and September I was able to show that they chose a different type of habitat.

Comparing Figures 4.4a and 4.4b which show the distribution of pairs that had established before and after 2 April, it can be seen that the earliest-arriving breeders tended to select coastal areas. While some late-comers could find unoccupied coastal territories, or fight for and win one within that area, many were forced inland to less preferable breeding territories; but I shall show later that in autumn these latter areas were preferred for foraging.

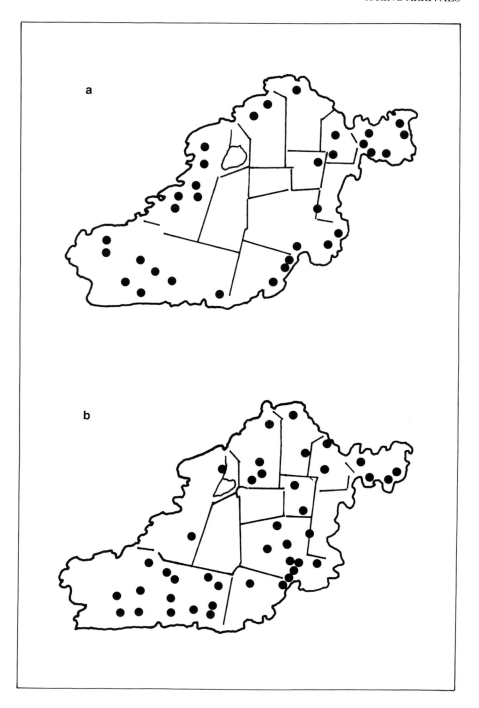

FIGURE 4.4 *Distribution of nest sites according to season: (a) nest sites of pairs established before 2 April 1951 and 1952; (b) nest sites of pairs established after 2 April 1951 and 1952.*

A further indication that the outer edge of the island was preferred was provided by the distribution of passage migrants shown in Figures 4.5a-e, where each spot indicates a mapped position of an individual migrant. Figure 4.5a shows that up to 26 March resting migrants were distributed around the edges of Skokholm. From 27 March to 15 April (Figures 4.5b and c), a period which includes the peak of spring migration, migrants were now resting in more inland areas. By 2 April an average of eight breeding pairs was established, and they and the unmated local Wheatears were still very mobile in their territories, generally forcing migrants inland. On some peak days, however, when 50-100 migrants might be on the island, the 'residents' appeared to be 'swamped' and were only occasionally aggressive to intruders; in these circumstances, a migrant might establish an individual territory for two or three days within the breeding territory of a local pair.

The distribution of Greenland Wheatears was very similar to that seen in the later stages of the migration of Wheatears of the typical race (Figure 4.5d), but I found that they were foraging to a greater extent in damp areas and among longer grasses where they might have been forced by the local population.

There was also a marked difference between the distribution of Wheatears on Skokholm on spring passage and their distribution on autumn passage (Figure 4.5e). The coastal zone, so much preferred in spring, was much less preferred in autumn and the island-born juveniles, the local adults and the migrants tended to gather on the heavily rabbit-grazed sward of common bent and Yorkshire fog, where the Wheatear breeding population had been rather low because of a shortage of rabbit and shearwater burrows: the rabbits that grazed the sward so closely burrowed in the dense bracken alongside.

In spring, therefore, Wheatears, even if they were only on passage and not potential breeders, preferred to rest in the cliff-top zones, where sea spray combined with the tunnelling of shearwaters and rabbits and the voracious appetite of the latter produced a breeding habitat for Wheatears which filled most of their requirements. In autumn, however, no longer dependent on burrows, they preferred a terrain over which it was easy to hop and which provided abundant food.

THE BEHAVIOUR OF MIGRANTS ON SPRING AND AUTUMN MIGRATION

I studied the behaviour of migrant Wheatears in spring and autumn both on Skokholm and Alderney, but chiefly on Alderney because Wheatears did not rest on the main island so I had no problems in trying to differentiate between local breeders and migrants.

Within hours of their arrival Wheatears had found a habitat which suited their method of hunting. When I arrived at my study area on Longis Common, Alderney, perhaps two or three hours after dawn, I often found that there had been a fall of Wheatears overnight and a dozen or more individuals would be foraging on the heavily grazed parts of the 20-ha common. If I then

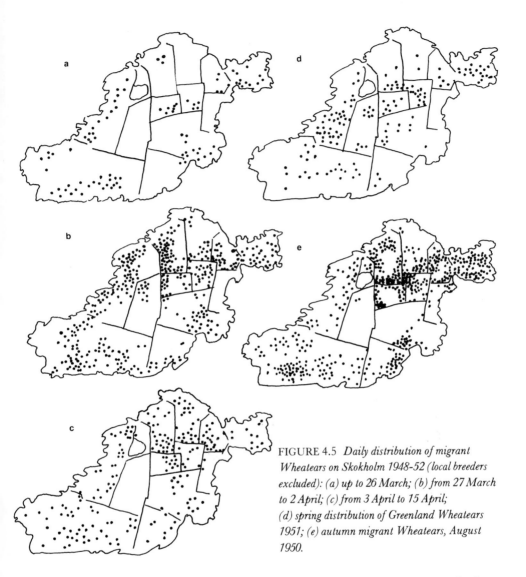

FIGURE 4.5 *Daily distribution of migrant Wheatears on Skokholm 1948-52 (local breeders excluded): (a) up to 26 March; (b) from 27 March to 2 April; (c) from 3 April to 15 April; (d) spring distribution of Greenland Wheatears 1951; (e) autumn migrant Wheatears, August 1950.*

walked across the island I might find other Wheatears where the grass had been grazed by rabbits, along well-trampled paths, where fences made good observation posts and, if the wind was blowing at about 30 mph (48 kph) or more, among the rocks of the foreshore. Only on Longis Common, however, where rabbits were abundant or where tethered ponies or cows, or folded sheep, had grazed the palatable plants hard, thus creating a lawn-like sward, did Wheatears gather consistently.

Since Wheatears migrate as individuals rather than in flocks (although many might well be stimulated to start a particular stage of their journey by the same factors), I was surprised that, so soon after dawn, they had managed to locate a resting place on an island 3 miles long and 1 mile wide (4.8 × 1.6 km) which had been used by generations of migrating Wheatears for resting and feeding on both spring and autumn passage. This was perhaps easier when they were flying high, but in daylight I occasionally saw them flying in low over the sea.

Later in the day, I often saw single Wheatears flying at a height of about 15-20 m over gorse, bracken and pasture in an undulating flight as they searched for a suitable habitat in which to rest and feed. As they came over the brow of the hills surrounding the common, often over my head, they flew down steeply towards the heavily grazed patches and joined any Wheatears already there.

In addition to the choice of habitat, a second important factor affected the distribution of migrant Wheatears both on Skokholm and on Alderney: wind strength and direction. Longis Common is bordered on three sides by hills which rise fairly steeply to about 15 m above the level of the common. If the wind was stronger than force 4 on the Beaufort scale (13-18 mph/21-29 kph) at the time of their arrival, I would find Wheatears grouped in the lee of the hills and there they, or some of them, would remain for several days. Initially the group might forage together, moving in one direction and then another for as much as 100 m, but eventually returning to the sheltered area from which they had started. A day or two later more Wheatears might arrive and, if the wind was blowing strongly from a different direction, they would localise themselves in the lee of another section of the encircling hills.

Thus, two groups of Wheatears could be localised on the common but in different parts and, unless the wind rose to force 6 or 7 (about 30 mph/48 kph) or more, there they remained. The groups might thin out a little as individuals continued their migration. When stronger winds blew from a direction — from the east or southeast — from which the hills gave little shelter, Wheatears which had been settled moved to the other side of the hill, or upwind to shelter provided by a seawall, or among the rocks of the foreshore. On the other hand, if the wind at the time of their arrival had been light, Wheatears might alight and settle anywhere where the grass was sufficiently short.

For the first day or two newly-arrived Wheatears fed rapidly; they were alert and held themselves erect (both the typical race and Greenland Wheatears), and often bobbed, fanned their tails momentarily and wing-flicked, all signs of nervousness presumably related to their unfamiliarity with a location new to them. After a further day or two their feeding rate slackened and they adopted a more relaxed posture, becoming erect again only when something caused them anxiety or excitement.

Draulans and van Vessem (1982), who studied Whinchats — which behave in a similar way to Wheatears on migration — in August, reported that individuals in larger groups fed faster than individuals in smaller groups and suggested that the rate of feeding was a function of group size. In their note they mentioned only briefly that numbers in their study area fluctuated, but they failed to notice that the larger numbers usually meant that there had been a fall of migrants which, being hungry, were feeding faster. Smaller numbers usually meant that some Whinchats had already satisfied their hunger and moved on, leaving behind those whose hunger was partially satisfied and which were no longer feeding as rapidly as they had when they first arrived. This is exactly the pattern followed by Wheatears when they are resting on migration. Fast feeding in this situation is a function of hunger and does not just result from the stimulus of larger numbers.

The tendency to group was a characteristic of newly-arrived Wheatears. When I first saw Greenland Wheatears grouping on Skokholm, I wondered if they did it because much of the preferred habitat had already been occupied by local breeders, or whether strong winds had driven then into sheltered spots. While habitat and wind do influence grouping, I became aware, particularly on Alderney, that whenever Wheatears of either race were unsettled, either because they had just landed in a new habitat or because they were disturbed by a passer-by, they tended to fly nearer to — even within 1 m or so of — other Wheatears. There, standing erect, they watched and waited. If danger passed they relaxed and resumed feeding, and gradually the distance between the individuals increased again until they were 3-5 m apart. On the other hand, if I, as a potential danger, was approaching slowly and directly towards them, first one and then another would fly off in different directions when I was 50-15 m from them.

When danger had passed they would return slowly, sometimes flying, sometimes hopping, to approximately the same area from which they had fled, and 30 minutes or so later they would be loosely grouped again: they sought confidence, as it were, in the closer proximity of other Wheatears — within the social distance. One can see the same behaviour in flocks of winter-

FIGURE 4.6 *Newly-arrived Wheatears grouping*

ing Starlings and Lapwings: when danger approaches slowly individuals in flocks closed up, and when danger comes too close, or within their 'escape distance' (Hediger 1964), they fly away.

Sandy platforms in front of rabbit burrows seemed another safe haven on spring and autumn passage and some Wheatears, when alarmed, flew and alighted on them for several seconds before flying further away. Others, when they had been on the common for three or four days, seemed particularly attracted to them and when not actually foraging they stood on the patches, looking around and preening. If they moved away to forage, in due course they returned to the same platform. Moreover, if, in spring, a male which had been localised for three or four days became aggressive, driving away other Wheatears, whether male or female, it often seemed to adopt one of the patches as its headquarters. Under these circumstances, does the platform and the burrow entrance represent a potential nest site or a place into which the Wheatear could escape? I would suggest the latter, since I never saw a Wheatear fly to its nest when alarmed.

On Alderney I had no help in the identification of individual Wheatears from rings, but it was often relatively easy to recognise individuals with reasonable certainty, for a few days at least, particularly if they were localised. Once I could recognise an individual which had an obvious plumage characteristic, such as a marked pectoral band or a particularly white supercilium, I could tell how long that individual rested and fed before continuing its journey. Some individuals stayed only a few hours. On autumn migration the longest stay I recorded was that of a male which was present for 16 days, while in spring a male was present for 11 days. The commonest length of stay in both seasons was between two and four days, although in late September or October it was apparently shorter.

For the first day or two after their arrival, individuals of the group would not be localised, although the group as a whole was: each hopped, erect and feeding fast, in more or less the same direction as other members of the group, sometimes striking out to one side or the other, even returning whence they had come for a few metres, but maintaining an individual distance of about 2-5 m. Since they were a group of individual birds without the flock cohesion of more social birds such as Starlings, and since one individual might forage for a short distance in one direction and a second head off in another, gaps were created in the group and after a minute or two another member of the group would often move into the gap. The thought sometimes occurred to me that a group of newly-arrived Wheatears was like a roomful of balloons being moved about by a gentle current of air, each balloon representing the individual distance of a Wheatear: when the balloons touched they bounced apart.

It is quite obvious that these Wheatears group, at least in times of anxiety, because they prefer to do so: they fly to join up with others, which presumably lessens each individual's anxiety. Within the group they maintain their individual distance.

So, we have a Wheatear which prefers not to see others too close (the individual distance) but seeks their company at a slightly greater distance, which I shall call the 'social distance'. Beyond that again, a Wheatear may show anxiety if it is 100 m or so from other Wheatears and it is alarmed — the 'lonely distance'. Wheatears usually fly from the lonely distance to put themselves and others in the social distance. As I shall show later, although Wheatears are generally thought to be relatively solitary birds, there is even a tendency for their nests to be grouped.

As individuals continued their migration, the group thinned out. Those that stayed longer tended to become localised and, being localised and less able or willing to retreat, they showed a greater tendency to attack others which came within their individual territory; the longer they stayed, the more aggressive they became. Thus, a mobile individual distance slowly changed over a period of two or three days into a localised individual territory.

By a Wheatear being localised I mean that it had become attached to a particular area on which it frequently foraged or stood and loafed and to which it regularly returned after going away for one reason or another. These areas were usually within that part of the common on which the group had originally established itself and they generally had a radius of roughly 25 m (say 0.25 ha), although if an individual had no neighbours it sometimes foraged as far away as 100 m before returning.

Localised females could be just as aggressive as males, maintaining their individual territories and attacking intruding males and females without fear or favour. Both sexes also showed a slight tendency to be more aggressive in spring than in autumn, and in both seasons they attacked other species such as Meadow Pipits and Linnets when they were within 1 m or so. Strangely, I have recorded them attacking Whinchats, which often grouped with them on migration.

It was at this stage, too, when individuals were becoming more aggressive, that, in autumn and spring, they began to use self-assertive displays in the

FIGURE 4.7 *Male bobbing and flashing tail*

defence of their individual territory. The first, and perhaps the most obvious, indication that an individual was a territory-holder was that, in either approaching an intruder or fleeing an enemy, its flight was undulating with its black and white tail fanned, thus clearly displaying its rump, the whiteness of which pulsated with the undulations. If it was standing in its territory the Wheatear bobbed when alarmed, and flashed its spread tail over its head; in the intervals between the 'flashing display' it stood erect or hopped towards an intruder in the 'advancing display' (which I used to call the 'dominance display'), calling 'tuc tuc' anxiety notes as it did so. (See also Chapter 6 on displays.)

A fourth characteristic of a Wheatear holding an individual territory on migration is the very quiet, warbling subsong, which at times was audible only a few metres away, although in winter quarters it is audible up to 100 m away (Tye, pers. comm.). I heard it more frequently in spring than in autumn. Since Wheatears became alarmed at my presence at distances of 15-50 m, and since latterly my hearing had become less acute, I am sure that I have under-recorded it. Subsong is used by Wheatears from a very early age. I heard a colour-ringed juvenile 30 days old and only just independent of its parents sing a five-second burst: perhaps my nearness stimulated it on this occasion, although I have heard subsong from two-month-old birds which were holding individual territories. On the other hand, Alan Tye tells me that he never heard first-winter birds or females sing in Senegal.

The fifth display, seen uncommonly on spring and autumn migration, is the zigzag flight, which is normally a self-assertive display and performed over the breeding territory. As with subsong, however, I have seen juveniles performing it occasionally when they were between three and four weeks old and before they were fully independent of their parents. Outside the breeding season none of these displays nor subsong is as well developed as when it is performed by territory-holders in the breeding season.

A sixth display, seen once performed by a migrant male which was holding an individual territory in spring, is the 'dancing display'. The male danced from side to side over a female (which was definitely not its mate) which was crouched in a slight depression in the ground. The entrances to holes and depressions in the ground had an importance to Wheatears not only as potential nest sites but also as an arena for some displays, a point to which I return in Chapter 6.

If, after all this show of aggression, the intruder did not retreat from his individual territory, the owner sometimes flew at it. If it still would not leave, the two would fight, trying to peck each other, sometimes flying up vertically for a metre or so in the 'head-to-head' fight.

Finally, there might be a change in a Wheatear's behaviour in relation to its habitat. Earlier in this chapter I showed that during spring migration on Skokholm the earliest arrivals rested on those areas with the greatest density of burrows, which were close to the cliffs, whereas in autumn they chose areas over which they could most easily hunt. On Alderney, sandy patches attracted

Wheatears in both spring and autumn. In spring, however, some migrant males hopped down the burrows as if investigating them as possible nest sites, although most of the holes the males entered were occupied by rabbits and had larger entrances than normally selected as nest sites by females. Furthermore, the investigations lasted only a few seconds and were possibly made by an inexperienced male with an imprecise image of what he was looking for.

So far I have described the behaviour of migrants which took several days to establish individual territories on Longis Common. Some males arriving on the common, however, behaved quite differently: within an hour or so of their arrival they were aggressively defending a territory of at least 0.25 ha. Instead of treating neighbouring or intruding males and females alike, they approached females less aggressively as if trying to entice them to stay. Their behaviour was more or less identical to the way in which unmated but 'resident' males on Skokholm behaved within a few hours of their arrival. It occurred to me that these individuals on Alderney, which were also displaying in other ways from time to time, were close to the end of their migration — that the drive to migrate was being overtaken by the drive to start the first stage of the breeding cycle.

If these newly-arrived migrants were really looking for a breeding territory, then they are likely to have been first-summer birds which had not previously bred: if they had nested the previous year, they would have returned more directly to their former territory. It also seems to follow that some late-arriving, first-summer birds do not necessarily stop migrating abruptly and establish a breeding territory at that point; they may occupy one or more potential territories in succession before they finally settle in to breed.

In summary, Wheatears on spring and autumn migration, and perhaps with a long distance still to travel, tend to group with other Wheatears, maintaining their individual distances, while they regain their fat reserves for the next stage of their journey. Those that stay for two or more days tend to become localised; whether male or female, they then defend an individual territory in which subsong and various displays may be performed. In contrast to this slow development of an individual territory by a migrating bird some Wheatears establish individual territories within a few hours of their arrival, and I suggest that, for these latter birds, the urge to migrate is more or less satisfied and that they are contemplating the establishment of a breeding territory.

UNMATED MALES

The main difference between the behaviour of passage males holding individual territories such as I recorded on Alderney and that of unmated local Skokholm males in potential breeding territories was that the latter warbled their subsong more frequently and isolated themselves in much larger territories. At the opening of Chapter 1, I described the typical behaviour of newly-arrived 'resident' males, which wandered widely and fed quite rapidly. They

FIGURE 4.8 *Enticement crouch*

were also learning where they could find food and shelter and possibly learning about potential nest sites. When their immediate hunger was satisfied, they spent more time sitting on tussocks, often with their tails fanned and looking around and warbling their loud subsong.

Sometimes a male showed no obvious signs of excitement, but occasionally picked up and manipulated a feather, a piece of grass or dead bracken for a few seconds before dropping it. Others made two or three visits in succession to burrows, but chiefly they spent their time hunting. One male, three days after it arrived and nine days before it was joined by a mate, suddenly picked up a piece of dead bracken, fiddled with it, flew to a hole in a bank and went in. After about 15 seconds he came out without it, hopped away from the hole, picked about five small pieces of grass, and returned with them to the same hole, which he entered after waiting about five seconds. Fifteen seconds later he emerged again and pecked at a bit of fluff, which blew away. At that point he seemed to lose interest in the nest and nest material and went on feeding. The male was a first-year bird and had not built a nest before.

As the days passed and more Wheatears arrived on the island, some to stay and breed and others for a short stop, the number of contacts between individuals increased. Loud subsong and the advancing display by a territory-holder were often sufficient to warn off an intruder, but very occasionally the owner would break into territorial song during intense disputes. If the intruder disappeared from sight, the owner often made short vertical flights to about 0.5 m. At first sight these short flights gave the impression that they were used for foraging, but, as with hovering (a behaviour pattern which appeared a little later in the breeding season) or the undulating flight, they were used chiefly for watching where intruders, both avian and human, had gone.

I could not always be certain of the first time that any unmated male territory-holder met an unmated female. On Alderney, in mid April 1980, a male I had been observing had been in occupation of a territory for four days and had attacked and driven out several intruders. I was too far away to hear if he was using subsong but assumed that he was. He had entered some burrows and remained in them for as long as 30 seconds, which was unusual for passage migrants. In this and many ways this male was behaving like a male establishing a breeding territory. A newly-arrived female, foraging, hopped slowly closer to the male. When they were about 20 m apart the male flew at the female, who took off before the male reached her and flew a few metres over a ridge. For two or three minutes the male continued this approach, flying up to her with an undulating flight but always alighting before he reached her; he did not attack her as he had males. She usually took off with tail fanned before he got to her, and the male, instead of chasing her, alighted. Finally, this female left the territory altogether and began a similar approach to another male, also holding a territory, and exactly the same behaviour was repeated. The closest that the male came to the female was about 1 m before she flew on. She remained in this area and continued this same behaviour for an hour, but finally moved away altogether. Neither male found a mate and both ultimately left the island.

This occurred on Alderney, but in 1948 on Skokholm I had recorded a very similar approach to a female by the male of pair W22. He had been foraging in a desultory way when a female arrived in his territory. He flew up in the song flight to about 5 m, then dropped to the ground, still singing, where he bowed with head low, tail depressed and fanned, and moved forward either by running or with short hops. The female moved about 100 m in two minutes, foraging, while close by her this male displayed and sang, apparently successfully because she remained in the territory. For a day or two after the female had joined the male she followed him around their territory, foraging energetically.

While a pair was formed on this occasion, the majority of my records, at this stage, show the male singing and displaying to females which entered the territory but hopped slowly through it and out the other side, unmoved by the owner.

On 5 April 1979, I watched a pair of Wheatears for most of the morning on Weeting Heath, Norfolk. The male, erect and alert, was quite clearly leading the female: he flew about 10-20 m ahead and alighted on ant- or molehills, and gradually the female hopped towards him. After a time he would fly on again, usually with undulating flight with the tail partially spread, and alight again. Within two or three seconds the female followed him. Now and then the situation was reversed and the male followed the female. Neither showed much excitement, just an occasional bob or flirt of the tail, and, rarely, the male flew at the female, passing just over her head in the 'pounce'. The one sign of anything unusual was the undulating flight, which was used in moments of slight excitement or when an individual needed the extra height to locate its mate or an intruder.

Once or twice the male fleetingly visited holes in the ground. Once he took off with the usual undulating flight and flashing white rump and fanned tail. When he alighted he waited for the female to catch up, then flew a further 20 m; this time he landed on a newly-excavated platform in front of a rabbit burrow, where he stood for a moment on the edge of the hole, bobbing and flashing his tail. Almost at once the female flew to him and went down the hole. As soon as she emerged, he flew off to another hole, where he bobbed and quivered his wings in the 'greeting' display. The female flew up beside him again and once more she went into the hole, in fact jumped in and out two or three times. Then, quite suddenly, the drive to look at holes seemed to be satisfied and the two continued foraging.

At this time the pair has the opportunity to become better acquainted with each other's habits, the limits of the territory and the habitats contained within them. As I shall show in the chapter on territory, Wheatears of both sexes are more closely tied to their territories than to their mates. There is sometimes a tendency among ornithologists to think that female birds have no territory of their own and that they just accept the boundaries of the male's breeding territory. It is forgotten that, in winter quarters and at resting places on migration, females of many species will hold individual territories. With the Wheatear, a single female can even hold a potential breeding territory against established pairs — at least for a few hours — as I shall show in the next chapter.

Once a pair has been formed and the territory established, the male's behaviour changes again: he follows his mate wherever she goes, at first keeping within a metre or two of her; after a week or so he becomes less attentive

FIGURE 4.9 *Female and male relaxing*

and just sits on a rise in the ground, or a tussock or a rock from which he can see her, and keeps a lookout for intruders or just preens or loafs. The female is now usually more energetic than the male, and the only time males were recorded as being the more energetic was when they were defending the territory or when they initiated one or other of the sexual displays.

A change in the vocalisation occurs when males are paired. Subsong is the characteristic utterance of an unmated male: this is a self-advertising song which is 'bursting with sex' and expresses the individual's sexuality, and which warns off other males and attracts females in the appropriate receptive condition. From this point it was heard only when occupying some of the sexual displays directed at the mate and, instead, the mated male sang 'territorial' song, of which the more aggressive version is a loud refrain with many variations, harsh notes and some mimicry. It is delivered from the ground, from a prominent position, or in song flight generally up to 7 m high but occasionally up to 20 m.

The fact that Wheatears rarely sing the territorial song before they are mated contrasts with the behaviour of many songbirds, which rely on territorial song until they are mated and then sing less often. For an open-country bird like the Wheatear, visual contact and display may be the main means by which territory is established, and the vocal role relates largely to readiness to mate and to the full establishment of territory. A quieter version of the territorial song, lacking the harsh notes and with less mimicry, was directed to the female. I shall discuss this point in Chapter 7.

Territorial song plays an important part in the defence of breeding territories. On Skokholm these were fairly large, ranging between 0.5 and 3.3 ha (1.8-8.0 acres) during 1948-53, and it was obviously impossible for a male to see all parts of his territory at the same time, particularly if it contained a number of rock outcrops. Thus the value of song, indicating his presence, and which could be heard over a distance of 200 m if the wind was not too strong or over 400 m in a calm, is obvious. In addition to intruding Wheatears, the close proximity of gulls, crows, Soay sheep and humans stimulated territorial song.

Song contests between neighbouring males are common at this period, often as a first stage in territory defence; the males perch on an eminence in their own territories and sing loudly, or they sing as they fly jerkily upwards, with fanned wings and tail, in song flight. In aggressive situations song phrases may be repeated at a rate of one phrase every four seconds, in contrast to the more usual rate of one phrase every 30 seconds.

Conversational song was also used in sexual circumstances when the male was following the female: it seemed quieter, less harsh and sometimes preceded one of the sexual displays.

When a Wheatear is paired, its vocabulary changes in another way. The usual anxiety note is a hard 'tuc', which may be uttered as a single note or repeated several times depending upon the intensity of the individual's anxiety. Once a Wheatear has established its breeding territory, however, it begins to use a 'weet' prefix, so that the complete anxiety note is then 'weet tuc tuc'. Wheatears used this note until the end of the breeding season and the pair had lost all contact with its young. It seems to be a note which signifies that an individual has a territory or family to defend. The 'weet' note was also used alone on occasions of extreme anxiety, particularly when the young were ready to leave the nest burrow or when they were just leaving.

INTERVALS BETWEEN ARRIVALS OF MATES

Males generally arrived on or passed through the island before females. For 17 pairs, I know the number of days that elapsed between the arrivals of each member of a pair: in three cases the male and female seemed to have arrived on the same day; in nine pairs the interval was between one and five days; and in five pairs the interval was between seven and eleven days. The male did not always arrive back before the female and this occasionally led to complications.

Wheatear X6875, a male nicknamed George, was hatched on Skokholm in 1947, and in 1948 and 1949 was mated with X6716. He was always one of the earliest males to return to the island and to his territory in the northwest corner. In spring 1950, X6716 failed to return and he mated with 001448, which became known as Margaret. She had been hatched on Skokholm in 1948 but had not been recorded in 1949. In 1951 Margaret returned to the territory she had held with George in 1950, but he had not arrived. She occupied it with a new (unringed) male for four days and the two were behaving as though they were mated. On the fifth day George returned and tried to re-establish himself, but throughout the first day the unringed male was vigorously attacking and chasing George: Margaret remained close to the unringed male in the typical posture of a newly-mated female and was not observed to attack George. On the following day there had been a change, and George was attacking the unringed male and now was closely attended by Margaret. On the next day a gale caused all three birds to leave the exposed territory. George and Margaret had moved to a sheltered but so far unoccupied area and continued as a pair, raising two broods.

In the adjoining territory a male, 10822, was mated with a first-summer female, which had lost one colour ring, and they produced a first brood successfully, but after that, as so often happens, the male 10822 disappeared. The female remained and eventually mated polygamously with George and produced a second brood. Both females defended their own territories against each other, and George defended a 'super' territory.

In 1952, when George would have been five years old, he failed to return to the island and so did the partly ringed female. Male 10822, however, returned and moved into Margaret's territory, and with her he successfully raised two broods.

The story illustrates several points. First, it is desirable for a male to return early to re-occupy the territory that it has known in previous years, a familiarity which seems to be of some benefit for pairs. The return to the same territory may also ensure that he meets the same partner, which again should be of some benefit to pairs. This story also shows that a female can return to and hold a territory that is known to her and which is not at that time occupied by a male; but I do not know what would happen if another pair tried to take it over.

I know of another female, 001447, the sister of 001448, both from the 'superfamily' W23, which held a territory for a day against the attacks of a resident pair. On the following day she left the area and the pair took over her territory. 001447 eventually mated with another male in another part of the island. In cases like these I do not know whether a female in possession of a territory accepts any male that comes into or wishes to occupy the territory. It is also interesting to note that male 10822 left a territory in which he had nested for one year and joined a female which had nested in her territory for three years. The move was, however, only a matter of 200 m into an adjoining area in which he might well have foraged occasionally in the past and which might have been accepted as an 'instinctive' territory, a point to which I return in the next chapter.

Chapter 5

THE USE OF SPACE

INDIVIDUAL DISTANCE AND TERRITORY

From the moment a juvenile Wheatear leaves its nest burrow at the age of about 16 days, it starts on a somewhat solitary life compared with young of most other passerine species, which stay together in family flocks after leaving their nests: a young Wheatear likes to see others of its own species — but at a distance! When the first juvenile leaves the nest burrow, often excitedly accompanied by one of its parents, it rushes to another burrow about 5-10 m away in which it localises itself for several hours, or for a day or two. There it waits to be fed, standing at the burrow entrance and begging vociferously whenever a parent appears. While it may join one of its siblings in another burrow, it is usually alone.

For the first day in occupation of its new home it may well withdraw into the burrow after being fed, emerging after a few minutes when it feels hungry again. Only rarely does the juvenile follow the adult into the open begging for food. Over the next three or four days it starts to wander a little, occupying another burrow for an hour or so and generally spending less time in hiding, feeding itself more, receiving less from its parents, but still meeting its siblings from time to time.

At three or four weeks old the juveniles start to show aggression. They begin to use the erect posture regularly towards siblings and occasionally towards trespassing adults (see Chapter 6). They chase siblings for two or three seconds, and once one even attacked a female from a neighbouring territory which had trespassed as she foraged for her own young; the attack caused the female to move away a little, but she did not retaliate. The dancing display was performed quite vigorously by a juvenile from one family I was watching. Other behaviour patterns showed links with adult defence of a breeding territory and included the occasional use by juveniles of various territorial and self-advertising displays and in particular the zigzag flight, normally used only by adults holding a breeding or non-breeding territory.

It may be argued that the use of such displays is only a matter of 'youthful exuberance', but I suggest that these juveniles were already exercising their aggressive drives to establish their individual distances or, to put it another way, they were developing a 'sense of territory'. In Chapter 15 I shall show how this develops once the young disperse and become independent of their parents and their siblings.

After I had been watching several broods on Skokholm in the late 1970s, what was slowly dawning on me after 25 years was that these young Wheatears were changing from what Hediger (1950) has called 'contact' animals (i.e. they preferred to be touching each other, as when they were in the nest) to 'individual' animals (where they preferred to stand alone), and that their individual distance was just then developing. In two previous papers (Conder 1949, 1956) I had defined the individual distance of Wheatears, both local breeders and migrants, as an area which had no visible boundaries, which moved with the bird, and into which no other individual could come. Morse (1980) redefined it as *the minimal distance tolerated without attack or retreat.* At that time, I had not thought of its origin or development. I knew that Wheatears in first-winter plumage, which replaces the juvenile plumage when they are about six weeks old, not only maintained individual distance in which they isolated themselves by retreating, but also, when they became localised in one place — and there is a tendency to do this — had to attack others which came too close, so that, by Noble's (1939) definition, they were defending or maintaining an individual territory.

FIGURE 5.1 *Aggressive male with tail spread searching for intruder*

It had taken me about 25 years to realise that the first indications of the Wheatear's territorial system, or of its territorial behaviour, became apparent almost immediately after the young birds left the nest burrow, and that in the following fortnight or so I would be watching the development of their territoriality. From that point Wheatears would always have some form of territorial tendency.

What is territory? None of the earlier writers on the behaviour which we call territorial — Altum in 1868, Moffat in 1903 or Eliot Howard in 1920, each of whom had described the behaviour of birds which fought and defended an area in which they nested, mated and fed — had defined what they meant by the word territory (actually, Altum never used the word anyway). Later, other ornithologists, notably Mayr (1935), Nice (1937) and Tinbergen (1936, 1939), each produced definitions which differed slightly and emphasised the functions that benefited principally the species they were studying.

In 1939, Noble submitted a paper on the role of dominance in the social life of birds in which he reviewed the functions of various types of territory, whether they were sexual and nesting territories, home ranges, occupied areas in which owners resented trespass, and so on. Noble's main subject throughout the paper concerned dominance particularly in relation to territory. Towards the end of a paragraph, he writes: 'In brief, *while territory is any defended area* [my italics], sexual and nesting territories are characterised by sexual or nesting activity, in contrast to a retreat which is occupied because it is familiar and defended because any newcomer is irritating to the resident.' Nowhere else in the paper does Noble use the words 'any defended area', not even in the concluding paragraph in which he summarises the main points of his thesis. Nowhere does he suggest that he has found the ideal definition of the term 'territory'. I have been driven to the conclusion that Noble, when he wrote that sentence, used it with several other sentences to extend and amplify the purport of that particular paragraph, and that he had no intention of producing a compromise definition of territory. If he had intended to produce a compromise definition, I am surprised that he did not praise the merits of the phrase and give the reasons for selecting it, and, finally, I should have expected him to have included the definition in the summary of his paper.

Nice (1943), faced with the dilemma of trying to establish a definition which reconciled the conflicting definitions already proposed by a number of authors but particularly Mayr (1935) and Tinbergen (1936, 1939), seized upon those words — indeed, defence was a common factor of the definitions of that time. Nevertheless, it seems quite extraordinary that Noble's words have for years been treated with little questioning, as if they expound a natural law.

In my view the emphasis on defence, and the treatment of Noble's definition for 50 years as though it were an unbreakable natural law, have been factors which, for many years, confined our thinking on territory to very narrow limits. I remember a sentence in Armstrong's book *The Wren* (1955) which stated, in referring to a female, that, although she was localised in an area, 'because she did not defend the area it could not be a territory'. I tend to think of that sentence as one of the missed chances of re-interpreting the theory of territory.

Another lost opportunity was exemplified by the definition of territory used by Hinde when introducing the *Ibis* symposium on this subject in 1956. He wrote that territory was any 'localised, defended area'; by introducing the word 'localised' he made the truth more difficult to reach. He wrote this in spite of the fact that two of the papers included in the symposium — Marler's on the territory and individual distance of the Chaffinch and my own paper on the territory of the Wheatear — referred to individual distance. Admittedly several other aspects of territory were omitted from his discussion, presumably because of lack of space; but sadly, in omitting individual distance, another opportunity of discovering whether birds, like other animals mentioned by Tinbergen in his book *Social Behaviour in Animals* (1953), had mobile territories was missed. So many students of territorial behaviour were seduced by the simplicity of the defence definition and happily discarded, as did Armstrong and Hinde, the bits of behaviour that, inconveniently, did not fit.

Another problem has been the obsession with aggression as though that were the only important drive. So few remembered that as long ago as 1943 Lack, in his book *The Life of the Robin*, wrote: 'It should be stressed that if this fighting is to result in a territory the retreat of the intruding robin is quite as important a part of the behaviour as attack by the owner.' (In the early stages of individual distance, before an individual Wheatear was localised, it generally retreated, but the more it became localised the more aggressive it became.)

Noble's definition was, therefore, despite his intentions, a compromise and, like many compromises, it solved some problems but left others unanswered: did a pair of Wheatears which nested a mile (1.6 km) or more from others along the narrow strip of heathland on the Pembrokeshire coast and which had no call to be aggressive to other Wheatears have a territory? Even on Skokholm, in years when the population was low and nests might be 500 m apart, pairs raised young successfully. In northeast Canada and in Greenland, where the nests are 500 m or more apart and aggression is reduced, pairs successfully raise broods. It is sometimes suggested that the presence of other pairs prevents parents from travelling too far in search of food, but in my experience on Skokholm between 1978 and 1983, when nests were further apart and Wheatears were hardly restricted by neighbours, they tended to restrict their own foraging for nestlings to a maximum distance of about 250 m from the nest.

One factor in particular has contributed to the continued acceptance of a compromise definition. Much research of the territorial systems of birds has been undertaken by people studying a species or its territorial behaviour for a PhD or as part of a university's research programme, which is sometimes dependent upon a three-year grant. It is only natural, in order to make the best use of the time available, that the ornithologist chooses a study area which supports a reasonably large population of the species in which he is interested so that he can observe and record the maximum amount of interaction. Thus, many studies tend to ignore low populations or the behaviour of pairs breeding in isolation, which, as in the case of the Wheatear, may be sufficiently different to throw additional light on behaviour.

In this respect, I originally studied Wheatears on Skokholm between 1947 and 1954 and again, intermittently, between 1978 and 1983 and I was lucky that, in the intervals, the island wardens annually censused the breeding population. Except for two years, the record of breeding populations from 1946 to date and for a few years prior to 1940 is, therefore, virtually complete (Appendix 5). This record shows that I was able to study the territorial systems of Wheatears when the population was highest — 38 pairs in 1951 and 1952 — and later when it was as low as 13 pairs.

Dissatisfaction with Noble's definition of territory has become more common in recent years, and several new definitions have been coined, but none has received general acceptance. Many authors still have a preference for defence-orientated definitions. Pitelka (1959), writing about the territory of the Pectoral Sandpiper, defined it as 'any exclusive area' which gives the pair use of an area without undue disturbance. Schoener (1968) also defined territory as an 'exclusive area'. Eibl-Eibesfeldt (1970) considered any space-associated intolerance as territoriality. Davies (1978) recognises a territory as 'wherever individual animals or groups are spaced out more than would be expected from a random occupation of suitable habitats'. Both Schoener's and Davies's definitions avoid the problem of overemphasising the defence aspects and the problem of implying motivation in the definition. Notwithstanding the current ornithological fashion of avoiding definitions which imply motivation, I shall define the territory of the Wheatear as *an area in which it isolates itself.*

As will become apparent later, both male and female own the territory in which they breed, and in spring the female's territory normally more or less coincides with that of the male but there may be some difference in the line of the boundary, which depends on the aggressiveness and support given by her mate. This explains the female Wren's behaviour described by Armstrong earlier and on which I commented above. In a few cases where the female Wheatear was localised in her territory before the male, the latter adopted the bulk of her territory to his. The process is seen again when a male mates polygynously with a second female and he adapts his boundaries to take in hers: the 'super' territory boundary is thus more or less aligned with those of his two females, but each of them defends her own territory against any neighbouring females.

In a sense, then, the Wheatear's breeding territory is a joint enterprise between equal partners in which, individually and jointly, the members of the pair isolate themselves in order to breed.

Finally, I think that, at least so far as the Wheatear is concerned, we have to forget our ingrained attitudes to territoriality and remember that a Wheatear, whether male or female, starts to isolate itself within its individual distance from the time that it leaves the nest burrow. It maintains that individual distance, or individual territory (the localised form of individual distance), by retreat or attack throughout the non-breeding season. On returning to the area in which it will breed, the individual distance or individual territory becomes much larger and, after the appropriate displays, male and female Wheatears form a pair.

Each maintains its own individual territory, which overlaps that of its mate. Male generally attacks male and female female. If one female is more aggressive than her neighbour, then it is likely that she will be able to penetrate more deeply into a neighbouring territory than her mate, who may be faced with a very aggressive male to which he has to concede ground.

At the end of the breeding season when the young have left, the adults revert to their own individual distance and go their own ways, perhaps to meet again, if they survive, next year, when the bond to that breeding territory will bring them together again: the bond to the territory is greater than the bond to the mate (Conder 1956). The fact that the female holds her own territory, rather than simply defending her mate's, becomes apparent in cases of polygyny.

Before leaving this section, I feel that it is important to consider the implications of the fact that territory is larger when a Wheatear first arrives, but decreases in size when other pairs establish themselves close by. If the population remains low, as in Greenland and Baffin Island, or even on Skokholm in the years 1978 to 1983, the territory remains large. Since in these cases the adults, from the moment they arrive, localise themselves within a territory of about 12-16 ha or range roughly 250 m from the centre of their territory and there are no neighbours to restrict them, I conclude that the size of this area is determined innately, and the term 'home range' would be an appropriate description of it. It is, of course, the breeding-season version of individual distance (so far as range is concerned) or individual area. Once population pressure builds up, pairs occupying their home ranges meet other pairs whose home ranges overlap with theirs. They fight, and ultimately establish a territory boundary.

Thus we arrive at a stage where we can say that the 'home range' of a Wheatear is the area in which it innately, or instinctively, isolates itself to breed, find food and so on, provided that the presence of other Wheatears does not restrict it largely to a defended area.

I also suggest that the tendency to enter an adjacent territory — or to intrude — is more a matter of a pair trying to enter that part of their home range which they had in fact lost to neighbours in a territorial battle.

BREEDING TERRITORY

In the previous section I concluded that the breeding territory of the Wheatear is best defined as *an area in which it isolates itself*. In this chapter I shall describe the characteristics of the breeding territories of Wheatears which, because pairs were now localised, were defended if another Wheatear, or some other species, intruded. Of course, if the breeding density of Wheatears is low and nests are widely spaced, then defence against other Wheatears is rarely needed.

In view of what I have said about the territory of Wheatears both in winter and in summer, it will be no surprise that the birds mate, display, nest, raise their young, obtain all their food and so on within their territories, which therefore tend to be large compared with those of some other passerine species. The average size of 99 first-brood territories on Skokholm in the years 1950-52 was 1.5 ha (3.8 acres), ranging from 0.48 ha (1.2 acres), in an area where the field of view was reduced by a number of rock outcrops, to 3.3 ha (8.1 acres) on the open plateau of the island; some parts of the largest territory were marshy, with grasses 10-20 cm high, or, on dry ground, were covered with mature bracken. This is the equivalent of 65 pairs per km². Territory size varied from year to year: in 1950, when the population was 26 pairs, mean territory size was 1.9 ha (4.6 acres), with a range of 1.1-2.8 ha (2.7-7.0 acres); in 1951, when the population had risen to 38 pairs, mean territory size had decreased to 1.3 ha (3.6 acres), ranging from 0.5 to 3.3 ha (1.2-8.1 acres).

The total area of the island occupied by Wheatears varied little, being 47.4 ha (117 acres) in 1950, 47.8 ha (118 acres) in 1951 and 57.9 ha (143 acres) in 1952. In 1951, the increased population was to a great extent accommodated by the contraction in territory size. Territory boundaries on flat ground changed from year to year, even if a territory had been held by the same pair in the previous year. In 1952, there was a colonisation of ground that had not been occupied by Wheatears previously, and which was again unoccupied

FIGURE 5.2 *Male in erect posture*

later when the population dropped, because the vegetation was tall and thick. When I returned to the island in 1978, the spread of the bracken had reduced the area suitable for Wheatears by about 4-5 ha but, even with a breeding population half that of the 1950s, Wheatears still tended to visit all areas of good Wheatear habitat though they did so less regularly.

Brooke (1979) studied Wheatears on Skokholm from 1973 to 1976, at a time when the breeding population had fallen to eight or nine pairs. This was the lowest recorded population since Lockley began the annual census in 1928, and was one-third to one-quarter the size of the population that I had been studying from 1947 to 1953. Brooke found that the average territory size was 2.9 ha, or 34 pairs per km².

During my visits to Skokholm between 1978 and 1983, I did not have the time to measure territory boundaries, but I measured distances between nests and compared them with figures for the earlier years. In 1950, the average distance between nests was 85 m, ranging from 24 to 135 m (n=42); in 1951, the average was 66 m and the range 16-146 m (n=42); and, in 1952, the figure was 62 m, ranging from 20 to 120 m (n=46). With a population of 14 pairs in 1978, the average distance between nests based on 17 measurements was 266 m (90-570 m). In the late 1970s on Skokholm, it became obvious that if a pair had no near neighbours the individuals would range up to about 250 m from the nest. Incidentally, the closest active nests mentioned in the literature are those recorded by Saxby (1874) in Shetland as being 6 in (15 cm) apart; on hatching, the families, totalling 16 birds, grouped together.

A survey of territory sizes recorded in the literature shows that the lower the density of the breeding population is the larger is the territory. Minimum territory size of the race *leucorrhoa* in Greenland was estimated as 12-16 ha (Nicholson 1930). In Baffin Island, Canada, Sutton and Parmelee (1954) recorded at least 800 m between each of four *leucorrhoa* nests. Tye (1980, 1982), in his study of the nominate race in Breckland, Norfolk, showed that the earliest-arriving Wheatears at first occupied large territories which might be up to four times the size they finally defended some three or four weeks later: the final size of 70 territories averaged 2.73 ha (range 1.92-6.68 ha), or 34 pairs per km² (range of 15-52 pairs per km²). In the northern Pennines he found that territory size was similar to that in the Brecks (1980). Smaller territories were found by Carlson and Moreno in Sweden, where ten contiguous territories averaged 1.2 ha (0.5-2.0 ha), but, even smaller than that, Ptushenko and Izomemtsev apparently found that in the Moscow region, USSR, territories ranged from 0.16 to 0.4 ha (*BWP*), sizes which seem to me incredibly small.

Territory tended to be larger on Skokholm in flat and open country and smaller in those parts of the island where the terrain was more irregular and broken by ridges and rock outcrops. Such ridges, rocks and banks, which were used as song posts, also formed visual boundaries between territories, at least for those periods when the pair was foraging on the lower parts of its territory; they reduced the chances of a pair seeing its neighbours continuously, which lessened territorial strife. In the flatter northern parts of the island the surface was unbroken except for earth and stone walls, and Wheatears could see each other and intruders more rapidly; even here some slight physical features regularly formed barriers, even though different individuals held territories in successive years.

In 1952, six pairs occupied just 4 ha (10 acres), giving an average territory size of 0.7 ha (1.7 acres). This area adjoined the cliffs over which of course there was no room for expansion and, consequently, it was an area where, in spite of the rocks, aggression was frequent. Ultimately, one pair moved its recently fledged young about 450 m through the territories of two other pairs, defending them fiercely against attacks by the owners: a fine example of juvenile distance or a mobile territory based on a brood of young (Jenkins 1944; Conder 1949).

The territory size of those pairs which had bred in previous years and which returned to the same locality, even to the same nest, to breed could also vary depending upon population density.

Second-brood Territories

Between 1948 and 1951, only 47% of first broods were followed by second broods, and most of the other adults tended to disperse. Consequently, second-brood populations were generally more widely spaced, were far less aggressive, and males sang less. Some males began to moult half way through second broods and from then on seemed far less energetic in their behaviour: some seemed to lose interest in the nestlings altogether. Earliest new feathers were seen in adult plumage on 1 July, so moult had presumably begun a few days before this. Females with second broods did not begin to moult until the end of the nestling stage.

As a consequence of the lack of song and aggression, I found the delineation of boundaries of second-brood territories almost impossible: minor battles were seen only when the young of first broods, which were slowly dispersing, trespassed close to second-brood nests.

The move of 450 m by brood W62 that I mentioned earlier and another move of 300 m by pair W18 are the only records I have of pairs making substantial changes of territory for a second brood. Many pairs changed their boundaries for second broods, usually enlarging them, or ranged more widely when foraging, but rather than a change in territory this might mean that the pair was able to make use of its complete 'home range'. Tye tells me that, in Breckland, he had one instance of a pair which shifted boundaries to an extent that the overlap was only 25-30% of the area of each territory.

Elsewhere, Thomas (1926) mentions a case in Sussex where a replacement nest was built 2 miles (3 km) from the first nest, which had been deserted. While one might expect to see Wheatears moving this sort of distance in the high Arctic, where their population is generally low, I believe that it is unusual for both of a pair to desert a territory — unless they take their young with them. As I pointed out in 1956, the individual's bond to territory, or its home range, is stronger than the bond between the pair, and recent observations confirm this.

Although so many of the Skokholm Wheatears moved away from the island after they had raised first broods, I had no evidence that a pair produced a second clutch away from the island or that Wheatears had come from elsewhere, such as Skomer 3 km (2 miles) away or the Pembrokeshire mainland, to produce a second brood on the island.

I have two records of females that changed their males for second-brood nests. In one case the male had disappeared, and in the other the male was still present but in heavy moult. In both cases the males and females were in adjoining territories.

Polygyny

Polygyny in Wheatears occurred uncommonly on Skokholm: I recorded two instances of first-brood polygyny when the population was low between 1978 and 1984 (Appendix 5), but none when it was high between 1948 and 1953. The first instance of first-brood polygyny that I knew of involved a male in 1980 which largely ignored his first mate and their nestlings and mated with a late-arriving female; the two nests were 35 m apart and in full view of each other. Antagonism from the second female increased as she began to lay. Both females apparently raised their nestlings with little help from the male. The second instance was in June 1981 when one male was servicing two nests, W81/06 and W81/09, 260 m apart. The habitat above Mad Bay was much favoured by nesting Wheatears (see Figure 2.5).

Second-brood polygyny seemed to arise because some males, for one reason or another, left their females and just disappeared. One male in moult was still in the territory when his former female mated with the neighbouring male. I have already told the story of George and Margaret (Chapter 4 and Appendix 6). In this case the disappearance of 10822 after the first brood fledged allowed George to have a polygynous, second-brood, relationship with 10822's former mate. Neither George nor the partly ringed female returned the following year and 10822 mated with Margaret (001448), which had, of course, been in the adjacent territory. Once again 10822 disappeared after the first brood fledged but 001448 remained in the territory until 25 July, when she was beginning to moult.

In mid July 1949, pair W63 had successfully raised a first brood and the female was incubating a second clutch. Her mate, 10292, was also attending another female in the adjacent territory which had juveniles out of the nest. No other male was present. His own female raised only two nestlings, the remainder dying in the nest. Positive evidence that this last instance was polygyny is slim but suggestive.

On Skokholm, the second-brood polygyny I recorded was always between members of neighbouring pairs; males ranged over the territories of both their females, but the females defended their territories against each other.

Brooke (1979) also recorded two instances of polygyny — presumably first-brood — on Skokholm and argues that his data support a graphical model of Orians (1969) which predicts that females select males on the basis of territory quality, and that polygamy will be more frequent in territories of higher quality. Unfortunately, Brooke's definition of high-quality or 'best' territory on Skokholm is not very helpful; he judges 'best' territories to be those to which Wheatears return first in spring, and in the same paper he says that the Wheatears that return first go to the best territories — which seems a rather circular argument that can hardly support Orians's model! In any case, I showed by colour-ringing that older birds usually return first and go to the territories they occupied the previous year (Conder 1956). Therefore, in contrast to Brooke, I saw no evidence to support Orians's hypothesis.

If first-brood polygyny is to succeed for Wheatears, the second female has to establish her own territory and the neighbouring male then has to mate with her. I have shown earlier that a territory is as much the female's as the male's, and any idea that the male encourages another female into his existing mate's territory is not in accord with the facts. Orians's model seems to be rather male-orientated and ignores the fact that the female has to select the territory and mate.

On Skokholm, leaving all the computer models aside, polygyny seems to arise when a male finds a territory-holding but unmated female nextdoor. Indeed, all my records show that the male came to the female rather than the other way around; he can get into her territory but his mate cannot. It is all very opportunistic.

Polygyny in Wheatears has been recorded elsewhere. Jenning (1954) reported two instances of bigamy involving nests which were 160 m and 350 m apart. Aro (1968) described two more cases on the Helsinki archipelago in which the nests of ringed males were 300 and 1,400 apart.

The literature records several cases of a third bird, usually a second male, coming to a nest and feeding the nestlings. For instance, Axell (1954) saw a second male helping to feed an exceptionally large brood of nine nestlings at Dungeness; but it was driven off about 10 m (which is not very far) whenever it was observed by the parents. Sutton and Parmelee (1954) also saw an extra male attending nestlings of Baffin Island. In other Wheatear species, first-brood young have been seen 'attending' second-brood nests. The amount of help given is not really known (Tye, pers. comm.).

As a general rule, breeding territories on Skokholm persisted until the young became independent of their parents, when many adults disappeared and almost certainly left the island. A few adults, however, held on to their territories, wholly or in part, until they finally left the island, and at least two remained in their territories until they completed the moult in August.

NON-BREEDING TERRITORIES

I have discussed non-breeding territories held by migrants resting on Skokholm and Alderney, both on spring and autumn migration, in Chapters

4, 15 and 16, and Tye in his article on the Northern Wheatear in volume 4 of *The Birds of Africa* (Keith *et al.*, in press) records that, even on migration stops in Africa, Wheatears establish individual territories temporarily.

Until recently, no extensive studies have been made of their behaviour in the winter quarters. My wife and I, staying in Kenya, found that the density of Northern Wheatears was not very high. They were defending territories against Pied and Isabelline Wheatears, and, although they retreated before the attacks of Isabellines, they drove off Pied Wheatears. Eggebrecht (1943) and Stresemann (1950) both recorded Northern Wheatears in their winter quarters driving away Pied Wheatears. Morel and Roux (1966) found that Northern Wheatears were very aggressive in defending their territories from other Northern Wheatears when wintering in Senegal.

The first point that emerges strongly and clearly from Tye's studies in an area of semi-arid thorn-scrub savanna in the Sahel of northern Senegal is that Northern Wheatears of both sexes hold individual territories which they defend intraspecifically and also against wheatears of other species. In Senegal and Zambia he estimates territory size as ranging between 2 ha and 4 ha, and in Kenya Leisler *et al.* (1983) state that the radius of the territory was about 90 m, giving a territory size of 2.5 ha. Territories were defended by the same individuals for long periods, mostly continuously, from November to January, when of course the first would be starting their journey northwards.

Tye was also studying the behaviour of Isabelline, which occupied the same habitat in winter. The two species defended mutually exclusive territories, although there were a few cases of overlap.

During the course of his stay, Tye also studied song from the Wheatear in its winter quarters. It was heard chiefly from males and only once from an adult female or first-winter bird. Song (loud subsong, following the terminology of *BWP*) was often loud and clearly audible over 100 m: in 245 minutes of timed observation of two males, Tye recorded 20 minutes (8.2%) of song. Quieter song was also uttered, although it may have been under-recorded. I know from my experience with first-winter birds on Skokholm that quiet subsong is so quiet that one had to be in a hide to be certain of hearing it, and, as I mention in Chapter 7, it may have been under-recorded on Skokholm.

I shall discuss song in winter quarters further in Chapter 7.

FIGURE 5.3 *Two males acting aggressively towards a third*

THE TERRITORIAL BEHAVIOUR OF JUVENILES

On Skokholm, juvenile Wheatears usually isolated themselves from their siblings in an individual area from the moment they left the nest burrows 15-18 days after they had hatched. From the nest burrow they hopped in one or more stages, taking temporary shelter in other burrows, a distance of between 10 and 70 m. Each juvenile took up residence in its own burrow, which might be 1-20 m from the burrows occupied by its siblings, although in the first day or two up to three sometimes shared a burrow.

The juveniles emerged from these burrows for increasingly long periods while they developed their hunting abilities and flying skills and generally learned more about their environment. Whenever alarmed, they escaped by rushing back to and down the burrow until, when about three weeks old, they escaped by flight, although only for about 30-100 m. Even though they might have wandered as much as 50 m from these burrows during the day, they tended to return to them to roost.

The attainment of independence was a mutual process in which the young wandered for longer distances from their burrows, which were the easiest places for the female to find them, fed themselves more often and solicited food from the female, or were fed by her less and less frequently. The parents, after their first brood, often began the preliminaries for a second brood, but sometimes fed the young until they were about a month old in the 'spare' intervals when seeking a nest site or even when building.

Unlike the young of some species such as Goldfinches, Greenfinches and Linnets, young Wheatears do not flock: they group as do Wheatears on migration and in these groups they isolate themselves by their individual distances. They maintain a minimum distance of about 2-4 m — sometimes more, sometimes less — between themselves and others of their own species. Although attacks by one on another are fairly common, actual resistance to an attack was not seen; the bird attacked either hopped or flew a short distance away.

From the time they become independent at the end of their fourth week, a further dispersal takes place. This is discussed in Chapter 15.

THE FUNCTION OF BREEDING TERRITORIES

Earlier I described the breeding territory of the Wheatear as an area in which individuals of a pair isolate themselves. Where population density is high and intrusion common, intruders are attacked and ultimately a boundary is established. Generally, males defend the territory against males and females against females. In a breeding territory individuals form pairs, display sexually, build their nests, raise their young and find all their food.

Another important element is that territory assists in continuing a bond between the two individuals which have been established as a pair the previous year. Both males and females showed a strong attachment to a territory in which they had previously nested; indeed, the evidence suggests that the bond to the territory is much stronger than the bond between the members of the pair (Conder 1956). Between 1948 and 1954, no pairs on Skokholm moved together at the beginning of the breeding season to an entirely new territory, although two pairs moved their young to new territories: one pair, already mentioned, moved 450 m and the other only 50-100 m and, in a sense, it might be argued (as I shall later) that they moved only to another part of their home range which had previously been occupied by another pair. All the evidence, based on colour-ringed individuals only, shows that individuals returned to the same territory — with the exceptions I have already mentioned — and, if their previous mates failed to return, they, whether male or female, accepted a new one in the old territory. I know of only one exception, where a male (10822) which had bred once before left his first territory to mate with a female (001448) in an adjoining territory which she was holding for her third year. (See the story of George and Margaret in Appendix 6).

Figure 5.4 indicates the fidelity of some long-lived Wheatears and shows the positions of all their nest sites. Female X6875 was the first mate of George from 1948 to 1951, and her eight nest sites are shown. Female WS884 was hatched on the island in 1947 and nested there from 1948 to 1953 inclusive. She had two mates over the years, and built a total of seven nests, of which two in succession were in the same burrow (I had removed the used nests for examination for parasites). X6948 was caught as a juvenile in 1947, but she also nested six times from 1948 to 1951.

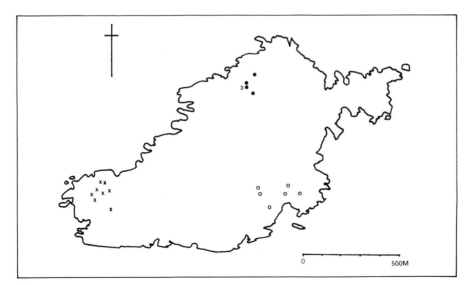

FIGURE 5.4 *Fidelity to territories of long-lived individuals:* o *marks 1948-53 nest sites of female WS884 (ringed as a nestling 1947);* x *marks nest sites of female X6875 (trapped as juvenile) when mated with X6716 in 1948-49 and sites in 1950-51 when mated with 001448;* ● *marks six nest sites, 1948-51, of female X6948 (ringed as nestling 1947), one used for three consecutive nests.*

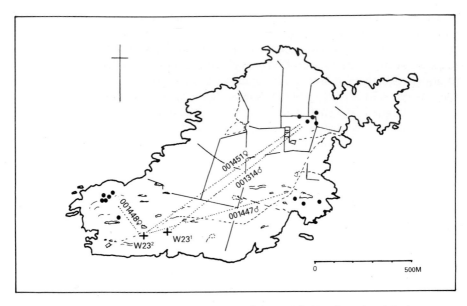

FIGURE 5.5 *Dispersal of juveniles from nests W23¹ and W23². Note longevity of the four offspring and incest between female 001314 from nest W23¹ and male 001451 from nest W23². Female 001448 mated first with X6875 in 1950, and in 1952 with 10822.*

Figure 5.5 indicates the territory fidelity of four moderately long-lived individuals, all of which were offspring of pair W23 in 1947. The map shows the location of nest sites. Note the incestuous relationship between 001451 and 001314.

On Skokholm, there was, even in the years of the highest population, room for further pairs in the less crowded localities of suitable, but not necessarily prime habitat.

Territory and Food

Wheatears breeding on Skokholm fed almost entirely in their breeding territories, which generally speaking seemed to provide an adequate supply of food. However, in 1984, a year of exceptional spring drought, the number of young leaving nests was apparently about 50% of normal (L. Gynn, pers. comm.). But this case was exceptional. In 1951 and 1952, when I was weighing nestlings daily, they sometimes did not put on as much weight each day as I expected, but whether this was because the parents were less efficient at foraging or because of a real food shortage in a particular territory I cannot say. Nevertheless, except in the years of drought, there is no evidence to show that nestlings dying in the nest (other than late-hatched runts) did so from a shortage of food in the natural environment.

From the data available, there appears to be little difference in the numbers of nestlings that fledged from territories larger or smaller than average. I have been unable to compare data from very small territories, say less than 0.8 ha (2 acres), with average or large territories because there were so few of them. In the years 1950-52, the average number of nestlings that left the nest successfully in 37 territories larger than average was 4.86, exactly the same as the average from nests in 37 territories smaller than average. On rather fewer data I have calculated the percentage of chicks hatched which successfully left the nest: in 17 territories smaller than average 94% of the nestlings hatched succeeded, and in 20 territories larger than average 96% successfully left the nest. There was therefore practically no difference on Skokholm.

From data which are available only for 1951, I have calculated the average maximum weight of nestlings from territories larger and smaller than average: from ten small territories it was 24.2 g, and from seven large territories it was 25.2 g (the difference is not statistically significant). Thus, there is no evidence at present to show that nestling Wheatears reared on Skokholm in larger territories fared better in the nest than birds from smaller territories. At the time of writing I have no information on the extent to which fledging weight affects a juvenile's chance of survival to the following year.

Competition for Better Feeding Grounds

Between 1948 and 1953, there seemed to be no evidence of competition for good feeding grounds or that pairs settled to breed more densely where food for the young was more abundant. That older and colour-ringed pairs or individuals returned so consistently for a further year or years to the same territory does indicate that the value of knowing a territory was greater than the desire to find a better feeding ground. Apart from the fact that Wheatears returned to the same territory year after year, a cogent objection to a view that food is of paramount importance in the territory is that the preferred feeding grounds were not the same as the preferred breeding grounds (see Figure 4.5).

RELATIONS WITH OTHER BIRDS

I think that the first point that one has to remember about Wheatears in relation to other bird species is that in the breeding season, when they hold a breeding territory, or in the non-breeding season, when they are in possession of an individual territory, they are very aggressive to a number of their neighbours. They tend to live in habitats where the number of breeding species per square kilometre is usually rather small and they seem to attack most of them.

Generally speaking their neighbours fell into three groups: raptors; species which were markedly larger than Wheatears and potential predators; and similar-sized species which in the breeding season might possibly have competed for food, nest material and so on. When breeding was over and before they migrated, Wheatears largely ignored the last group. On the other hand, Alan Tye tells me that Wheatears attack pipits and larks in their winter quarters.

As one might expect, raptors caused Wheatears some anxiety and falcons, particularly the swift-flying species, caused most. The commonest raptor on

Skokholm in my study period was the Buzzard, which attempted to nest in most years. Over much of the island, Buzzards tend to fly higher than 150 m in the breeding season in order to reduce mobbing by the large gull population. A Buzzard at that height worried Wheatears only a little: they would pause for only a few seconds in whatever they were doing. In another part of the world, Baffin Island, Wheatears showed a similar lack of fear of Rough-legged Buzzards (Sutton and Parmelee 1954), and in Senegal in winter little notice was taken of vultures and Buzzards (Tye, pers. comm.).

Kestrels did not breed on Skokholm, but were nevertheless a common visitor which regularly hunted over the island. On Longis Common, Alderney, particularly in late summer, and after a good breeding season, as many as three Kestrels could be hovering at 20-30 m over the 20-ha common on which about 30 migrant Wheatears might be foraging and some holding their individual territories. Yet Wheatears generally ignored them. Furthermore, Kestrels generally ignored Wheatears, except once when, in September, a Wheatear was chased twice in quick succession for about five seconds over Longis Common by a male, which did not get very close; the Wheatear circled around and landed in its individual territory after both attacks. Oddly enough the Kestrel, which had been flying at a height of about 1 m across the common before it encountered the Wheatear, then continued at the same height and pounced on a horse dropping, on which it tried to balance as though it was holding prey, beating its wings to keep its balance; then, after a few seconds, it moved on and pounced on a second horse dropping about 10 cm away and held on to it again for a few seconds before flying off. These droppings were at least a month old and grey-brown in colour, and no longer attracted dungflies. They were dry and being moved slightly by the wind, so it is possible that the Kestrel mistook the dropping for a mouse, particularly as the Kestrel was flying directly towards the sun.

The story was, however, very different, whether on Skokholm or Alderney, when one of the fast-flying raptors appeared on the scene. In the late 1940s and early 1950s, Peregrines and Merlins were not uncommon and they bred along the Pembrokeshire coast and on some of the other islands. In the late 1970s and 1980s Peregrines bred very irregularly on Alderney, but since then they have been occasional visitors with Merlins and, less commonly still, Hobbies. When any of these falcons arrived, Wheatears scattered in alarm, and in the breeding season they perched and called a constant stream of high-pitched 'weet-weet-weet . . .' or 'see-see-see . . .' notes for as long as three minutes, which indicated maximum alarm. I have no records of Merlins taking Wheatears on Skokholm, but the latter had plenty of room to escape. Casement (1979) records that a Merlin which rested on one weather ship for 24 hours caught and ate three Wheatears which were also resting there. On the same ship five months later, another Merlin also caught a Wheatear.

Sparrowhawks, too, appeared from time to time both on Skokholm and latterly over Longis Common, Alderney. They hunted chiefly the wooded valleys but several times a day they swept out, fast and low, over the common, causing panic among the Wheatears, Meadow Pipits, Linnets, Skylarks and any other small birds that might be feeding there. The Sparrowhawk has the habit of flying within a few inches of the ground and sometimes gliding with wings outstretched and legs dangling and almost touching the ground. In this way, particularly if any dead stems of plants such as ragwort had survived the winter winds and were still standing, it could approach potential prey without being seen. One spring morning, using this technique, a Sparrowhawk surprised a Wheatear which just had time to dive into the wide mouth of a rabbit burrow before the hawk alighted and ran in too: the hawk emerged after about 15 seconds, and after another pause left the area; three minutes later the Wheatear emerged, and after looking around continued foraging.

In other parts of the Wheatear's world range it is taken or disturbed by falcons. In Greenland, the presence of two Gyrfalcons and their families stopped Wheatears from feeding their young. Nicholson (1930) describes how, the moment a procession of four falcons came over, the cock slipped into a convenient rock crevice, from which it emerged very cautiously three minutes later. He was just able to dash down into the glen after a good look around before the falcons came over again.

On islands in the Red Sea, Sooty Falcons, which feed their growing young on autumn migrants, also take Wheatears. Sherif Baha El Din (1984) records that the remains of hundreds of wheatears, not only the Northern Wheatear but other *Oenanthe* species as well, were found at the plucking points of Sooty Falcons.

Similarly, in winter quarters in West Africa, Kestrels, which catch many birds in winter, and Red-necked Falcons frequently chase Wheatears, which respond to the presence of a falcon in the vicinity by crouching on the ground or dropping from a higher perch; the Red-necked Falcon, an efficient bird predator, usually stimulated Wheatears in the open to fly to cover (Tye 1984).

Another predator which is known to have taken juvenile Wheatears is the Little Owl. Three or four pairs nested on Skokholm, usually in rabbit burrows in the earth and stone walls and sometimes among the boulders down the cliffs. In the breeding season one of its main items of prey was Storm Petrels, which often nested in the same type of holes as owls; juvenile Wheatears, either inquisitively or to take refuge, sometimes entered these holes and were killed and eaten. On 10 June 1949, we found a burrow with an adult and two nestling Little Owls surrounded by 36 Storm Petrel remains and Wheatear feathers. Wheatear feathers were also found outside the burrows and occasionally in pellets. One pellet contained the colour-ringed legs of a juvenile which had been ringed as a nestling nine-and-a-half weeks earlier and had been seen alive when four-and-a-half weeks old.

A second group of birds which made Wheatears sit up and take notice included Carrion Crows, Herring, Lesser Black-backed and Great Black-backed Gulls and sometimes even Oystercatchers. The big gulls were continually flying over the island, and whenever one flew directly overhead a Wheatear would watch it. When the nest contained eggs or young, males in particular would call alarm notes and sometimes sing. Individual Great Black-backed Gulls, which had developed the habit of wandering around those parts of the island which had a high density of burrows and of looking into them, sometimes successfully caught a young rabbit which, incautiously, might have been snoozing in the sun at the burrow entrance. These gulls, which had individual hunting habits (Conder 1953a), were also successful at catching Manx Shearwaters which were sitting close enough to the burrow entrance to be seized, or which had been calling and had thereby attracted the attention of the gull. Noisy broods of Wheatears whose nests were in burrows in the thrift peat were occasionally dug out and the nestlings eaten. Sometimes the whole brood was caught, but once the nestlings were mobile some managed to escape and, being colour-ringed, were recorded again in other parts of the island. Wheatears were obviously not the main prey of the Great Black-back but suffered by nesting in an area where many of the more usual prey of these gulls had their burrows.

The last of the large birds which sometimes caused Wheatears alarm were Oystercatchers, not because they were in any way predatory, but because their shrill calls as they flew, sometimes in the butterfly flight, around their territory stimulated the Wheatears' anxiety and they, too, called and bobbed in alarm. I have no evidence to suggest that Wheatears had any fear of Oyster-catchers.

It is understandable that Wheatears should be alarmed by birds of prey and the larger birds, whether they were real or adventitious predators. It is also understandable that Wheatears should attack neighbouring Wheatears which might be intruding on territories or competing in other ways, but it was less easy to understand why they were aggressive to so many species which were quite dissimilar in appearance and in their habitat requirements and whose only problem seemed to be that they were close to hand and easily attacked.

Since Meadow Pipits were one of the commoner passerines on the island, it was not surprising that, when Wheatears were aggressive to species other than their own, Meadow Pipits featured most commonly, being attacked by both males and females particularly during the so-called fertile periods of females (egg-laying and incubation periods) and to a lesser extent when the young were in the nest. All my records are of attacks on Meadow Pipits which were within 20 m either of the nest or of the aggressor itself. Wheatears in fact did not fly such long distances to attack Meadow Pipits, or other species for that matter, as they did when attacking intruding Wheatears. I have no records of Meadow Pipits retaliating during these periods, but later in the summer — June or July — Meadow Pipits with broods regularly attacked young, but independent, Wheatears which always retreated.

I recorded little interaction between Rock Pipits and Wheatears, which may have been because Rock Pipits are larger birds than Wheatears or because their breeding territories overlapped with the breeding territories of Wheatears only around the edge of the island. I did, however, record two female Wheatears frequently attacking Rock Pipits during the egg-laying period: one female was particularly aggressive and harried a Rock Pipit which was collecting nest material, but, in spite of being larger and in spite of being involved in nesting, it did not retaliate but retreated. Wheatears in Senegal also attack Tawny Pipits, which Tye (1984) speculated were competitors for similar prey.

Starlings are somewhat larger than Wheatears, but they were also threatened when Wheatears were nesting. A female Wheatear was collecting food by digging into the grass roots when a Starling flew at her, displaced her, and began digging in the same area; the female moved a few centimetres and turned and threatened the Starling. On another occasion, a male Wheatear flew at a Starling which came too close to the nest and then assumed the threat posture when the Starling did not move. On Alderney, an immature Wheatear was pecking quite vigorously at ants on an anthill. A Starling appeared, walked up to the anthill, and the Wheatear withdrew a few metres; after feeding for two minutes the Starling withdrew, and the Wheatear hopped back and began feeding again, at which point the Starling rushed back. This behaviour was repeated three times until a second Starling appeared and the Wheatear left.

In the 1980s, when Blackbirds had begun to nest on Skokholm, I saw nesting female Wheatears flying at Blackbirds, but they did not press home an attack and the Blackbirds did not move.

Other species which I have recorded as being attacked when Wheatears were either in a breeding territory or in an individual territory are Skylarks, Spotted and Pied Flycatchers (particularly when foraging on the ground), Linnets on Alderney, and Willow Warblers on Alderney on autumn migration and foraging in the corymbs of ragwort. Tye (1984) recorded several interactions between Great Grey and Woodchat Shrikes and Wheatears, which seemed to be associated with competition for food or foraging space.

In the Arctic, Wheatears attacked Snow Buntings (Nicholson 1930) and Lapland Buntings (Sutton and Parmelee 1954).

Chapter 6

THE WHEATEAR'S DISPLAYS

A Wheatear has a varied range of spectacular displays and songs by which it stimulates or reacts to its mate, intruders or predators. In the early part of the breeding season when establishing its breeding territory, a male Wheatear, warning off competing males and trying to attract a mate, makes itself as conspicuous as possible both by the postures it adopts and by its songs and calls. Postures and song are often linked in the sense that each tends to accentuate the effect of the other. On Skokholm, a Wheatear's voice can be heard further away than its postures can be seen; on calm days, I have heard a Wheatear singing as much as 400 m away.

Postures express simple states such as contentment, as when a Wheatear, having fed vigorously, sits on a tussock in the sun with its head sunk on its shoulders and its body in a more or less horizontal position, sunbasking, and apparently looking thoroughly relaxed. Then, with varied postures and song, it shows antagonism to intruders or advertises its presence in a territory, and demonstrates its paired status to other males and females which are within sight and sound. Finally it courts its mate, expresses its sexual readiness and thereafter helps to maintain the pair bond. Some of the displays, particularly those used in self-advertisement, are used in several different circumstances, whether expressing anxiety, sexuality or antagonism.

As with song, the colour of the Wheatear's plumage conveys to other birds its own specific identity and, in the breeding season, its sex. Then, because the plumage of each individual varies (obvious if you look hard enough), colour will aid in the recognition of mates.

As I mentioned when introducing the Wheatear, the colours of its plumage have a dual purpose. When Wheatears are in their primary breeding habitat — the arctic-alpine zone — the striking colours help to hide the males in that broken ground. When displayed by various postures and movements of the body and accompanied by songs and calls, however, the blacks and whites and the sandy-buffs of the throat and breast become fascinating and stimulating to their mates or threatening to rivals and intruders.

Postures are a language. When relaxed, Wheatears stand in a hunched posture and others, with no pressing function to perform, lie on their belly in the sun in the lee of a grass tussock, and apparently doze. Occasionally they sun-bask, when the feathers around the head appear ruffled and the wings are outstretched. This posture indicates to other Wheatears that this individual has no worries, that there are no enemies nearby; and, through social facilitation, it can have the same effect on neighbouring Wheatears, which, seeing this individual in this relaxed position, will behave in the same way. As soon as a cause for anxiety appears, however, the Wheatear will draw itself up into the erect posture and others will see and react.

Erect Posture

The next series of postures I shall describe emphasise these points very well: if a Wheatear does not heed the message given by a neighbour, it can be in for serious trouble.

Probably because birds fear man, we see the erect posture most commonly: it signifies alertness, anxiety, uncertainty or lack of confidence. It is usually the first indication that a Wheatear's contentment has been disturbed. The moment a Wheatear alights in its territory, even after a short flight, or in a new area after migration, it adopts this erect posture with head, neck and body stretched up, the back making an angle of about 70-80° to the horizontal, continually looking around, perhaps on the lookout for danger or for companions. If it has moved only to a different part of its territory it may show alertness at first, but it soon relaxes and continues foraging with its body in a more normal posture; but if the Wheatear has stopped, say on Alderney, for a short rest on migration and is on entirely new ground from which it can derive no confidence from previous experience, it is likely to keep its erect and rather anxious posture for a day or two and relax only slowly as it becomes accustomed to its new surroundings.

Bobbing

If a Wheatear becomes more anxious or excited, it begins to bob, bringing the fore part of its body down to the horizontal, flicking open its wings, lifting and flashing open its tail and, because the wings have been flicked open, also exposing the white rump. When very excited, the bird bobs every second or so: the breast comes down lower still, with wings flashing over the back so high that they almost touch and in so doing showing the black and grey underwing pattern; the tail is lifted and fanned outwards, showing the black and white pattern over the bird's head.

Bobbing is a common display which expresses a form of excitement and is used in a variety of situations, usually accompanied by the 'tuc tuc' call notes but by the 'weet tuc tuc' when mated. The intensity with which the notes are uttered matches that with which the Wheatear bobs: when only mildly disturbed the male hardly bobs at all and just flicks its wings out sideways.

The presence of an intruder, a predator such as a crow or a gull, or humans close to the nest was the most frequent stimulus to bobbing, but Wheatears also used it at rabbits, sheep or goats when they came close. The display was fairly mild during egg-laying, but became more intense as the days passed and it reached a frenzied climax when young were leaving the nest. After that, the intensity would decline until the young became independent. Bobbing was also used in association with the advancing display in its various forms (which originally I called the dominance display, after Armstrong, 1942) when intruders were in the territory.

Bobbing was a display frequently used by the female: she lacks many of the displays used by the male and many of the colour patterns that go with them.

The Significance of the White Rump and Tail Pattern

The most extreme form of bobbing is very similar to the bobbing and bowing by which juveniles solicit food from parents, except that in soliciting food the posture is held for a second or so and the wings, instead of being flicked out momentarily are held outwards and quivered in the manner in which, with slight variations, females solicit copulation from their mates.

One of the actions which has prime significance in a wide range of displays is the exposure of the rump by fanning the tail out sideways and drooping the wings a fraction. In my studies, when almost any excitement other than purely sexual excitement affected the Wheatear, whether it was warbling before the female arrived, or in other displays such as in the moth flight, the zigzag flight, song flight or dancing display, whether it was about to attack an intruder, fleeing from danger, or approaching its mate in one of the self-advertising displays, the tail was fanned and the white rump exposed. The action was used when the Wheatear was in its most highly excited state; and it was used at the very first sign of any excitement. The male, however, when it was approaching the nest or in some of the strictly sexual displays in the presence of the female, kept its tail closed and thereby reduced its conspicuousness. Similarly, a female normally kept hers closed when approaching the nest or in some of the sexual displays.

Sometimes males, and occasionally females, flew in such a way that the patch became even more conspicuous: when, for instance, the pair had seen an intruder of whose exact location they were not quite certain, and flew in the direction in which they presumed the intruder to be. Then, even females used the undulating flight, in which the tail was fanned spasmodically and the white rump given yet more prominence.

To summarise, the white rump and black and white tail pattern are exposed in self-advertisement display, in aggressive situations, and when fleeing an enemy (when it may act as an alarm to neighbours). On the other hand, it is not exposed when the adults are approaching the nest — unless of course they are alarmed — or during sexual displays.

SELF-ADVERTISING DISPLAYS

So far I have described displays and postures which are used throughout the year by Common Wheatears when moved by general excitement such as anxiety, fear, sex or aggression. During the breeding season, in particular, they are associated with other displays, sometimes antagonistic and sometimes sexual; while the basic displays continue to express a heightened excitement, the additional displays signify the kind of excitement.

In this section I shall describe the self-advertising displays used in aggressive or sexual circumstances, in the order in which they generally occur if a persistent intruder declines to retreat after its initial interception and a variety of displays, until the owner and the intruders actually fight.

Both sexes of the Common Wheatear defend their breeding territory. Which bird undertakes the defence usually depends upon the sex of the intruder; females usually attack only females, but males will attack females as well as males — and even the bigger Greenland Wheatears were displayed at and occasionally attacked when resting on the island.

Self-advertising Displays in Aggressive Circumstances
The male accompanies his self-advertising display by a warbling subsong when he is unmated; when mated, he usually sings the loud primary song. When he is mated and expressing his presence to neighbouring or intruding males, then his song is loud and tends to be harsh, but, when just expressing his presence to his mate who may be incubating in a nest burrow, then the song tends to be quieter and more musical, lacking the scratchy notes. When either mate uses subsong to the other, they are 'being sexy'.

While the self-advertising displays are a sort of basic virility posturing, they are alternated with other displays which indicate the real motivation of the display, and the posturings, by the energy that is put into them, also show the strength of the drive moving them.

The Advancing Display
When an intruder has alighted and the owner has failed in his first attempts to dislodge it, males, particularly, display with increasing use of limbs and colour patches, with more songs and song flight until, if all the displays, threats and bluff fail, a fight ensues in which the participants try to peck each other.

In the simplest form of the advancing display one of the pair, depending upon the sex of the intruder, in the erect posture will hop or run-hop, often pecking at the ground and sometimes quite energetically, about 2-3 m from the intruder. The owner edges slightly towards the intruder, which usually retreats. Sometimes, by hopping a little to the left or right, it tries to find a way around the owner, but almost always the owner covers the intruder's action. Finally, after two to three minutes, the intruder usually retreats, but others are more determined and one encounter lasted 14 minutes.

Their movements reminded me of the children's game 'fox and geese' in which four geese may move only forward on the chess board in their endeavour to trap the fox, which can move in any direction. The owning Wheatear seems to move towards the intruder with the inevitability of skilfully handled geese.

FIGURE 6.1 *Male in advancing display*

The advancing display is not particularly aggressive and was recorded most frequently in spring when Wheatears were establishing their territories; it often occurred on territory boundaries. This display is known to some ornithologists as 'parallel walking'. It was also seen in late summer when Wheatears were holding individual territories.

The Flashing Display

If the close presence of the owner in the advancing display fails to cause the intruder to retreat from the territory, aggressive actions become more intense. Movements of the head, wings and tail used in the flashing display add emphasis to the owner's actions.

The erect posture is again the basic element of the display. The male usually faces the intruder, but sometimes faces away from it, with beak pointing well up and exposing the sandy-buff throat. The tail is depressed and fanned, exposing the black and white tail pattern; every two or three seconds it is further depressed and further flashed out, and at the same moment the wings are flashed upwards momentarily, sometimes well above the head, which then accentuates the patterns of grey and black both on top of the wing and on the wing lining. If the two birds are standing sideways to each other, the displaying bird occasionally lifts the nearer edge of the fanned tail so that the underside is slightly exposed to the intruder. During the performance of the display, which may go on for about 15 minutes, the owner either sings the warbling subsong, the primary song or a series of twangy notes, sometimes using imitations of other birds.

Females were not recorded using the flashing display, presumably because they lack some of the black and white patterns (except on the tail).

A crouching variation of this display occasionally follows as flashing becomes more intense: the body is held horizontally rather than erect, and the head is held up, showing off the throat. The wings and tail are flashed as before. This posture is very reminiscent of a Robin's aggressive display. The circumstances in which this variation was seen did not apparently differ from those of the more erect version.

I saw one particularly spectacular version of the display on 27 March 1949, when three males were involved. At 17.45 GMT, two males with their tails fully fanned were singing, one of them loudly and continuously, from Little Bay Point; they were usually about 3 m but sometimes as little as 1 m apart. A female was about 20 m away and apparently feeding. The situation became more hectic when a third male appeared. The first two turned their attention to the third, which appeared to ignore them and carried on feeding, even when the first two were less than 1 m away. The first two males adopted exaggerated postures of the flashing display: flashing wings and tail every two or three seconds, beak high, and so on. They followed the third and, as they moved forward, lowered their heads so that they were almost crouching, a behaviour pattern usually known as the 'rodent run' and very similar to the rodent run of Blackbirds (Snow 1958). They hopped and ran in the advancing display about the same distance as they did when foraging, then stopped, often on top of a tussock, and continued with the flashing display. Sometimes, instead of running, they flew the same distance at 5-10 cm above the ground. Often they ran ahead of the third male, which showed him the fanned tail feathers (although this action was more noticeable when the first two males were displaying at each other). The first male twice flew at the third, which

FIGURE 6.2 *Male in flashing display*

moved about 200 m and continued feeding. On the second occasion, two or three minutes later, the third male left the area altogether.

The fight on 27 March lends support to my belief that the instinctive size of a Wheatear's territory has a radius of about 250 m centred on the area where it localised itself when it arrived. In dense populations, territories reduced by aggressive neighbours would have a radius shorter than the instinctive radius: they would have lost part of their instinctive territory to a more recent arrival. This means that one Wheatear owns a territory which is part of his instinctive territory, while the rest is within the instinctive territory of a neighbour. Hence the prolonged battle between the two males, each of which 'owned' in slightly different ways this piece of ground, and hence the reason why both turned on the third male.

After the departure of the third male, the two remaining males continued displaying to each other. Twice the second male flew up in the song flight, and the first male at one point used a displacement activity — or incongruous activity: it suddenly bent down and pecked at the ground as though picking up nest material, but the whole movement was very stiff. Almost the whole time the first male was displaying, it flicked its wings above its head for a fraction of a second, so fast that I could only just see it.

In the flashing display the Wheatear demonstrates a variety of colours and patterns: the intruder, faced by an owning and displaying male, will see the black and white facial mask, perhaps the black interior of the beak, the sandy-buff chin and throat on the stretched-up neck, between the owner's legs the black and white tail, and finally the black and grey pattern on the wing linings. No matter which way the owning male turns or what position he adopts, one or more of these colour patterns becomes visible. On some occasions when the male was erect and stretched, or else crouched and facing the intruder, a black line appeared to run from the top of the bill to the tip of the tail in a more or less straight line through the ear-coverts, wing-coverts, flight feathers and on to the tail.

As happens with so many species when showing aggression to intruders, Wheatears try to display from a higher point than that occupied by their antagonists. On Skokholm, the higher point in a territory might be an anthill or a thrift tussock, but a wall or a rock outcrop could give an even more advantageous position. Sometimes males sang and used the flashing display when about 100 m apart, either from a perch or in song flight — and song in windless conditions could be heard from a distance of 400 m (although such conditions were rare). These more remote types of flashing display coupled with song flights rarely led to big battles or confrontations, and after a minute, provided the other made no further aggressive moves, the owner returned to its nesting activities.

On the other hand, when the contestants were 1-5 m apart in the territory of one of them, the posturings, songs and calls often became intense and, if the intruder had not retreated — which it often did by hopping away slowly and feeding as it went — a number of other displays and activities might follow.

Flashing Display by Male Owners to Intruding Females
As a rule males do not display aggressively at intruding females, which are usually attacked by female owners. Exceptions occur, however: mated males whose females may be incubating approach intruding females with the full flashing display, as well as with the bowing display, song flights and so on. Male territory-holders also use the flashing display, but in a less exaggerated form, in support of their own females which are displaying at intruding females. Finally, if the intruding female does not remove herself after being displayed at by the owning female, the male will often attack her until she retreats.

SONG FLIGHT

If one of the participants in the flashing display does not retreat, the owning male may behave in a variety of ways which would ultimately lead either to the retreat of the intruder or to a re-adjustment in a territory boundary. The song flight is another common way by which the Common Wheatear makes itself very conspicuous to mate or competitors: it is another way of projecting more effectively the message conveyed by song.

The male, which has perhaps been singing while perched on a tussock or a rock, will fly upwards, head to wind, with neck stretched upwards, wings beating in bursts lasting about a second, followed by a pause, which produces a very jerky upward flight. The tail is fanned throughout, but at intervals it is flicked outwards even further. The flight path varies too: the male appears to fly up sometimes almost vertically and at other times at an angle of about 50° to a height of about 5-10 m, and rarely up to 18 m; the average height of all song flights I recorded was 7 m (see figure 6.3). At the top the male sang one or more refrains as it continued its jerky flight, after which it dived back, with tail still fully fanned, either to its previous perch or to another.

Although many males sang only one refrain from the summit and returned at once to the ground, others moved around rather like a Whitethroat at the apex of its song flight, singing and fluttering their wings in bursts, still with their head reaching upwards and their wide-spread tail hanging down, dancing up and down before returning to the ground. Occasionally, instead of returning to the same perch or one nearby, they travelled up to 100 m, still in the song flight, and then landed in another part of their territory; or, having descended to within 0.5 m of the ground, they flew to other parts of the territory to join a mate or threaten another intruder. In another version, the male flew up in a rather gentle climb with wings held almost horizontally, hardly beating at all, as though in the moth flight, and sometimes gliding with tail fanned; then it changed into the normal song flight pattern.

Except perhaps during an exceptionally intense encounter with intruding Wheatears, song flight was not used until the pair was formed. Very rarely, however, first-winter Wheatears when holding individual territories used the song flight; although I was too far away to hear what type of song was used, it

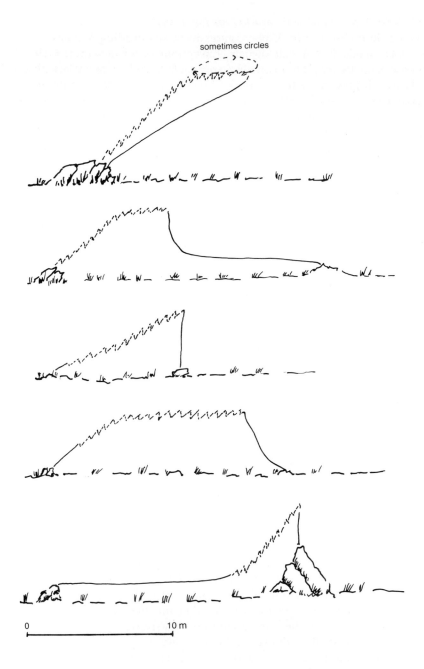

sometimes circles

0   10 m

FIGURE 6.3 *Some patterns followed by Wheatears in song flight; some males return to same perch, particularly after song flight has been used for the quiet conversational song, but when the loud territorial song is used male may travel up to 100 m. Song uttered in jerky ascending flight shown by the broken line, and the unbroken line shows the dive back to a perch.*

was probably subsong since first-winter birds sang it occasionally when they were perched on walls in their individual territories.

The song refrains used in song flights, some musical and some harsh and scratchy, I shall describe in the next chapter, but they are very varied and depend largely upon the target of the display, whether at human or avian predators, at intruding Wheatears or to a mate. While the vertical flight may be used only to sing a single refrain for no apparent reason (perhaps *joie de vivre!*), it can also be used as an element in a sequence of displays which may have a territorial or sexual basis and other elements in the sequence can influence the performance of the song flight; a song flight may look like a mixture of the song flight and the butterfly flight.

Wheatears flew at various heights in ways which slightly resembled song flight but which were not displays. I have already mentioned hovering as a method of observing potential predators or enemies. Wheatears flew up as high as 18 m to catch an insect, after which they returned to earth. I discuss flycatching later, in the chapter on food and feeding.

Butterfly Flight
The butterfly or bat flight, or impeded flight (Armstrong 1942), in which the wings are beaten slowly and the tail fanned, was often used in association with the song flight, usually at a persistent intruder but occasionally in sexual encounters: the male, during his approach to the mate before copulation, and once after he had chased his mates, switched into travelling song flight with slow wingbeats, and finally he performed the dancing display over her. I recorded the butterfly flight less frequently than the moth flight. Nor did I record the butterfly flight as a display on its own in the way that it is used by Greenfinches and many other species as varied as Goldfinches, Oystercatchers, Black-tailed Godwits, and Puffins and other auks.

FIGURE 6.4 *Song flight*

Other displays which were used in conjunction with or in sequence with song flight were the moth flight and the zigzag flight, both of which are described later in this chapter. Song flight was also used before and after aggressive chases, the flashing and the bow. In territorial disputes, the male often used the flashing display when it returned to its perch from the song flight. When song flight was used in sexual circumstances, males also performed other sexual displays such as the zigzag flight before landing.

The Dancing Display

The dancing display, or trench display, is one of the better-known Wheatear displays and has been described by a number of authors (Selous 1901; Lloyd 1933; Pettit and Butt 1950; Monk 1950; Edwards *et al.* 1950; Conder 1950). When I first saw it, and before I had read all the literature, I called it 'the sabre dance' because the fast tempo of the dance reminded me of a piece of music of that name composed by Katchachurian. All three names embody characteristics of the display: it gives the impression of a dance; the bird danced over is in some form of depression in the ground; and the movements of the display are very rapid. It is performed as a single display or in a sequence of displays, chiefly in sexual but occasionally in aggressive circumstances, by both sexes but chiefly by the male. I have also recorded it from juveniles as young as 24 days old when still dependent on their parents.

FIGURE 6.5 *Dancing display*

When members of a pair dance, the male usually dances over the female: I have never seen a female dance over her mate, although she dances over intruders, not only Wheatears but also Rock and Meadow Pipits. On Skokholm, one essential element of the display was that the second bird was either in the entrance to a burrow, in a depression in the ground or in an angle between some vertical object, such as a rock, a wire-netting fence and so on, and the ground. The dancing bird then fluttered and glided, usually singing the subsong, about 0.5-1 m from side to side between 15 and 60 cm above the bird in the depression, and moving so fast that it appeared to be just a bundle of feathers. When the mate or the intruder was in an angle between the rock and the ground, the dancer fluttered between the ground and a point on the rock above the second bird. Sometimes the male danced more over the hole than over the female, even though she was within 5 cm or so of the centre, but he had his head pointed towards the female. Rarely, the male danced over an empty depression with his mates 1 m away.

The bird displayed at is given the maximum exposure of the underwing and tail patterns, all reinforced by movement and the warbling subsong. This rather frenzied display lasts only about ten seconds before the displaying bird changes to some other display or flies off. Once, an intruding male, dancing at a female, gradually subsided 2-3 cm in front of her, where he adopted a posture with wings spread and fluttering, tail up and fanned and head up, his beak was open thus showing the black mouth lining, and he was singing. At this point the female pecked at his beak, and a few seconds later her mate appeared and the intruder flew off. I cannot say whether this was an attempt by a strange male to 'steal' a copulation, as suggested by Carlson et al. (1985), but if so it was the only instance of which I was aware.

Occasionally, males were recorded as standing and resting after this display, with their crown and throat feathers fluffed out slightly, tail moving up and down and the body still in a somewhat horizontal position.

The dancing display is therefore used in antagonistic and sexual displays, but chiefly between the pair. Some of the circumstances that led to the display were similar to those that led to the pounce; for instance, the female might be building rather slowly and the male danced over her, when in similar circumstances he might well have pounced on her. On other occasions the male danced over the female after she had given the greeting display, but he flew to her only after she had stopped greeting, perhaps soliciting copulation. During the incubation period, a male danced over his mate after she had performed the greeting display, and after the dance he chased after her closely. In virtually all these incidents the elements were sexual.

By contrast, the dance was regularly used by an owning male over intruding males, including, once, a Greenland Wheatear, and even by intruding males over a male territory-owner. There are several records of owning males dancing over their mates in the close presence of intruding males (perhaps an incongruous activity in these cases), or after song contests with other males in which the song flight was used, or when an intruding female was present and was not being attacked by the owning female.

Although this display was used in both antagonistic and sexual circumstances, the hole-in-the-ground element was essential, and in a number of different types of display we shall see that a hole in the ground (not just the nest burrow) has an especial importance for Wheatears.

Of particular interest was the occasion on which a 24-day-old juvenile used this display. With its siblings the performer had left its nest seven or eight days earlier and the brood was still partially dependent upon its parents. The members of this brood had already begun to separate and to show a tendency to establish temporary individual territories, but occasionally they came together again. In the incident, on 10 June 1980, one juvenile was investigating a wall, once flying to the top but chiefly hunting along the base. It came to the entrance of a large rabbit burrow and went in, and as it did so one of its siblings flew up and alighted close by. Suddenly it began to dance across the first juvenile for about ten seconds with wings spread wide and tail spread, fluttering and almost gliding across the front of the hole in which the first juvenile stood framed against the darkness of the hole. At the end of that display the second juvenile stood very erect with beak pointing upwards in the posture used in the advancing display (although it had no coloured throat to demonstrate), which it held for about 30 seconds before it moved away. So it would seem that, even at this age, holes or depressions in the ground may have a significance for juveniles other than as an escape route.

When they are three or four weeks old, juveniles of the same brood, still partly dependent on their parents, display fairly regularly. I have seen them bobbing, performing the dancing display, the zigzag flight and the advancing display, and attempting copulation. Except for copulation, all these displays are self-advertising and their target is their siblings. I point out later that juveniles, as soon as they become independent, begin to disperse and become temporarily localised, establishing individual territories. I suggest that the use of these self-advertising displays confirms the existence of whatever is a 'territorial feeding', or a 'low-intensity sexuality', even while juveniles are still dependent upon their parents.

The Zigzag Flight

In the zigzag flight, which shows some similarity to the dancing display and which lasts about five seconds, the male Wheatear, with its tail fanned showing the white rump, flies fast about 0.5-1 m above the ground, but occasionally as high as 3 m in a very erratic flight, twisting and turning from side to side. Sometimes it makes several circuits roughly centred on some point, which might be the nest, the female, along or up and down a wall, around a patch of stinging nettles, a dock or some other distinctive group of plants or near an intruder. Generally the male does not sing during this flight, although it may have been singing or calling immediately before and occasionally it begins to sing and call as soon as it stops and, perhaps, begins another display.

Males use this display in a variety of circumstances, either when threatening intruders, or even when intruders are just present, but most frequently with their own mate. On Skokholm, they showed much individuality in the frequency with which they used it; male W55 quite commonly flew in the zigzag flight to the nest while it was still being built or in its vicinity. Sometimes, apparently, it is used as a displacement activity when, for instance, the female has not been building very energetically, a circumstance in which most commonly the male would actually pounce on the female or occasionally use the dancing display. Another male approached his newly-arrived mate with the zigzag flight. Another, which had been using the flashing display at an intruder, zigzagged to his mate, which went into the greeting display as he approached. I also observed an unmated female using this display as it defended a territory against a pair, and the same female also sang during this encounter — unusual behaviour for a female under normal circumstances. Only occasionally were mated females recorded using short zigzag flights, during which I heard something which resembled subsong; unfortunately, I was uncertain of the circumstances which preceded the flight.

Finally, it is also used by non-breeding Wheatears. On autumn migration on Longis Common, Alderney, several Wheatears which had arrived about three days previously were attacking and chasing each other as they became more widely spaced, and one performed the zigzag flight for about five seconds over a rabbit warren which had a fairly dense growth of ragwort over it; later in the morning it performed twice more. These individuals were

FIGURE 6.6 *Zigzag flight*

migrants and the display might well have been related to the displaying bird's individual territory. I have other instances when first-winter birds on migration alternate normal flight with the zigzag flight.

Juveniles used the zigzag flight quite commonly when they are between three and four weeks old, before they are fully independent and when they are making longer and longer flights. Sometimes a juvenile will begin a normal flight around the parents' territory, gradually flying faster and ultimately swinging into the zigzag flight.

Like the dancing display, the zigzag flight appears to be another self-advertising display which can be used in sexual circumstances and in relation to territory, sometimes in clearly aggressive circumstances.

THE DISTRACTION DISPLAY

A distraction-lure type display, when a bird flutters or shuffles along the ground giving the impression to a human observer that it is disabled, has been recorded for many species but particularly for waders. During the 35 years or so that I have been watching Wheatears, when I have visited some 300 nests, checking clutch or brood size and temporarily removing the eggs or nestlings for measuring or weighing of about 100, I have never seen Wheatears performing the distraction-lure display. I have, however, found one reference to this behaviour in the literature.

Meinertzhagen (1938) recorded, that in Afghanistan, a pair of Common Wheatears which had nested under a pile of stones feigned being wounded and fluttered off when they returned to the pile and found that Meinertzhagen had demolished it and taken the eggs. Unfortunately, he did not give a complete description of the display, but he did comment that it was strange that the pair displayed after the nest site had been destroyed and the eggs taken!

AGGRESSIVE DISPLAY

Intruder Interception

The first stage in territorial defence is the flight by one of the owners towards an intruder, whether the latter is still flying into the territory or whether it has already alighted within it. The simplest response by the intruder is to retreat out of the territory without attempting to contest ownership. Sometimes it has to be chased only a few metres before it retreats; sometimes it may fly to the edge of the territory, and occasionally, having started to retreat, it may circle around and the whole performance begin again. In most chases both birds fly swiftly in a level flight but with the tail spread, exposing its pattern and white rump.

In the early stages of an encounter with an intruder, another type of flight may be seen. When the owning male of a large territory on Skokholm was apparently uncertain of the exact location of the intruder, it would fly in the undulating or 'bouncy' flight, with tail fanned and rump exposed part way to the area in which it last saw the bird. The undulating flight, like the rapid raising and lowering of a long-necked wader's head, seems to help the owner locate the intruder more precisely. In Chapter 1, I described other circumstances in which this flight was used.

Later in the year, when they were involved with second-brood nests, both males and females at times chased juveniles, presumably from other first-brood nests, to distances of up to 50 m.

If the interception by the owner fails, then a range of self-advertising and aggressive displays is performed.

The Flying Attack or Pounce

The chief feature of this display is that one of the pair, usually the male, flies at its mate and passes only about 5 cm above her head, gliding the last 20-30 cm with wings outstretched and calling a harsh rattle which sounds like 'tetete-tete' as it passes over. Very occasionally he lands momentarily on her back and then flies on. Originally I called this behaviour the 'pounce', because it greatly resembles the behaviour of the male Song Sparrow in similar circumstances described by Nice (1937).

In my studies, the commonest situation that led to the male pouncing on the female was her slowness either at searching for nest sites or at searching for and carrying material to the nest. It is quite usual for the female to visit nest holes or build energetically for a period, then she will forage, preen for a time or stand around loafing, preening or pecking at the ground in a rather desultory fashion. After she had been inactive for about 30 minutes to an hour and showed no sign of restarting, however, the male sometimes flew at her in this way and this usually stimulated her to continue whatever had been her primary activity at that stage. Sometimes the female crouched as the male

FIGURE 6.7 *Flying attack, or pounce*

passed over her, but she also jumped about 30 cm into the air or even appeared to jump over him, and there were times when I was unclear what had stimulated the male to pounce on the female.

Both members of the pair also flew in this way at intruders, which often flew away.

All in all, if I dare to imply a motive for this display, there appears to be an element of anger, irritation or, indeed, frustration involved, rather than aggression of the type used against intruders. Perhaps the basic stimulus to the display is that the bird attacked has not done something which the circumstances suggested that it should, whether to continue building in the case of a mate or, if an intruder, to retreat.

Threat Posture

In contrast to some of the other crouching postures, when the body is largely held horizontal, in the threat posture the Wheatear bows so deeply that its chin appears to touch the ground and the tail is held up almost vertically (I recorded that it was sometimes fanned and sometimes not). The wings are not used. The posture is held for three to five seconds.

The threat posture is performed chiefly by the male, presumably because the main feature displayed is the black and white head pattern and sometimes the black gape. Normally it is used in an early stage of an interception of an intruder, sometimes the very first attack, and less commonly after the owner has been indulging in the flashing display. The owner usually faces the intruder, but sometimes stands sideways on. Sometimes it is stationary when displaying, but it also moves, either by shuffling forward, head lowered, or by swinging around so that it seems to be using its chin as a pivot. Sometimes the owner attacks the intruder immediately after the display.

On the two occasions that I recorded the tail as fanned, the display was performed by a female: once when she was attacked by a male, and once when two females were involved in a boundary dispute.

Only once have I seen migrant Wheatears use the threat posture. In September 1986, a first-winter bird was foraging and another flew at it. The first bird threatened by adopting this posture as the second approached, and it turned away at about 2 m.

Another form of threat was delivered by a female to a juvenile which appeared on a rock 5-8 cm above her head. Her body was lowered on flexed legs, her wings were almost fully extended upwards, showing the underwing pattern, and she called with her gape wide open and pointing towards the intruding juvenile. She showed an aggressive response to the sudden appearance of this juvenile, which was soliciting food although it was independent.

Incongruous Activities

If neither contestant retreats and the self-advertising displays fail to drive out the intruder, and the owner's tendency to expel the intruder from his territory remains unsatisfied, the action may be varied in two ways: the owner may use displacement or incongruous activities, or it may fight. In the first case, one or other begins plucking at the vegetation, picking up loose feathers, pecking hard at the ground or even entering burrows in a form of displacement activity of which there are many examples recorded in the literature (see, for example, Armstrong 1942). Sometimes contestants peck at the ground in intervals of fighting, but basically these activities act as safeguards which lessen the chance of actual fighting in which one of the individuals might be damaged.

Most territorial activity occurs in early spring before the eggs are laid, so that pseudo-nest-building activities are an appropriate switch of behaviour to make. Such pseudo-nest-building occurred when both members of a pair were present and threatening one or both of an intruding pair.

Sometimes it seems that the presence of the intruder directly causes this action but, as in the pounce, the failure of a mate to react appropriately is another cause. For instance, a male has been attacking an intruding male but gave up for a time, and the owning male pecked vigorously at the vegetation. In another instance, the intruder was a female and the owning female did not persist in her attacks; and this time it was the male that pecked at the ground and entered burrows as though they were nest sites. A male attacking his own mate when she does not persist in her attacks on a female intruder, and *vice versa*, is not uncommon either.

Throughout this chapter, I have described incidents in which it would appear that the failure of one member of a pair to do what it should be doing in a particular situation creates dissatisfaction in its mate, which then behaves incongruously in an apparently aggressive fashion towards its partner. In the section on flying attacks, or the pounce, I described how, time and time again, the male pounced on the female when she was not building very energetically during the nest-building period. In next section on fighting I describe how female W61 attacked her mate physically (not just flying close over his head) after he had given up attacking the larger Greenland Wheatear which had established its individual territory within their breeding territory. Female Goldfinches with attentive males incubate for longer periods than those whose males are less attentive and sing less (Conder 1948a).

Each member of the Wheatear pair appears to have distinct roles when breeding. A male has chiefly to defend the territory by singing and displaying, to maintain the pair bond and to ensure a safe feeding area free from competition from other Wheatears. A female's role is to build the nest and later to incubate, brood the young and so on.

This means that an intruding male in the territory releases one or more antagonistic displays by the owning male. Normally the female watches male-to-male battles from a short distance but does not take part. Her aggressive drive is also stimulated by the presence of the intruding male, but her drive to attack the intruder is satisfied if her mate attacks and drives him out. After all, as I pointed out in Chapter 5, it is not only his territory but hers.

Similarly, in the nest-building period the male's drive to build a nest (which in fact he occasionally does) is usually satisfied when the female is building. If the female is slow or her drive is not so strong as his, however, the male, by pouncing on the female, satisfies his own drive to see her building and, once she begins again, all is well.

FIGHTING

While the displays which I have just described, as well as the incongruous activities, may well be sufficient to establish which bird is territory-owner at a particular locality, they are also a prelude to fighting — the ultimate action in territory defence.

In a typical incident on 18 April 1949, male W61 was displaying at a larger and more brightly coloured, male, Greenland Wheatear. Very little ground was given and the males stood 1 m or less apart throughout the action. W61, the owning male, sang at the beginning, but later both called 'weet' and 'tuc' notes. Both males initiated attacks, but the Greenlander was more often the aggressor. They flew at each other and, because neither retreated, they fought for about five seconds at a time, pecking at each other, sometimes on the ground, springing into the air for about 0.5 m in an aerial battle. While they were in the air their positions changed the whole time, but when they landed they invariably landed on the correct side — the ground they were defending.

The battle went on for some hours: the Greenlander had established its individual territory not only within the territory of W61 but within 20 m of the burrow in which that pair was building its nest. Throughout this performance, female W61 had been feeding or apparently feeding 1 m or so from her mate: she was obviously interested in what was going on. In the next day or so, when her own mate failed to attack she virtually gave up nest-building in order to attack him: another incident in which the non-executive partner behaved incongruously when her mate failed to react adequately to an intruder.

For a fight between males of the typical race to continue as long as this is exceptional, and normally one of the protagonists will retreat after a few seconds, usually less than ten. While they last the battles are fierce, with each bird trying to peck at the other; sometimes one is prostrate with its back on the ground, wings outstretched and tail spread, while the other is on top pecking it. If the fight persists the contestants fly up as high as 3 m, still endeavouring to peck each other.

Greenland Wheatears on Skokholm may have fought more fiercely since they were localised in comparatively small individual territories and there was less space in which to retreat. Furthermore, the local birds are smaller and may be more easily dominated. The local birds were often nest-building or involved in some other urgent breeding activity so that, as well as the drive to expel the intruding Greenlander, they might have a conflicting drive to return to nesting activities.

Although male Wheatears fight only as a last resort, fighting between females is commoner than between males, perhaps because they lack the

FIGURE 6.8 *Females fighting*

important colour patterns displayed by males in aggressive circumstances. Battles between females, once started, also tend to be long drawn-out and fierce. On 10 April 1949, female W56 was intruding into the territory of an unringed female, and a male was also present. During the early part of this encounter, both females used the erect posture in the advancing display. Then the owning female made a series of flying attacks at female W56, which flew up about 1 m in a two-second flight in which she circled around and landed close to the spot from which she had taken off.

These attacks continued for several minutes; the owning female always seemed to initiate them, and between the attacks there were gaps of about ten seconds during which the females either looked at each other or pecked at the ground in another incongruous activity. This became very intense at one point when both pecked at the grass, pulling it up and mandibulating it in a very marked manner. At one stage, the unringed female carried the material about 2 m before she dropped it into a depression in the ground. When female W56 started pecking vigorously at a disturbed thrift clump, the owner immediately followed suit at another clump about 1 m away; the intruder hopped off, and

109

the owner immediately moved to the same tuft and pecked hard at it.

Finally the two flew at each other again. There was another battle, each holding on to the other with its claws and pecking at the other vigorously; one female was on its back on the ground. Eventually they separated and female W56 flew off south, followed, but not closely, by the owning female. One male was in the vicinity, and I had the impression that it was male W56. He had kept close to the two females while they were fighting and occasionally flew at one of them, but I could not determine which.

The three points of interest in this incident are, first, that, lacking the black and white patterns of the male, the females did not use the flashing display but moved from the erect posture, through the advancing display and incongruous activities, to actual fighting. Second, it was interesting to see the owning female leave the tuft at which she had been pecking in order to peck at the tuft at which the intruder had been pecking. It almost seemed as if, by pecking at the same spot as her antagonist, she was beginning to lose the control imposed by the incongruous activity; indeed, a few seconds later she did physically attack the intruder, which fought back for a time.

The third point relates to the length and the fierceness of the encounter, and I would argue again that these two factors are indicative of a female trying to move into what was her instinctive territory but which she and her mate had lost when the neighbours had established their breeding territory.

SEXUAL DISPLAYS

Turning now to those displays whose nature is entirely sexual, the enticement crouch is adopted by a male as it approaches a potential mate. This can lead .o the sexual chase and other displays which either lead up to or are connected with copulation. It should, however, be remembered that interspersed with those displays may be any of the self-advertising displays, particularly the erect posture, flashing display and the song flight, the song, delivered either in the song flight or from some high perch, lacks the louder and harsher notes of aggressive song and is therefore more musical — to my ears — and softer. The quiet, warbling subsong which the male has used when unmated but holding a territory is now used only in sexual circumstances.

The Enticement Crouch

In contrast to the crouching posture which males adopt in some versions of the flashing display and which I described in the previous chapter, in the enticement crouch the legs are held at about 20-25° to the horizontal, the beak is held low and the throat partially obscured. The tail and the hind part of the body are held down, and the tail is depressed (and thus less obvious to the female) and slightly fanned but not so widely as in aggressive displays: indeed, the white rump and tail pattern are hardly exposed at all during sexual displays (see figure 2.8, p. 64).

The enticement display is not commonly observed, and an incident on 13

April was fairly typical. A male had been warbling the subsong and was apparently unmated. A female eventually appeared, and the male sang and flew up in the song flight to about 5 m with wings beating very slowly and deeply as if in the butterfly flight. It then returned to the ground, where it postured with tail depressed and slightly fanned, head forward bowing, and, in that posture, walked and hopped forward a few paces, stopped, moved and postured again. While the female continued foraging, covering about 100 m in two minutes, the male performed more song flights and, after one, dived back at the female, returned to the bowing position with head low, and pecked twice at the ground. As she moved, the male moved too, keeping to one side of her and between 0.5 and 3 m from her. She appeared to ignore him, except to call harsh scratchy notes when the male dived at her. During the whole of this display, the male was singing and calling.

This was clearly a male trying to entice a female to stay, and there are some resemblances, to the behaviour that I described in Chapter 4 when, in the spring of 1979, two males used the same display to entice a female to stay with them. The action of pecking at the ground is interesting, and perhaps indicative of a conflict between the drive to attack an intruder and that to attract a mate: normally an intruding female would have been attacked by an unmated male as though she were a male.

Sexual Chase
In the sexual chase, which was rather uncommon and seemed to precede what I presumed was copulation, the female apparently flies fast enough to keep ahead of the male but never leaves the territory, whereas intruding females attacked by owning males try to escape and fly out of the territory. Normally these chases only lasted up to 30 seconds and often less, but on 3 June 1981 a pair chased very energetically for 35 minutes.

The spring of 1981 had been exceptionally wet and cold, and the pair I had under observation was very late with its nesting and may have lost its first clutch: in normal years the first juveniles would have left the nest by this time, but the pair was building very rapidly. Both members of the pair were visiting the nest site and both stayed inside for about two minutes. The male emerged first, followed by the female, which hopped away foraging for 5 minutes. She flew back to the male, which was still near the nest and which then chased her for the next 35 minutes with breaks of one or two seconds but very occasionally up to 15 seconds. The female ended the first chase by diving into a hole in the ground which was not the nest burrow, and the male flew in directly after her. They remained down for about 15 seconds, and I thought that in these circumstances they might have copulated, although I could see nothing. After the second and third chases, when the female landed close to the wall, the male danced over her in a display which seemed a natural successor to the energetic sexual chase.

The chase was very clearly of a sexual nature. At no time did the male posture aggressively in the pauses of the flight; he did not stand very erect nor was his tail ever fanned or his wings flicked upwards. The course followed by the pair was very circumscribed and, from where I was watching, it appeared triangular or a figure of eight, chiefly over a wall junction of three fields, but, perhaps more significantly, also over the nest site. The flight was low, 25-50 cm above the ground, rising to pass low over the walls and then dropping down on the other side, giving the flight a switchback appearance.

The female never tried to fly away from this area or to escape from the male: she was always fairly close to the nest, and only once towards the end of the chase did she lead the male about 20 m from it. In the early part of the chase the two flew for 20-30 seconds without a break, but 30 minutes later the flight lasted only two or three seconds before they alighted and rested. They were rarely more than a body length apart; even when the female turned a corner or looped over a wall, the male was able to keep close to her. Both birds had their tails fanned, displaying the white rump and black and white tail pattern. When on the ground, 1-2 m apart, the female regularly used the greeting display. Sometimes she turned sideways to him with her tail closed and held up at about 45° (soliciting copulation?). The wings were stretched out and quivering, tips curved down slightly, and sometimes flicked out further. Once or twice I noticed that, when the male was to one side of her, she lifted the wing furthest from him slightly higher so that it might have been more visible to him. In these circumstances the female appeared to be soliciting copulation.

As the male took off to restart the chase, the female often went into the greeting display for a fraction of a second. In one way she seemed to be enticing him to the chase, but she nearly always took off fractionally after him so that from the beginning of each chase they were very close together. Towards the end, when the chases were very short, she greeted him and half-flew along the ground on the tips of her toes.

At one point, the neighbouring female was in the middle of the area over which they were chasing and at times within 2-3 m of the nest site, but the pair ignored her entirely. She actually joined in four of the chases, following the male, but significantly, she was always 1 m or more behind; either she could not or she did not wish to get as close to the male as the male was to his mate. Finally she disappeared; but she had not been chased away — the two owners were still too involved with themselves!

The Greeting Display

The greeting display is a strictly sexual display used by the male or the female only when they meet. Almost always one or the other is moving towards its mate, often in the vicinity of the nest, as for instance when they are searching for a nest site or when the female is building, and, very frequently, when they are feeding young of both first and second broods. The bird to which the display is directed, usually the male, occasionally greets in return but often just flicks out its wings.

Most of my records show that the female generally initiated the display, in which the breast is lowered so that the body is almost horizontal but the head and neck stretched up. The tail is lifted slightly but not fanned; and the wings are outstretched, slightly drooped and shivered sideways very rapidly. The legs are fairly upright so that I had the impression that the bird was on tiptoe, particularly when it moved forward a few centimetres. Occasionally the displaying bird shuffles its wings immediately before the display. It usually faces its mate but is sometimes in front, turning its head back to look at its mate.

Normally the two greet for five to 15 seconds, but in one intense session the female was recorded as wing-shivering with pauses of a second for two minutes, sometimes hopping towards the male and sometimes away from him. After two minutes the male, having apparently shown no interest in the female, flew off about 20 m and was followed by the female, which, after alighting about 5 m away, moved towards him with quivering wings. This time the male turned towards the female and attempted to copulate with her, but the attempt was thwarted when another female appeared and the male, unusually, left his own mate to attack the intruder. It would appear that the first female was using the greeting display to solicit copulation, but there are one or two odd points. It is, for instance, unusual for the pair to copulate in such an open position and not in a hole or depression. It is also rather extraordinary that the male was so easily diverted by the intruding female, since, in the sexual chase described earlier, an intruding female joined the chase and was entirely ignored. It would appear that the male was not so prepared for copulation as the female. The female called a variety of high-pitched, rather quiet squeaky notes, and quiet piping or warbling notes.

While the greeting display is used in the early part of the nesting cycle — sometimes not in a fully developed form — it occurs even more frequently once both parents are feeding their young (of either brood) and when they

FIGURE 6.9 *Greeting display*

meet regularly at the nest entrance: circumstances which are typical for the initiation of this display.

One further element could be important, too. In the above incident both male and female were bringing food to the nest, and the greeting display does have a resemblance to the posture by which the juveniles solicit food. It is for consideration whether the greeting display is a relic of courtship feeding.

Courtship Feeding

I never saw a male pass food to his mate and the latter then eat it, which is what I consider real courtship feeding to be. On the other hand, very occasionally the female solicited food from the male which she then took into the nest and presumably fed to the nestlings. On 30 June, pair W82 was feeding its second brood and the male flew to the nest entrance with a caterpillar. The female came out of the nest and began the greeting display at the male, which moved away; still greeting, she moved up towards him, and again he moved away; again she moved up and this time he dropped the caterpillar, which she picked up and took into the nest. Similar behaviour was recorded from another pair in 1949, but, so far as I was concerned, I saw no sign of courtship feeding on Skokholm. Nevertheless, there may be a link between these rare cases of females soliciting food from males for the nestlings and perhaps the earlier behaviour pattern of courtship feeding.

Moth Flight

The moth flight, in which the wingbeats are shallow and fast, is uncommon and is usually a sequel to the greeting display, but can also be a prelude. Either adult uses it, but only over comparatively short distances and for only a few seconds at a time. When I was close enough to hear, either sex occasionally called quietly or used the quiet warbling subsong.

The flight used by either parent to carry away faecal sacs from the nest appears identical, except of course that the circumstances are different and the moth flight is sometimes accompanied by song.

Copulation

One of the extraordinary things about my study is that only once during the 35 or so years that I have been looking at them, with varying degrees of intensity, have I seen Wheatears copulating: in fact, I have seen juveniles mock-copulating when three or four weeks old and before they became independent of their parents more often that I have seen copulation performed by adults. J. C. S. Robinson, who helped me on Skokholm, saw adults copulating on one further occasion. Circumstances often suggested that a pair was copulating, but the birds were in a depression or hollow in the ground or in a burrow entrance and, although the fluttering of raised wings could be seen, the bodies could not.

My only sighting was on 20 April, when female W55 was hunting alone and the male approached in a low flight. She crouched, wings held clear of the body but not extended so fully as in the greeting display, the head low with open beak and the tail raised. The male alighted and adopted a similar posture, except that his wings were quivering very rapidly. They copulated. The female threw her head up vertically, held her wings clear of the body for an instant, motionless at first, then quivered them; her head slowly drooped. Meanwhile, the male hopped 60 cm away to a rock, where he raised and then lowered his tail very slowly for about a minute. Finally he shook his feathers very thoroughly, paused for an instant and then flew away. The female had already started moving about, trembling very violently and making no pretence at feeding.

What I recorded on 3 May 1979 on Skokholm was more typical of what I usually saw. Both adults had been hopping towards the nest, the female quite slowly. The male was motionless, not even bobbing. Then I noticed that he was leaning forward a little, and finally he took off towards the female in a hovering flight — almost in a combination of the normal flight, the dancing display and the flying attack. The female moved into a hollow in the ground into which she was followed by her mate. All I could see was the fluttering of wings, which lasted for about ten seconds until the male came out but immediately dived back again. He repeated this three times before he emerged and flew back to the point from which they had started.

What I suspected of being copulation, such as the incident just described, was usually seen only during the nest-building and egg-laying periods up to the laying of the third egg. Both sexes solicited copulation, but the male more often than the female. The displays by both sexes which preceded it were varied, and there was no set pattern; but the female was almost always in a some form of depression or deliberately moved into one. Quite often the male had been giving subsong quietly as he had been watching the female feeding or gathering nest material.

At its simplest, the male might fly direct to the hollow in which the female was standing, and the fluttering of wings suggested that they were copulating. Quite often, however, he used one of the display flights in his approach such as the moth flight, the zigzag flight, the butterfly flight (uncommonly), the sexual chase (once or twice) or the forward-leaning posture. When the male was within 0.5 m of her the female usually greeted, to which the male sometimes responded with the same display or danced over her. Following presumed copulation, both male and female might continue greeting before they moved apart; or the female might use a posture which I did not see commonly, with head stretched up, wings slightly drooped and tail fanned; or the male dance or sing.

When the female solicited copulation, the male, whose basic function at this time was to accompany or attend the female and to defend the territory, was always singing. She flew to within 5 cm of him and greeted, or adopted the ecstatic posture in the hollow, after which presumed copulation followed.

There were instances when I thought the circumstances suggested that the birds had copulated, but they were entirely out of sight in a burrow. On these occasions they were more excited than when they were going into a burrow with material or searching for nest sites. Furthermore, the pair rarely entered a burrow together at this stage — except apparently to copulate.

Mock Copulation by Juveniles

From the day that the juveniles leave the nest burrow until they become independent at about a month old, they are gradually beginning to separate, but so long as their parents are prepared to feed them they remain loosely grouped. Occasionally, when one juvenile encounters another and conditions are appropriate, the two will perform one or other of the displays. I have already described how one juvenile danced over a sibling in exactly the external situation in which the dancing display occurs between adults.

Occasionally, a juvenile would mount another as though copulating but not attempting to join cloacas. At first I thought that they might have been trying to reconstruct the situations they had known in the nest, but they had been out of the burrow about a week and out of the nest a few days before that; also, the position was a copulatory position more than anything else. When this behaviour occurred, the young had usually been standing together in front of a burrow for several minutes, sometimes showing excitement by bobbing slightly without moving their wings. Then one ran at another and jumped on its back for about a second before the other dislodged it. On another occasion, a juvenile close to a second juvenile leaned forward: it was not soliciting copulation, not was it greeting or anything like that. The first had, by chance, adopted a posture which in adults could lead to copulation, and the sibling responded by mounting the first and staying there for two or three seconds looking down. None of the participants fluttered its wings.

Precocious behaviour, including copulation, by young birds has been recorded from a number of different bird species, and indeed from other classes of animals. Armstrong (1942) comments that these actions are without direct sexual reference until the organism has reached the phase in which the actions function in due sequence in the whole sexual pattern, and Hinde (1956) also questions the applicability of the word 'sexual' in the circumstances. The young Wheatears are developing a whole range of innate behaviour patterns whether they are strictly appropriate or not, and when one sees another in, for instance, the entrance of a burrow one of the reactions which would be appropriate to that situation later in life is triggered off.

Nevertheless, in view of the fact that within a few days, by when juveniles are independent, they establish their individual territories and perform this and other displays, I wondered if they were also developing some low-intensity sexuality. Must there not be some internal tendency that permits a juvenile to react to external stimuli in a way which would be appropriate when it is adult in the breeding season? If that internal situation was absent, why did they react more or less appropriately to the external situation? Leaning forward is a posture that must occur on countless occasions and is usually ignored.

Chapter 7

SONGS AND CALLS

INTRODUCTION

For the first few weeks of watching Wheatears I had difficulty in deciding whether some sounds they made were what we usually call songs or whether they were calls. The boundary between them seemed imprecise. The males has a wide and varied range of vocalisations of his own, some of which are harsh and scratchy; they consist only of a note or two and they sound like a not very mellifluous call. Furthermore, he can hear the songs and calls of other species, as well as mechanical noises from human artifacts, and immediately incorporate them into his range of vocalisations. Wheatears also make much use of subsong, a vocalisation whose significance is usually underestimated.

Thorpe (1961) argues that we usually understand song as 'a series of notes generally of more than one type uttered in succession and so related as to form a recognisable sequence or pattern in time'. He states that song is primarily under the control of the sex hormones and is generally concerned with the reproductive cycle. Call notes, he says, are concerned with the coordination of the behaviour of all other members of the species (young, the flock and family companions), mostly in situations which are not primarily sexual, and are rather more concerned with maintenance activities — feeding, flocking, migration and responses to predators.

Catchpole (1985) defines songs as 'long, complex and produced only by males' and points out that calls are used in quite specific contexts; they may affect the behaviour of the receiving individual and they may be thought of as expressing a tendency to behave in a certain way. As I have shown in Chapter 5 and shall describe later in this chapter, female Wheatears also sing.

Because male Wheatears on Skokholm tended to keep a rather long distance between themselves and people — as well as living in a rather windy habitat — and because some of their vocalisations were uttered very softly, I could not always be certain that I heard every note or every section of the song a Wheatear uttered. Furthermore, most of this study, and certainly the most intensive parts of it, was undertaken before portable tape recorders became generally available. In 1978, when I did use one to tape what Wheatears called to each other when they met at the nest, I recorded a quiet call which I had heard only once previously — about 30 years before — in nest W67 when I was lying in a hide only 30 cm from the adults.

When my hearing was acute I tried to make as many phonetic transcriptions of songs and calls as was feasible. It was not easy to decide what selection of English vowels and consonants most clearly approximated to some elements of song uttered by a Wheatear, particularly those which were scratchy or vibrant like the calls of many of the chats. In any case, the modern recording equipment and the use of sonagrams has shown that birds do not use consonants.

The only guide to bird phonetics that was easily available in those days was included in the introduction of *The Handbook of British Birds* (Witherby *et al.* 1938). North's (1950) paper was published when I had already been using the other system for three years, so I did not change. During the course of this study, I wrote down several hundred different transcriptions of what I thought a song or call sounded like, but I could not be confident that anyone else would hear and transcribe the song in the same way. So, while part of my descriptions is based on my phonetic transcriptions, I am grateful to the Editors of *BWP* for permission to use the sonagrams from that work, which has enabled me to illustrate a few of the different forms of songs and calls.

SUBSONG

I intend to begin my description of the Wheatear's vocalisations with subsong, since that type of self-advertising display is used throughout the year, by both sexes and by juveniles as young as 30 days old. Unlike many other songbirds, Wheatears use subsong when they first occupy their breeding territory and 'territorial song' only in intense battles with intruders or when mated.

In the past, there has been some argument as to what types of quiet song are covered by the term subsong, which was proposed by Nicholson (1927) to distinguish songs 'which are so inwardly or faintly delivered that they do not carry anywhere near the distance over which a birds is capable of making itself heard' from the full or true song consisting of a 'flow or pattern of notes warbled or delivered more or less at the top of a bird's voice'. Since then, others, including Lister (1953) and Thorpe and Pilcher (1958), have discussed subsong in general terms and tried to define it. Various terms have been proposed instead of subsong, such as 'secondary song', 'quiet song', 'inward song', 'strangled song' and so on, which, in their turn, have been used to label different forms of subsong.

Fifty years after his first attempt, Nicholson (1977) now lumps most of the varying characteristics together and defines subsong as 'a useful general term to denote quiet forms of song. More specifically it refers to a quiet, extended warbling in which fragments of the territorial song may be heard, often with imitations of other species.' This definition avoids suggesting a function for subsong, on which there is still no general consensus. For the purposes of this book, however, it is a useful starting point.

A Wheatear has two forms of subsong, but because of their quietness they are difficult to separate. I considered that the commoner was what *BWP* called 'quiet subsong'; it is a rambling warble of which the individual notes are often uttered so rapidly that I had some difficulty in picking them out and in writing them down phonetically. Some elements are harsh and scratchy, some squeaky, and others are sweet piping notes; some are so quiet that, sitting 20-30 m away, I could see the bill moving but could hear nothing: sometimes mimicry of other species is included. This seems to be more often used in sexual situations throughout the breeding season by male and female. It is also the version used by juveniles, as well as by males before the arrival of females.

The second type of subsong, called 'loud subsong' in *BWP* is louder and apparently harsher, with more scratchy notes; the singer also mimics other birds. On a still day, I heard subsong from a male 200 m away across a valley; the sound of the higher-pitched notes carried further than those which were harsh and scratchy. Panov (1974) called it 'Battle Song', and I considered that it was used chiefly in aggressive situations.

Others who have heard subsong from Wheatears have described it as a low inward warbling, reminiscent of Skylark song (Nicholson and Koch 1936). In South Africa, Borrett and Jackson (1970) stated that once or twice the subsong sounded like the full version of song but chiefly it sounded warbler-like.

Males usually sang the quieter form of subsong when no intruders were present; this changed and gained in strength as an intruder or enemy approached. Occasionally the 'weet' alarm call was incorporated, not, I think, as an expression of alarm but as another element in the Wheatear's vocabulary (the urgency and volume which normally accompanied the 'weet' call was missing. Mimicry of various species, which I shall discuss in greater detail later, was also included in the louder type of subsong. Some imitations sounded perfect; among the voices I detected were those of Goldfinch, Swallow, the 'bree' of a juvenile Wheatear's location call, Pied Wagtail, Linnet, Dunlin, Meadow Pipit and Song Thrush; other notes and phrases sounded like the song of a Blackbird, the trill of a Skylark and a Yellowhammer call.

Subsong is used by male Wheatears, and by females to a lesser extent, when occupying their individual territories in winter quarters: I heard them in subsong in Kenya, as did Borrett and Jackson (1970) in South Africa. M.F.M. Meiklejohn's report of a Wheatear singing in its winter quarters (Witherby *et al.* 1938) does not state whether the Wheatear was singing the loud, territorial song or subsong, but I should have expected the latter since I know of no references to Wheatears in what *BWP* called 'territorial' song outside the breeding season or their normal breeding range. Subsong was used by adults, juveniles as young as 30 days old and first-winter birds when on migration and in possession of an individual territory for a few days.

Once male Wheatears arrived on their breeding grounds on Skokholm they wandered around their newly-established territories, feeding and warbling their quiet subsong. They did not use the loud territorial song or the 'weet' alarm until they were mated, or unless they became involved in an energetic battle with neighbouring males when they might even burst into full loud song. Intruders of both sexes were apparently listening to the subsong of owners. After pair-formation males generally sang the louder territorial song, but subsong was still used to accompany some of the sexual displays. The male also used subsong when standing near the nest or when visiting it during nest-building or incubation; sometimes the female emerged from the burrow after the male had been subsinging at the entrance. It was probably used during copulation, but I could never be certain. The pair continued to sing it when they were involved with second broods and repeat layings. One female sang it while being held in the hand and ringed.

Occasionally, females warbled the subsong loudly and vigorously in territorial disputes with other females and, like the males, incorporated the vibrant, harsh and scratchy notes in these songs. On 13 April 1949, a female, so far unmated, was defending a territory against a pair of which the female was doing most of the attacking. For an hour I watched the unmated female as she resisted attacks from the owning female by adopting the erect posture of the dominance, or advancing, display, and singing the subsong loudly. They actually fought four or five times in this period, pecking at each other with bodies touching, and were still fighting after the sun had set. Next day the lone female had left. She then lost a ring, and it was several weeks before I caught her on another part of the island where she and her mate were nesting. This incident was interesting, not only because the female was subsinging vigorously but because she was apparently trying to establish and defend an individual territory alone, very much as an unmated male would do when it first arrived.

Most often females sang when in the presence of a male, and particularly during the performance of the greeting display. As a rule they sang for only ten or twenty seconds at a time, and rarely as long as males. When I was watching nest W67 from a hide directly over it, I could see and hear the female singing a quiet subsong while she was on the eggs, in answer to her mate which was warbling softly at the entrance; if I had been observing from my usual distance, I could not have heard that song.

In late summer, after the males had moulted and were holding individual territories before they migrated, they were heard in subsong. Juveniles as young as 30 days and those which had not yet moulted into their first-winter plumage used it while they were maintaining their individual territories. Indeed, it was the type of song used all the year round by both sexes of all ages, particularly when they became localised for a time.

Thorpe and Pilcher (1958) considered that the subsong of the Chaffinch had no specific communicatory function, a view held for the subsong of Robins by Lack (1965) and for Great Tits by Jellis (1977); but it was thought to have resulted from low sexual motivation. I would emphasise, however, that whenever I heard a male or female Wheatear in subsong they were holding either a breeding territory or a small individual territory in winter or summer and that subsong had a self-advertising function just like the full song of other species.

Snow's (1958) study of the Blackbird showed that, while some types of subsong, such as subdued or strangled song, had apparently no communicatory function, others had, particularly when they were associated with some of the displays. The pattern of song and subsong of Blackbirds is very similar to that of Wheatears: for instance, older male Blackbirds do not sing full song until they are mated. There are other similarities: the dominance or advancing display of Wheatears is similar to an equivalent display by Blackbirds.

Whereas Willow Warblers and many other migrants returning to their breeding territories sing the full song to advertise themselves as territory-owning males lacking mates, the Wheatear, in exactly the same situation, uses subsong. If the full song of some species at this time and in these circumstances communicates some message to their neighbours, it seems clear to me, therefore, that a Wheatear's subsong used in exactly the same circumstances also communicates some message.

I rather favour the idea that subsong is chiefly an expression of an individual Wheatear's sexuality, whether of high or low intensity, or, if one was ever to dare to use the term, that it is the sexual portion of the self that is advertised. The quieter and more musical form is used by both sexes, in and out of the breeding season, in sexual displays when holding an individual territory and so on, whereas the harsher scratchy form is used in a defence function when another Wheatear intrudes into a breeding territory.

Thorpe (1985) seems to suggest that subsong is transformed at the beginning of the breeding season into true song. I would consider this unlikely in the case of the Wheatear, because subsong remains important for the pair and is used in parallel with full or territorial song.

Subsong, and particularly the situations in which male and female Wheatears use it, needs more investigation, and the interpretation of it may well reveal much about the true nature of territory.

TERRITORIAL OR LOUD SONG

Territorial song is the best-known of the songs of the Wheatear, and for a short description it is difficult to better that of Tucker to be found in *The Handbook* (Witherby *et al.* 1938). He writes: 'Song superior to other British Chats and delivered with great gusto, is short, pleasantly modulated warble in which melodious rather lark-like notes are mingled with harsh, creaky and rattling sounds; to some extent imitative.'

The circumstances that stimulate song also affect its quality and pattern, its sweetness, its scratchiness, whether the sections or song phrases are repeated more than once in song flight, and so on. Some males repeat certain songs or even sections of song before the phrase is changed, sometimes abruptly and on other occasions with only slight variations.

I would appear to have more records of the loud, territorial song being used in aggression, hostility or alarm rather than in sexual circumstances, which is, perhaps, understandable since I was often the cause of aggression. On the other hand, I also had the impression that, even when they were not conscious of my presence, Wheatears more often sang the loud song aggressively, i.e. with the scratchy elements, than sexually, and that they more often used subsong to their mates than the softer, sweeter, shorter, version of full song.

I never heard the female singing the full loud song as used by the male. Her song was the subsong.

The basic territorial song pattern used in aggressive situations usually consists of up to three sections. The first has one or two notes, the second has two or three — but up to ten — rapidly repeated notes, and the third, which tends to be quieter, is often a repetition of the first section: for example, in the song phrase 'zee zee widdle ee', the first and third sections consist of the high-pitched 'zee' or 'ee' notes.

Figures 7.1a to 7.1c are sonagrams giving examples of territorial song (reproduced here by kind permission of the Editors of *BWP*). Figure 7.1a illustrates a typical, bright and brisk, self-advertising song, the kind of refrain that is used when perched or when in song flight, or both. It consists of a number of fairly high-pitched notes, with little of the harshness or scratchiness that is associated with the songs of other chats such as the Whinchat, Stonechat, Black Redstart and so on.

In Figure 7.1b, however, the singer, at the end of a number of high-pitched notes, begins to use the harsh and scratchy notes and the sound stretches across a number of frequencies. Figure 7.1c shows a yet harsher and less musical type of song with the sound spread over a number of frequencies, particularly the low ones. It is also possible to see the bisyllabic nature of the refrain, with the three-note introduction and the very harsh conclusion. This song phrase is used on the ground or in flight, usually in aggressive situations.

As one can see from the time-scale at the bottom of the sonagram, each phrase tends to be quite short, although each may be repeated with variations as many as ten times in one song burst.

The sudden appearance of a Peregrine or a human near a nest with well-grown nestlings could drive Wheatears into a frenzy such that the first section of the song often consisted of rapidly repeated, high-pitched elements sounding like 'see see see', which were recognisable as variations on the 'weet' alarm call, followed by the scratchy second and third sections of the song.

Thus far I have been using sonagrams to demonstrate one aspect of territorial song used in song contest with other Wheatears, or in the presence of Little Owls, other birds of prey and also man. Particularly in the close presence of people, the song could become even harsher, more guttural (one might even say that a male singing these song phrases had an excessively sore throat or syrinx!) A fine example is shown in Figure 7.1d, when the singer is in the flashing or advancing display either against another Wheatear or a human intruder: every time the male flashed out its wings, or flicked out further the

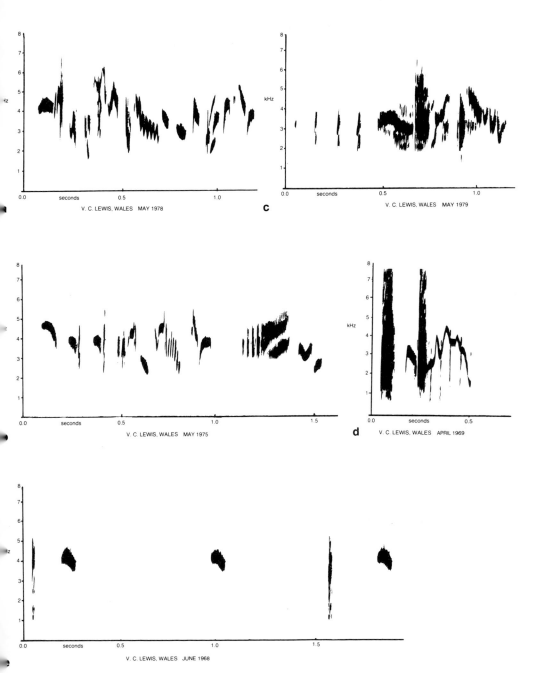

FIGURE 7.1 *Sonograms of Wheatear vocalisations: (a-c) loud territorial song showing increasing harshness of the song; (d) very harsh short song often accompanying flashing display; (e) call notes 'tuc weet weet tuc weet'.* (Recordings by V.C. Lewis and sonograms by kind permission of the Editors of Birds of the Western Palearctic.)

123

already spread tail, it would use this harsh and vibrant aspect of its song. It is an extraordinary range of sounds for a bird and it was originally thought to be part mechanical, but, having listened to Lewis's recording, I am now convinced that it is entirely vocal.

It was with this last type of song, in particular, that I had difficulty in deciding whether the sounds were songs or calls, or indeed mono- or polysyllabic. Their form was more like a call, but they were used in song flight and in other circumstances in which song was commonly heard. Finally, I concluded that they functioned as songs.

Recently, Joan Hall-Craggs played me a number of tapes made by several recordists of the territorial songs of Wheatears from different parts of Britain. I was astonished by the variety, not just by the variations in the refrains, but in particular by the variety of the 'scratchy' notes — the clicks and the 'snaps, crackles and pops' that are usually so quiet that one is rarely close enough to pick them up. Some bird-song experts wonder whether these sounds are mechanical — made by wings or tail. I have never seen a Wheatear using its wings in such a way that it would make sounds like those I heard on the recordings, and I remain convinced that all these sounds are vocal.

The Common Wheatear can produce a remarkable range of sounds which have yet to be fully explored. Later I shall be discussing mimicry by Wheatears: again this emphasises the Wheatear's enormous vocal capacity, all to express a limited range of drives or tendencies.

CONVERSATIONAL SONGS

Songs sung by males in the presence of their mates are not noticeably different from those used in aggressive circumstances. They have the same basic structure and are clearly recognisable as Wheatear song, but they are quieter, more musical, and usually lack the harsh, shingly and vibrant notes. Occasionally males mimic other species. I regularly noted sexual song as consisting of quiet conversational notes, particularly when the males had flown to within 10 m of the nest or had even hopped to the entrance and sang. I was normally unable to hear any response from the female unless I was in the hide immediately over nest W67, when I saw and heard the female answer with very quiet subsong. The female often left the nest after the male had been singing quietly at the entrance. I know of no sonagrams of this type of song, since Wheatears do not sing it when disturbed.

RATE OF SONG

Wheatears were not particularly early risers: in mid June I heard the first song on Skokholm at about 05.30 GMT, when the sun rose in west Wales at 04.00 GMT. But nothing might be heard in the next two hours. I got the impression that at this time of the year the Wheatears generally became active at about 05.00 GMT. In the evening in April the last songs were heard a few minutes before or after sunset, at about 19.15 GMT. (The length of the working day as recorded automatically is described in Chapter 12). Occasionally, songs from Wheatears were heard in darkness, but these were probably from individuals disturbed by Manx Shearwaters from the longer grasses where they roosted.

Frequent strong winds coupled with the long alarm-distance of Wheatears made the counting of their songs difficult: while a single refrain usually lasted a second or so, the refrain could be repeated up to ten times, or ramble with several phrases flowing into one another. On 26 April I recorded male W9, when singing to his mate, singing up to 14 times a minute, which meant that the phrase was repeated at three-second intervals for five minutes. This was an exceptionally high rate, but on three further occasions I recorded him singing 15 phrases a minute.

I found that the most satisfactory way of indicating the volume of song was to show the average number of minutes in a 50-minute observation period during which some form of song was recorded. Such figures that I have indicate that song reached a peak during egg-laying, when females were staying longer and longer on the eggs, and a second and higher peak just before the eggs hatched. The mean number of minutes in which song was recorded was just over eight during the 50-minute period (16%), ranging from no song at all to song in 27 minutes (54%) during which 141 phrases were sung. As soon as the eggs had hatched, the average number of songs per minute dropped to two per 50-minute period, and with few exceptions males sang very little during the last ten days or so of the nestling period, although some males became more vociferous when the nestlings were about to leave and when parental anxiety was reaching its peak. Within a day or two of the juveniles leaving the nest, and if the male had not begun to moult and was otherwise in a condition to recommence nesting, he started to sing again.

The rate of song from individual males varied. Weather also affected them: severe cold in late March and early April was recorded as inhibiting song. It

FIGURE 7.2 *Song flight*

was also clear that wind strengths of forces 4, 5 or above on the Beaufort scale had a similar effect. On the other hand, it was obvious that on days with little or no wind some males sang regularly, and it was on such a day in May that male W77 sang 141 refrains in 27 minutes.

In terms of volume of sound, Wheatears have a very large range. On the one hand I could be within 20 m of a male whose beak was moving but I was unable to hear anything, and I have records of hearing Wheatears singing on windless days when they were 500 and 450 m from me.

SONG MIMICRY

As long ago as 1874, Saxby wrote in his *The Birds of Shetland* 'I am greatly perplexed that no allusion whatever is made to the marvellous power possessed by the Wheatear of imitating the notes of other species.' He went on to describe a Wheatear mimicking a Ringed Plover and then a Lapwing. Finally, he gave a list of ten birds that he had heard being mimicked by Wheatears. Surprisingly, nobody else seems to have recorded Wheatears mimicking the songs and calls of other species, so that in text-book descriptions of the species' song we find such phrases as 'to some extent imitative', although Armstrong (1973) goes some way to putting the Wheatear in its proper rank. A number of other Wheatears, including the Red-breasted Wheatear (Ebbut 1967) and Capped Wheatear (Vincent 1947), as well as other chats such as Whinchat (Thielcke 1976), are known to be imitative. John Buxton (1950) lists 14 species whose songs and calls have been recorded in Redstart song. Armstrong lists a number of other chats recorded as mimicking other bird species.

Table 7.1 lists those bird species and subspecies whose songs and calls I have heard mimicked on Skokholm, the species recorded by Saxby in Shetland and the status of these species on Skokholm in spring during the period based on Conder (1953b). Table 7.2 summarises the numbers of mimicked species according to the status and shows, as one would expect, that those species which bred on or visited the island regularly were the species most frequently copied, although the number of species which visited on less than ten occasions per month during spring migration and were yet mimicked was surprisingly high.

Although at least 43 bird species were recorded as being mimicked by Wheatears, more than one call or phrase of song was recorded so that the total of mimicked vocalisations heard was much greater. Wheatears occasionally produced a phrase of song which was not part of their normal repertoire and which sounded like a bird call but one unknown to me, and I wondered if it was the voice of a species which the Wheatear had encountered in its winter quarters in Africa.

I heard two non-avian sounds from Wheatears. One was a copy of the squeal of a young rabbit when frightened. The other was the squeak of a pulley on a flagpole attached to the Observatory buildings which was used

TABLE 7.1: *A list of the species and subspecies whose songs and calls have been mimicked by Wheatears, including those species recorded by Saxby (1874) in Shetland and by Williamson (1965) on Fair Isle*. The third column shows the spring status of these species on Skokholm based on Conder (1953b)*

	Shetland	*Skokholm*
Buzzard		Breeds
Oystercatcher	Saxby	Breeds
Ringed Plover	Saxby	Irregular visitor
Golden Plover	Saxby	Irregular visitor
Lapwing	Saxby	Breeds
Dunlin		Regular visitor
Snipe		Regular visitor
Whimbrel		Regular visitor
Curlew		Regular visitor
Redshank	Saxby	Irregular*
Green Sandpiper		Autumn visitor
Common Sandpiper		Occasional
Turnstone		Regular*
Herring Gull	Saxby	Breeds
Swift		Regular visitor
Short-toed Lark		Vagrant
Skylark	Saxby	Breeds
Sand Martin		Regular visitor
Swallow		Regular visitor
House Martin		Regular visitor
Meadow Pipit		Breeds
Rock Pipit		Breeds
Yellow Wagtail		Irregular visitor
White Wagtail		Occasional visitor
Pied Wagtail		Occasional visitor
Wren		Occasional visitor
Robin		Occasional visitor
Black Redstart		Occasional visitor
Ring Ouzel		Irregular visitor
Blackbird		Irregular visitor
Song Thrush		Irregular visitor
Willow Warbler		Regular visitor
Blue Tit		Irregular visitor
Chough		Irregular visitor
Starling		Breeds
House Sparrow	Saxby	Vagrant
Chaffinch		Irregular visitor
Greenfinch		Irregular visitor
Goldfinch		Occasional visitor
Linnet		Occasional visitor
Twite	Saxby	Not recorded
Yellowhammer		Irregular visitor
Corn Bunting	Saxby	Not recorded
Non-British Birds	? species	

TABLE 7.2: *Numbers of bird species mimicked on Skokholm according to status, based on Conder (1953b)*

Breeders	9	
Regular visitors	10	Recorded between 21 and 31 times a month
Occasional visitors	7	Recorded between 11 and 20 times a month
Irregular visitors	12	Recorded between 1 and 10 times a month
Vagrants	2	Not recorded annually
Autumn visitors	1	Not recorded by Observatory in spring

three or four times a day; a Wheatear nesting about 200 m from the building incorporated this sound into his song.

The possibility that Wheatears were copying African bird calls which they had heard two or three months earlier poses the question: how quickly do Wheatears learn imitations and for how long do they retain these vocalisations in their repertoire? I have evidence that they can reproduce perfect copies of other birds' calls within hours, even within seconds, of hearing a specific call. On one occasion, I was walking back to the Observatory with the Assistant Warden when we stopped to look at and listen to a Wheatear and a Meadow Pipit about 3 m apart on a wall. They were calling alarm notes very vigorously and presumably had young. When the pipit flew off, the Wheatear continued calling Meadow Pipit alarm notes so perfectly that neither of us would have been able to distinguish the copy from the genuine if we had not actually seen and heard the Wheatear from a very short distance. The Wheatear had obviously heard Meadow Pipits before, but there can be little doubt that in this incident it was copying this particular individual and using these notes in its own territorial defence call.

A Short-toed Lark was recorded on Skokholm for the first time between 9 and 13 April 1952. On 10 April, Wheatear W145, in whose territory the Short-toed Lark had largely localised itself, sang the 'prrit' note of the lark, which is rather harsh and pebbly. No doubt Wheatears could have heard the species somewhere on migration, but, clearly, W145 was reacting to the Skokholm bird. I have records of Wheatears mimicking Goldfinches and Pied and White Wagtails within 24 hours of the appearance of these species or subspecies on the island.

Not all imitations were of the same degree of excellence as that of the Meadow Pipit's alarm notes; occasionally I noted that imitations were very Linnet-like or very lark-like — in fact, they revealed every gradation between those which were reminiscent of a particular song or call through to perfect mimicry. Copies, as we have seen, could be picked up very quickly and were then perfect or near so, but, as time went on, the reproduction sometimes deteriorated: for instance, a male Wheatear singing Goldfinch notes on 30 April was not mimicking that species as well as he had on 23 April when I first heard him doing so. On the other hand, early in March, before Swallows had been recorded by the Observatory staff, Wheatears were using Swallow call notes and must have remembered them possibly from a previous season six months before or from hearing them in their winter quarters. Armstrong (1973) mentions that many species of bird have been recorded singing songs which they had heard months earlier.

When using mimicry, Wheatears might use imitations of up to four different species in a single burst of song. For instance, on 21 March, a male in subsong sang the calls of Pied Wagtail, Meadow Pipit, Skylark and notes very similar to a Blackbird. On 3 April, another male incorporated Meadow Pipit, Swallow, Skylark and wagtail 'chizzick' notes in his song. On 25 March, a third Wheatear sang Goldfinch twitter, twitter of Swallow, then another call which I failed to identify although I had heard it before. It would seem that the choice of imitation used was more or less haphazard.

Mimicry was used, in subsong and full song, chiefly where territorial defence was involved and only occasionally when the male was singing the quieter and rather more sexually motivated, bond-maintaining song to the female. I could detect no other function than that of increasing the already varied repertoire of song phrases of individual Wheatears, and I have the impression — but little concrete evidence — that the most energetic singer, which may have the largest repertoire, may also have been more successful in maintaining a larger territory.

Thorpe (1985) says that the tendency for birds to mimic a wide variety of sounds can probably be regarded as a special development of the ability to learn songs from their parents or their conspecifics. Thorpe quotes Marshall (1950), who says that most mimics are strongly territorial, and this is certainly true of the Wheatear.

FIGURE 7.3 *Male in flashing display*

Thorpe also mentions that there is some evidence that variety in bird song is attractive to females and that mimicry may simply be a way for the male to increase the size of his repertoire. I have pointed out elsewhere that Wheatears which sang more often tended to have larger territories. Another suggestion is that mimicry is important in interspecific territoriality, since some species mimic competitors and may use their mimicked song to this end and mimic mainly local breeding birds. So far as the Wheatear is concerned, it is perhaps relevant to point out that mimicry is used chiefly in territorial song; Skokholm Wheatears, however, did not use the territorial song of local species but mostly their alarm notes.

Another feature of birds that mimic well is, apparently, that their songs have a loose sense of overall pattern, a point which I made about Wheatear's songs at the beginning of this chapter.

CALLS

At this point I am going to describe the Wheatear's calls, which help to co-ordinate the behaviour of relatives and neighbours and which are concerned primarily with intruders, predators and relationships within the family.

'Tuc'

The commonest call heard from adults and older juveniles and first-winter Wheatears is the 'tuc' note, which expresses excitement, anxiety or alarm. It can be uttered very quietly as a single element, but much more commonly as 'tuc-tuc' and with intervals of two or three seconds between each double note or rather loudly at a much faster rate. The 'tuc' call shows on the sonagram

FIGURE 7.4 *Sunning Wheatear spotting possible cause for alarm*

(Figure 7.1e) as a single bar across a range of 1-5 kHz (the other blob on that figure is the 'weet' call). A Wheatear calls it in a wide variety of circumstances when the level of excitement is low, as for instance in low-intensity chases between individuals, whether males or females. It is used by one or both contestants in the more intense territorial disputes, in intervals of song contests, in the gaps in song phrases both when the bird is on the ground or in flight, and in fights with intruders when the owning male may be calling or singing other notes and phrases. It is also used by the male when the female is searching for nest sites or when she is building, when he visits the burrow entrance and the female is down the burrow, and later when the young have hatched and he is bringing food. It is also used whenever a raptor, a human or a dog is in the vicinity of the bird itself or the nest. Any situation which generates any excitement, whether sexual, aggressive or fear, is the basis for a Wheatear to use this call.

In mid July, when adults are moulting, the 'tuc' notes began to sound like 'tch', or the rather unformed call of the juveniles when they had just become independent.

The 'tuc' note used by Greenland Wheatears sounds flatter, more like 'tac' rather than the 'tuc' of the call of the Skokholm birds, and is rather similar to the call of the Skokholm juveniles.

In West Africa, Tye (in press b) states that this call is used in similar circumstances and is the usual alarm note given in response to invasion of the territory by humans and some species of birds, especially other wheatears, shrikes and occasionally Tawny Pipits. It is also used very commonly in intra- and interspecific aggressive interactions.

'Weet'

In some text-books the general alarm of the Wheatear is described as 'weet-tuc-tuc'. In my experience, the 'weet' or 'yeet' note was generally used only in the breeding season when the Wheatear was established in a territory and was mated. At low-intensity anxiety it prefixed the 'tuc-tuc' notes and the whole call sounded like 'weet-tuc-tuc'. As the nesting cycle progressed, the intensity and the loudness of the calls increased: the greater a Wheatear's anxiety or fear related to the adults, nest, eggs or young, the louder and more rapidly the calls were uttered; when a predator or human was at or near a nest containing well-grown nestlings, the calls were often so frenzied that the 'tuc' notes were omitted and the adults called a continuous stream of 'weet' notes, which seemed then to sound like 'see-see-see'. They might at that point also be incorporated as a prefix to a song pattern.

The 'weet-tuc-tuc' notes are used in a wider variety of situations than the 'tuc' notes alone, particularly in intervals of song, song flights (when song expresses increased anxiety at the presence of a human intruder), in the intervals of displays of a territorial nature such as the zigzag flight, dancing display and so on — in fact, whenever excitement, whether aggressive or from fear, becomes intense.

The 'weet' note carries well and I heard it when I was more than 200 m from the male in a force 2 wind (6-11 kph/4-7 mph).

The different effect of the 'weet' and 'tuc' notes was shown when nestlings were leaving the nest burrow. While they were actually moving to other burrows the female called 'tuc' notes, but the young came out into the open; as soon as she used the 'weet' notes they dived back underground.

The rates at which the 'weet' note are uttered vary and occasionally seem to be used excessively. I was watching a pair of Wheatears from about 100 m as they were building a nest a fortnight before the first egg was laid, and the male began calling 'weet' vigorously. Four minutes after I had arrived, I counted 86 calls in one minute, which was a slower rate than it had called when I first arrived. During the next four minutes it called at about the same rate until it flew off. It perched on a rock, where it was not bobbing but only sitting and watching for long periods, occasionally preening and shaking its feathers as it called. The female showed no sign of anxiety. This individual seemed to be over-reacting compared with the behaviour of other males in this situation.

Towards the end of June 1949, some males 'lost their voices' and then seemed to call a note which sounded like 'chip' and which, although quiet and insignificant, was used in the same circumstances as the 'weet' note. As I mentioned before, males tended to moult about ten days before females and, at that point, seemed to lose interest in the nest and territory: perhaps this change in voice was related to some physiological change taking place in the male.

In rather the same way that before the male was mated it might use the loud, territorial song in intense territorial battles, so 'weet' might also be used in similar circumstances. Interestingly enough, Wheatears in possession of individual territories in winter quarters also used this note when, for instance, a Great Grey Shrike declined to leave a Wheatear's individual territory (Tye, in press b).

'Tetetetete'

This call expresses anger or hostility: it is a rapid rattle or repetition of five or six elements which I transcribed as 'tetetetete', each element of which is hard and has a vibrant or pebbly quality. Typically the male uttered this call as he passed 1-2 cm over the female's head in the pounce after she had been inactive for 30 minutes or so. At first I was uncertain as to which of the pair called, so close together were they. Both male and female, however, used the pounce, against other species such as Black Redstart, Little Owl and Spotted Flycatcher, when clearly it was the aggressor that called. I heard this call only in the breeding season. The range of circumstances in which the pounce was used is described in Chapter 6.

The Vibrant 'Bree' Call

The vibrant 'bree' call is, to my ears, identical to the contact or location note of juvenile Wheatears. By adults it was used in March, April and early May (last date 23 May), whereas juveniles first called this note in the nest but chiefly from late May until early August. Tye (in press) states that a similar call was heard from three birds in Senegal which were in either adult female or first-winter plumage; he also says that this call is uttered if an intruding Isabelline Wheatear attempts to displace a Northern Wheatear from the latter's individual territory.

Both sexes called it when they were together, often when they were idling, and it may have functioned as a contact or location note with sexual overtones. Occasionally it was used in the intervals of displays which led up to presumed copulation.

On the other hand, a male occasionally used it in the prefix to song delivered either in song flight or from the ground, and also in subsong when unmated and when intruders were present. These were the only circumstances in which aggression was possibly involved.

ADULT NEST CALLS

Only by being very close to the nest was I able to hear the very quiet notes which the female, and occasionally the male, used to nestlings, particularly when they were only a few days old and had not heard the female approach down the nest burrow. I described it as being a strangled call which lasted less than a second and which can easily be imitated by placing the teeth against pursed lips and drawing in air quickly — a sound sometimes used by people when caressing a cat.

Once the nestlings were able to see an adult approaching down the burrow and gaped, the call was less frequently heard. I last heard female W67 use it when the nestlings were ten days old, on which day they showed fear for the first time: in response to the adults' 'weet' and 'tuc' calls the nestlings had turned from the light at the nest entrance and were crouched in the back of the nest when the female came down the burrow; since they continued to hide their heads, she used this call and immediately the nestlings turned around and solicited food.

THE CALLS OF THE YOUNG

During my observation of nestlings W67 through a tunnel into the nest from a hide over the nest, I was able to listen to the nestlings from a distance of about 25 cm. I heard the very first thin 'ee-ee-ee . . .' food calls when they were three days old. On the fourth day, the whole brood stretched up when a parent was about to feed them and called more loudly. Throughout the nestling period, this high-pitched note was used, although it changed its character, whenever the nestlings were fed by their parents.

When the young were eleven days old, I detected a new note which was staccato and had the vibrant quality characteristic of so many Wheatear calls; this I wrote down as 'tchi' or 'tchie'. At ten days old the nestlings began a number of new activities; they climbed on the rim of the nest to defecate for the first time and they cowered in fear. By the twelfth day this vibrant call had developed into 'tchee' or 'tzee', which I sometimes also wrote down as a nasal 'ing', 'zweng', or 'zwang' in an attempt to convey that peculiar, vibrant, quality.

Although nestlings used this call in addition to the high-pitched food call, it was a location or contact note which, more or less, said 'I am here' and which was frequently used after the young had left the nest, split up into their separate burrows and were still being fed by their parents. The twelfth day had also seen major changes in the nestlings' behaviour, and, in particular, some of the nestlings left the nest for the first time and began to move up and down the burrow.

On the thirteenth day I heard the first attempt at the alarm note. It was hard and unmusical and sounded like 'tchuk', which days later had changed into a more recognisable 'tuc' or 'tac'. The call seemed flatter than the 'tuc' of an adult and more like the 'tac' of Greenland Wheatear.

On the fourteenth day one or more of nestlings W67 began to make a sneezing or snorting noise, and while doing so their bodies shook. This noise was heard only over a short period of about 24 hours and I have no idea of its significance, if any.

By the fifteenth day the nestlings had a basic vocabulary of three calls and their variations: (1) 'see-see-see' appetitive calls, still used when being fed; (2) 'tzee' or 'bree' vibrant contact or location call; and (3) 'tuc' or 'tac' alarm call.

The food calls were used only as long as the juveniles were fed by their parents. The 'tuc' calls were used throughout the individual's life. The 'tzee' or 'bree' vibrant call, with varying degrees and changes of pitch, was also used chiefly while they were dependent upon their parents, although this call, or one so similar to it that I was unable to distinguish between them, was also given by the adults in early spring. These vibrant notes were used, rarely, by juveniles after they were independent, and then only in inappropriate circumstances as, for instance, when an eight-week-old colour-ringed individual which had been independent for a month and had already moulted into first-winter plumage solicited food from an adult by using this call.

Chapter 8

FOOD AND FEEDING

While I was studying the Wheatear, I collected information in a rather anecdotal fashion on the different ways in which the birds caught their prey in various habitats on Skokholm, on the Pembrokeshire cliff tops, Weeting Heath, Alderney and elsewhere. From time to time, on Skokholm and Alderney, I measured their actual feeding rates under different weather conditions and at different seasons. On Skokholm, too, I examined the fauna of the grassland over which Wheatears were foraging and, using 10 cm and 0.5 m squares, I recorded such arthropods as I could see in each square. I identified them usually to Order and occasionally to species. While I recorded any species of insect which I saw being caught and eaten, I collected virtually no quantitative data on the Wheatear's food items.

One of the best studies of the foraging behaviour and food of the Wheatear in Britain was undertaken by Alan Tye in Breckland, Norfolk. This was part of a larger study in 1976 and 1977 of the Wheatear's breeding biology and population size (Tye 1980) and its social organisation and feeding habits (Tye 1982). Cornwallis (1975) studied the comparative ecology of eleven species of wheatear (genus *Oenanthe*) in southwest Iran; Brooke (1979, 1981) studied Wheatear feeding methods on Skokholm; Moreno (1984) examined the strategies used by Wheatears when searching for food in north Spain; and Carlson, Moreno and others, in a number of papers, examined the feeding methods of Wheatears in Sweden.

I shall be describing certain aspects of these studies, and particularly those by Tye and Cornwallis, in relation to my own observations.

THE WHEATEAR'S METHOD OF HUNTING

Hop and Peck
The usual foraging method of a Wheatear on the short turf of its typical habitat is to hop forward a few paces (a hop-move) then peck down at its prey or, if it fails to see anything when it stops, to look around and after a few seconds hop on. If it is hungry, a Wheatear may run, rather than hop, or use an odd combination of running and hopping known as an asymmetrical hop. 'Hop

and peck' is the method of foraging that Cornwallis (1975), and Tye (1982) after him, call 'dash and jab'. I think, however, that the latter is a rather misleading term: while Wheatears occasionally do make a 'dash' after an insect in a prominent position and jab at it, their typical hop and peck is more sedate. *The Handbook*'s description (Witherby *et al.* 1938) as 'moving on the ground in quick succession of long hops, frequently halting . . .' remains a very good one. Moreno (1984) calls this method of foraging 'running ground gleaning'.

Occasionally, a Wheatear would fly after prey for 0.5 m or so only about 5 cm above ground. In fact, only 4.6% of the moves made when Wheatears were foraging in this way utilised short flights (see Feeding Rates).

Scooping
Quite often a Wheatear caught a slow-flying insect less than 5 cm above the ground by running or hopping rapidly towards it, with head lowered and beak open, and thus scooped the insect up.

Jumping
In the 1970s, I was looking at a print by Gould of a Wheatear displaying. In the background Gould had depicted two Wheatears apparently jumping 2-3 cm into the air without opening their wings, a behaviour pattern which, in spite of having watched Wheatears for many years, I had never seen. About ten years later, I was again watching a family of recently fledged but still dependent juveniles on Skokholm. Among other activities they were pecking in exploratory fashion at various objects near them, and occasionally they found something which they ate. Suddenly one, copied later by a second, started to jump 1-2 cm into the air without opening its wings, exactly as Gould had recorded many years before; presumably they were jumping to catch low-

FIGURE 8.1 *Scooping*

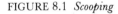

flying insects. Since then I have seen one or two adults hunting in this way, but, even though Gould knew this behaviour well enough to illustrate it, jumping without using wings is not a common method of hunting and is not mentioned by Cornwallis or Tye.

Digging

In addition to pecking at prey on or among grass blades at the end of a series of hops, a Wheatear, like other members of the thrush family, also uses its beak to dig, sometimes very vigorously, into sand, loose soil, plant debris and so on to find food. It hops to the point where it has seen a movement, and stands there displacing sand or soil by sideways flicks of its beak — just like a Blackbird displacing leaf litter — until it finds the prey it is seeking, when it will often pick up four or five items, probably ants, before moving on. Wheatears use this method to take termites in Senegal (Tye, pers. comm.).

Perch and Pounce

Where grass is less heavily grazed and taller, making hopping difficult or impossible, and where 50 cm-high plants unpalatable to rabbits such as ragwort or thistles provide perches, Wheatears use a different technique which Cornwallis and Tye called 'perch and pounce.' Moreno (1984) called it 'perch to ground sallying'. They perch on the corymbs of ragwort, scan the vegetation and then flutter or hover over it, before finally dropping down to take an insect.

Since most of Skokholm's vegetation was grazed short by the large rabbit population, and winter gales had flattened ragwort and other plants on which Wheatears had perched and pounced the previous season, this hunting method was uncommon in the breeding season; in fact I have only one record of a pair, whose nest was close to a patch of bracken, perching on fronds and then dropping to the ground to pick up insects — rather like a hunting Stonechat. On the cliff-top zone of the Pembrokeshire coastline, however, Wheatears occasionally used it in the breeding season. In Breckland, Tye found that 'perch and pounce' was uncommon (used on less than 2% of all attempts to catch prey) and confined to times and places where fairly high concentrations of large prey were available, or in tall vegetation and often after heavy rain.

Perch and pounce was more commonly recorded on Skokholm in late summer, when the plants had regrown. In July, a juvenile apparently found this method rewarding and spent several days localised near the Observatory buildings, where it perched on hogweed and dropped on the grass for food; occasionally it sought insects in the umbels of the plant itself. Oddly enough, Wheatears on Skokholm did not seem to make much use of the dry-stone walls to look for insects on the ground, although Moreno (1984) records that they did in north Spain: they sometimes used them as a take-off point to catch flying insects. Of course foraging Wheatears when alarmed often flew to a perch, such as a wall or rock outcrop, bushes or wires of one kind or another elsewhere, which gave them a good vista. When, eventually, they relaxed, their attention might well be attracted by an insect on the ground, to which they would then fly: in this case the use of a wall for hunting was fortuitous.

In Senegal in winter, while studying Northern and Isabelline Wheatears and their relationship to Great Grey and Woodchat Shrikes, Tye (1984) showed that for both wheatear species there was a significant positive correlation between the height of the perch and the distance to the prey: the higher the perch, the greater the distance they flew for prey. Of the perches used, however, 80% were less than 1 m high. Alan Tye tells me that 'perch and pounce' is much more common in Senegal, but he is not clear why, although it could be the greater availability of suitable perches.

On Longis Common, Alderney, autumn migrants occasionally perched on ragwort, bramble and fence posts to look over the longer grasses, and from these they fluttered low until they spotted their prey and dropped into the grass after it. They avoided perching on flimsy species such as sea radish or wild cabbage. At this time of year, too, migrant Whinchats appeared in the same habitat and they regularly grouped with the more numerous Wheatears. The latter often copied the Whinchats and perched on adjacent ragwort corymbs and hunted like them, even though other Wheatears would be hopping and pecking close by.

Since Wheatears generally prefer wide vistas, did they make use of higher perches for hunting in late summer, or at other times for that matter, because plant density had reduced the vista and increased the danger of being surprised by an enemy, or because it was an optimum method of foraging? Since Wheatears survive and bring up young without using the perch and pounce method of hunting, it is not vital to them. If tall plants were an essential element of the breeding habitat, then Wheatears would not nest on Skokholm or other gale-swept habitats. It sometimes seems to me that researchers, in searching to test computer models, often leave out of their calculations some fairly vital elements of behaviour.

Hovering

As I explained earlier in this book, Wheatears hover principally to observe potential enemies or intruders, although a hovering Wheatear sometimes spots an insect and drops down to catch it. Of course, when using the perch and pounce method it flutters and hovers over the long grasses, but hovering at 3 m or more is not a normal method of hunting. Nevertheless, two males and a female which were on passage through Skokholm frequently hovered over the muddy edge of the North Pond. One in particular would fly from the grassy edge, hover at about 5 cm above the wet mud, dip down and pick up an insect from the surface, dropping its legs for a second or so, and the continuing to hover; usually its legs were dangling as it moved to a different area. All the same, the action suggested that the Wheatear had discovered that the mud would not support its weight and that, to catch the prey in an unusual place, it had to adopt another technique which was really a variant of 'perch and pounce.'

FIGURE 8.2 *Aerial sallying, or flycatching*

Aerial Sallying

The Wheatear is very adept in catching prey which are flying only a few centimetres above ground or as high as 20 m. The method it uses depends on the type of prey and the height at which it is flying, which, in its turn, often depends on the strength of the wind: few insects fly high in strong winds.

Low-altitude chases usually occur when a Wheatear is hopping and pecking, sees an insect flying only a few centimetres above ground and flies after it for 1 m or so: if it fails to make a catch, it returns to hopping. Bees and butterflies such as meadow browns, large and small whites and small coppers are all chased in this way, not always successfully. These low aerial chases can be quite spectacular when the intended prey is an active 'dodger': one Wheatear chased a bumble bee for about 30 seconds, twisting and turning in its flight, but failed to make a kill.

Typically a Wheatear climbs steadily and directly towards a prey which it has marked, but, if the angle of climb is steep, the flight can become laboured. In a way the upward flight after an insect resembles the song flight, but it is a direct climb whereas in the song flight the Wheatear beats its wings intermittently and thus climbs jerkily. Bumble bees, bluebottles and slow-flying beetles are usual targets. Juveniles as young as 22 days, which had only just begun to use flight as an escape reaction, caught flying prey in this way. Slow-moving and very conspicuous march flies such as *Bibio johannis* or *B. marci*, which were quite common in spring on calm days, were not seen to be taken, and Tye found that they were ignored in Breckland.

The habitats in which I watched Wheatears on Skokholm were less sheltered than those on Alderney. Consequently attempts to catch insects higher than 1 m or so were less common and, in any case, Wheatears rarely flew higher than 6-7 m there. In Breckland, more sheltered than either Skokholm or Alderney, Tye regarded aerial sallying as uncommon early in the season, but more frequent later on when flying prey such as bumble bees and adult grasshoppers were more common; even here wind speed might have been a factor.

On calm, rather humid days in August, Wheatears climbed to about 20 m on Longis Common, to catch bumble bees or ants when they were swarming.

Flying ants attracted many species: Starlings hunted them on the ground as well as in the air, circling around in a somewhat ungainly fashion; above them would be Swallows and occasionally late Swifts; higher still would be gulls, both Black-headed and Herring. Like Starlings, Wheatears fed on ants on the ground but quite often flew up to take a single insect, after which they returned to the ground or occasionally to the power wires. It was apparent that, before taking off, Wheatears had seen and marked the flying ant which was their target, unlike Starlings which circled around continuously among the flying ants, apparently snapping up several prey items.

Wheatears rarely picked up a number of food items in quick succession — except of course, when they found ants on an anthill! In this respect particularly, their method of hunting resembled that of a Spotted or Pied Flycatcher, whereas a Starling's method was more like that of a Swallow: they were able to eat one insect after another without alighting.

Swarming ants led to one unusual encounter between a Wheatear and a Swallow, two species that I did not normally expect to see competing for food; nor, indeed, did I expect aggression between them. Wheatears were making use of banks of spoil alongside a trench on the common as lookout posts, and from there flying up to about 10 m and chasing flying ants. Suddenly a Swallow dived at a Wheatear standing on top of the spoil; the latter flew off and dived into the bottom of the trench, which was about 1 m deep. The Swallow then flew down the length of the trench and drove out the Wheatear, which again alighted on the spoil. The Swallow did not press the attack further. I had difficulty in understanding what stimulated the Swallow to behave in this way.

FEEDING RATES

On both Skokholm and Alderney, I tried to find out the rate at which Wheatears were hopping and feeding by counting the number of hops in each series of moves — or hop-moves. Table 8.1 shows that, in nearly 70% of the hop-moves, Wheatears hopped between two and five times before they stopped to peck down at an insect or to look around. The mean number of hops in each hop-move for all Wheatears studied was 3.8 (S.D. = ± 2.8, range 1-c.30 hops, n = 515). If I omitted hop-moves of more than ten paces, which comprised 3.1% of the total, the mean dropped to 3.2 hops (S.D. = ± 1.6 hops, n = 479). I was interested to see that Tye, in his Breckland study, found that the move length of his Wheatear population was 4.3 hops per move, which, incidentally, he calculated as equivalent to a distance of 0.69 m ± 0.04 m. I never measured the distance covered by hops, but Tye calculated that the mean of 19 hop-moves was 16.0 cm ± 0.5 cm (range 13.33-20 cm).

Tye also calculated that when a Wheatear was standing still it could scan an area of 50 cm^2 and suggested that it might be able to scan as much as 0.59 cm^2 for large insects. He further suggested that Wheatears chose a hop-move length that resulted in no overlap in scanned areas.

TABLE 8.1: *The mean number of hops by Wheatears in each hop-move*

	NUMBER OF HOPS										Pecks-down	Flies	Total hop-moves
	1	2	3	4	5	6	7	8	9	10+			
All ♂♂	29	44	25	21	18	8	5	6	—	8	37	2	164
				(Mean = 3.7; S.D. ± 3.0)									
Adult ♀♀	17	54	48	26	21	8	2	4	—	5	30	9	185
				(Mean = 3.4; S.D. ± 1.9)									
Autumn migrant ♂♂ & ♀♀	14	25	35	24	34	18	3	8	1	4	59	12	166
				(Mean = 4.1; S.D. ± 2.2)									
Totals	60	123	108	71	73	34	10	18	1	17*	126	23	515
			(Mean = 3.8; S.D. ± 2.8; n = 515)										

* = 10, 10, 10, 10, 10, 10, 10, 11, 12, 15, 20, 20, 20, 20 & 30.

I also tried to discover, using the small amount of data that I had available, whether males and females during the breeding season, or first-winter birds on autumn migration, showed any great differences in the lengths of their moves when hunting. The mean of 164 hop-moves by off-duty males was 3.7 hops per move (S.D. = ± 3.0 hops), while the mean of 185 moves by females, when foraging between spells of incubation, was 3.4 hops (S.D. = ± 1.9 hops); whereas, of 168 hop-moves by first-winter birds of both sexes, the mean was slightly greater at 4.1 hops (S.D. = ± 2.2 hops). As one might have expected, the females, which were hunting at the end of a spell of incubation, were making shorter moves before taking or looking for food, whereas the first-winter birds, which were holding individual territories and at a stage in their lives when they had some days to put on weight before they started to disperse or migrate, did not need to hunt so rapidly. Nevertheless, it was obvious that, generally, the hopping rates of the three groups were little different.

Success Rate
Looking first at the rates at which males and females foraged, for off-duty males I had 44 observations of Wheatears hopping and pecking which showed the number of pecks-down per minute, and for 29 of those I also knew the number of hop-moves per minute that preceded the peck-down. On average, off-duty males made one hop-move every 5.9 seconds (S.D. = ± 2.6, n = 29), or about ten a minute. For incubating females which had just left the eggs for a feeding break, the figure was one hop-move every 5.7 seconds (S.D. = ± 4.2, n = 185), or ten-and-a-half a minute: although the difference was very slight, the female moved more frequently than the male. Similarly, the pecking rate for males was one peck every 15.7 seconds (S.D. = ± 1.9 seconds, n = 44), which is just about four times a minute; whereas the females' pecking rate was slightly faster at a peck every 11.7 seconds (S.D. = ± 11.9, n = 188), or about five pecks-down a minute. The mean rate for all males and females was 4.8 pecks-down a minute.

Tye, looking at this situation from another angle, found that the number of moves for males and non-incubating females was also about ten pauses a minute, but for an incubating female the pauses were shorter and she made more hop-moves a minute, which is. approximately what I discovered on Skokholm. Tye states that 'in addition to searching a larger area per unit of time, and despite having shorter pauses, incubating females found more food per unit area searched than at other stages'.

By assuming that all pecks-down resulted in the capture of a food item and stating the pecks-down as a percentage of the number of hop-moves, it appears that the success rate of females was greater at 50.9% (S.D. = ± 27.1, n = 187) than that of males, whose success rate was 39.8% (S.D. = ± 21.5%, n = 29). These results were to be expected, since almost all the females observed were breeding females which had just left their eggs after a spell of incubation: they had only a third of the daylight hours in which to hunt since they were incubating for the other two-thirds of the day, while most males were off-duty birds whose role at this time was defending or guarding a territory.

The mean rate at which all those Wheatears studied pecked down, whether on Skokholm or Alderney, and including first-winter birds — some of which had been bred on Skokholm — was 7.4 pecks a minute (n = 261). It was clear again that off-duty females foraged at a faster rate than off-duty males before breeding or first-winter birds during rests on migration. The mean rate for females was 9.8 pecks-down a minute (S.D. = ± 2.5, n = 122), and for the others 5.1 pecks-down a minute (S.D. = ± 2.7, n = 139).

When discussing, in Chapter 13, the rates at which the parents brought food to their nestlings it is clear that weather, and rain in particular, reduced the number of visits to the nest with food. In early spring, wet and windy conditions also reduced the feeding rate of Wheatears: on 10 April, when the wind strength was about 56 kph/35 mph with some sun and heavy showers, the rate had dropped to 3 pecks a minute and the peck-down rate of breeding females also dropped to 6.6 pecks a minute compared with the mean of 9.8 pecks a minute. At the other extreme, some of the fastest peck-down rates occurred when the temperature was very high: when a thermocouple placed among the grasses showed 17-18°C, the peck-down rate varied between 11 and 19 pecks a minute.

Tye concluded that Wheatears using dash and jab, or hop and peck, chose a move-length relative to search radius that reduced the distance moved to a minimum consistent with avoiding searching the ground more than once, and which maximised scanning rate for a given pause length and for running speed: i.e., if pause length and running speed are such and such, the birds chose a move-length that maximised the search rate at those values.

GROUND SURVEY

Having discovered a little about the hopping, pecking and success rates of a hunting Wheatear, I tried to find out what food items were available to the birds by inspecting the vegetation over which they had been hopping and pecking.

I used either decimetre squares (10 × 10 cm) or half-metre squares (50 × 50 cm) placed more or less at random in 20-50 (but occasionally up to 100) positions within the area in which a Wheatear had been hunting and counted all the invertebrates I could see among the grasses in ten seconds. With the aid of a thermocouple placed in the vegetation, I also read off the ambient temperature.

From 29 March until 5 April, there seemed to be few food items available: the temperature among the grasses was between 8°C and 9°C and the ground was still damp. In 100 decimetre squares I found only springtails (Collembola) and mites (Acari). The mean number of these animals per square was 2.8 (S.D. ± 0.5), and of the 277 individuals 89% were springtails. It would appear, therefore, that the earliest residents which returned to the island at the end of March could claim the territories they had occupied the previous year but food in them was in very short supply, a situation which Tye also found in the Brecks.

From 11 April to 1 May, more animals became visible, including the ants *Myrmica scabrinoides* and *Lasius flavus* and a few chalcid wasps. Diptera were represented by gall midges (Cecidomyiidae) and march flies (Bibionidae). Spiders also emerged, including those of the genera *Erigone* (Linyphiidae) and *Lycosa* (Lycosidae) as well as *Xystichus erraticus* (Thomisidae). Finally, I found some true bugs (Hemiptera).

In this period, the mean number of arthropods I saw in 374 decimetre squares was 1.1 individuals (S.D. = ± 0.4), at a time when the thermocouple recorded temperatures among the vegetation of between 11°C and 18°C. Springtail numbers were again much reduced; in fact, with four exceptions their numbers had dropped to less than half what they had been in late March and early April. On 17 April 1950, with 0.8 cm of rain recorded, on 29 April, with 0.3 cm, on 30 April, with 0.6 cm, and on 1 May, when the vegetation was still very wet after rain, springtails had, however, increased markedly: in 90 decimetre squares on those four days the mean number of springtails was 1.3 per square, and apart from one mite no other arthropods were recorded. In Chapter 12, I describe the rates at which parents brought food to the nest and how the number of visits with food declined on wet days. These checks on the number of obvious prey items in the hunting areas gave a further hint that Wheatears might well have to work much harder to find sufficient food for themselves or their families in wet weather.

There was little change in the period 2-9 May, although the mean number of prey items had dropped to 0.7 per square (S.D. = ± 0.7, n = 209 squares); springtails had disappeared again, but other arthropods had increased.

Since I had been finding so little in the decimetre squares in the previous period, I decided that, between 10 and 24 May, I would also check what I could find in a smaller number of squares with half-metre sides in addition to the decimetre squares. Taking the decimetre squares first, the mean number of animals in 60 squares was 0.2 (S.D. = ± 0.1); and the fact that it was as high as this was again due to a rainy day on 24 May, when in six squares the mean number of individual springtails rose to 12.5 per square. On the other hand, when I looked at 90 half-metre squares, the mean number of prey items was 1.1 (S.D. = ± 0.5) per square. Although drier weather had driven the springtails into hiding, the other insects and spiders were still present and in addition beetles (Coleoptera) were beginning to appear.

Between 10 and 24 September 1950, I repeated this survey on nine days, on eight of which I searched for possible prey items in 50 decimetre squares and on one, only in 20 squares. The mean number of prey items was slightly larger than in April at 1.2 per square, but daily variation was much greater, ranging from 0.7 to 1.9 items per square. The figures were probably depressed owing to the wet, windy and overcast conditions which predominated during these 15 days (with some rain falling on 14 of them). This wet weather almost certainly accounts for the high numbers of springtails at a season when I would have expected them to have been less obvious, and for the low numbers of ants.

The species composition had also changed with the season. Small Nematoceran flies, which I had not seen in spring, were often quite common, and representatives of the Orders Coleoptera (Staphylinidae and Chryso-melidae) and Hemiptera, as well as spiders of the genus *Lycosa* and mites, were also slightly more obvious than they had been in spring. The Hymenoptera were represented only by ants, and the four species known on the island — *Myrmica scabrinoides*, *Lasius alienus*, *L. flavus*, and *L. niger* — were recorded, of which the commonest were *M. scabrinoides* and *L. flavus*.

In this search, I also compared the size of the possible prey items with that of an ant: 13% were the same size, 2% were larger and 85% were smaller.

More recently, and in another part of the British Isles, Tye (1982) carried out a major study of the social organisation and feeding of the Wheatear in Breckland, Norfolk/Suffolk. To examine their food, he captured invertebrates in habitats utilised by foraging Wheatears by using a vacuum pump mounted on a back pack with a flexible tube leading from the pump to a rigid circular sampling head in which a fine nylon net was mounted. Since some inverte-brates avoided capture by this machine, either because they were quick enough to jump out of the way or because they could burrow quickly and deeply into the vegetation and thus resist being sucked into the trap, Tye developed a visual sampling method similar to my quadrat technique which, while it did not estimate the absolute numbers of all the invertebrates present, produced results which were comparable between areas and dates; these enabled him to estimate actual availability of invertebrates important to Wheatears.

Flying insects, which were taken by Wheatears, were also difficult to count. Tye therefore counted and identified, at least to their Orders, all those flying less than 1.5 m high over a line 15 m in length in 30-minute periods.

He also collected faeces when time permitted and, although the quantities collected were insufficient for comparative work, he was able to provide a list of prey types with an estimate of abundance. The nestlings produced faeces when handled and they also regurgitated pellets, both of which Tye collected.

Tye grouped the invertebrates which were collected, or examined, by the first two methods into two main categories: (1) distasteful animals or those capable of rapid movement, which he called 'dodgers'; and (2) all other invertebrates, which he called 'sitting ducks'.

The first category consisted of Homoptera, both hoppers and aphids etc; all Diptera except Bibionidae (march flies) and Tipulidae (crane flies), which are slow to take off and slow in flight; of the Hymenoptera, Chalcidae, Ichneu-monidae, Cynipidae, Braconidae etc, most of which were small wasps, some parasitic, some gall wasps, and so on. Tye shows that bees and wasps were chased and caught in flight and their potentially dangerous sting removed. Distasteful insects such as cinnabar moths and their caterpillars were never observed to be eaten.

145

'Sitting ducks' also consisted of some aphids (Homoptera); Heteroptera (more plant-sucking bugs); march flies and crane flies; some Coleoptera (beetles); Dermaptera (earwigs); Orthoptera (grasshoppers), which in spite of being able to jump remained conspicuous after landing and were often chased by Wheatears; Thysanoptera (thrips); Acari (mites); Araneae (spiders); various Arthropods, including Chilopoda, Diploda, and Isopoda; and Mollusca. This category also consisted of insect larvae, mainly of Lepidoptera (butterflies and moths) and Coleoptera; and Hymenoptera, including ants, wasps and bees.

Ground sampling showed that there were peaks in abundance of the 'dodgers', particularly of the larger animals, in May, June and also September, owing mainly to the abundance of small flies and hoppers. Most of the 'sitting ducks' peaked in April and May, then declined to low levels for the rest of the season. Ants were by far the most numerous component of the 'sitting ducks' for most of the season, but owing to their larger size beetles and grasshoppers were also of major importance to Wheatears. Two factors which probably mitigated the effects of the fall in the availability of 'sitting ducks' in June, July and August were the abundance in July of large aerial insects, of which the most important were bees, and the increase in mean prey size caused largely by the growth of grasshopper nymphs. Ladybirds were not eaten. Some butterflies were eaten, but the blues (Lycaenidae) and skippers (Hesperidae) were fast-flying and difficult to catch, as also were large Diptera.

Cornwallis (1975) studied prey availability to eleven different species of wheatear in Iran by pitfall trapping and examination of the fauna of 100 m transects. He states that in all the areas investigated the prey available on bare ground where wheatears, including *O. oenanthe*, mainly foraged was broadly similar, and consisted chiefly of worker ants (Formicidae) with smaller numbers of Coleoptera, Orthoptera, Homoptera and Araneidae mostly of 2-10 mm in length. Rarely eaten by Wheatears in southwest Iran were molluscs (Gastropoda) and some plant seeds. Gut analyses showed that, roughly speaking, prey taken by wheatears were similar in taxonomic composition and size distribution to the arthropods available in the environment. 'This is consistent with the hypothesis that wheatears eat any prey item that they encounter as long as it is large enough to be worth fetching and not too large to handle.'

FOOD

Vincent (1988), reviewing published descriptions of Wheatear food, has described the species as chiefly insectivorous, although it also eats spiders, other small invertebrates and molluscs, and has occasionally been recorded eating berries.

In view of the Wheatear's almost circumpolar breeding distribution and its winter distribution south of the Sahara, it is not surprising that a very large number of food items has been recorded: I do not propose to list them here.

From the summaries by Vincent, it is obvious that beetles, and particularly Curculionidae (weevils), figure most prominently in the diet of an adult Wheatear, followed closely by ants (Formicidae). Beetles and ants appear in almost every list of Wheatear food. Also appearing commonly are the larvae of butterflies and moths (Lepidoptera), bees and wasps (Apoidea) and grasshoppers (Orthoptera). In the diet of young, Vincent again shows that the larvae of Lepidoptera and Tipulidae are particularly important, as are, to a lesser extent, spiders.

In Breckland, Tye analysed 72 faecal samples from nestlings. Those from nestlings less than four days old contained only small animals less than 6 mm in length. Faeces from older nestlings contained bees, large flies and quantities of fragments of a bettle *Isomira*. The only large prey found in faeces from nestlings younger than five days was caterpillars. Faecal remains suggested that the major components of the nestling diet were Coleoptera, Orthoptera and Aranea, but that Diptera and Lepidoptera larvae were also important.

In pellets regurgitated by juveniles, he found mostly *Bombus* species, the remains of caterpillars of peacock butterflies *Inachis io*, mites *Erythraeus phalangoides* in three pellets, grasshopper remains in one pellet and large beetles.

Summarising, Tye states that, taking into account faeces, pellets and direct observation, the most numerous prey in the diet of nestling Wheatears were grasshoppers, ants, beetles, spiders, bees, caterpillars and large flies. The prey which contributed most to energy intake were grasshoppers, bees, caterpillars and large flies, except in the case of the very young nestlings when smaller prey (small beetles, spiders, grasshopper nymphs and ants) were probably more important.

I was able to recognise in the field only the larger prey items taken by Wheatears. At times, however, I could guess fairly reliably that a Wheatear, if it was pecking at an ant heap or where ants were common, was eating ants, and I was able to check from the contents of a few faeces what some of the ant species were.

Faeces were easily come by: close watching revealed the spots where they were dropped, and it was then easy to collect them. Generally speaking, faeces gave only a partial indication of what a Wheatear was eating, but, since only the hard parts of the prey pass through the gut and are evacuated in an identifiable form, and since the soft prey items are either digested or are too mashed up in the process to be recognisable, they are not a wholly reliable way of assessing the Wheatear's food intake. Furthermore, what does survive the passage through the gut may be misleading. Were the vegetable remains, such as bits of moss and grass, found in some faeces from Skokholm's Wheatears deliberately eaten as part of their diet, or were they accidentally plucked when the Wheatear was pecking at some very small prey such as a springtail or an ant? Since these vegetable remains had passed through the gut undigested, they would seem to have little nutritional value to Wheatears and were possibly the only clues that the soft-bodied springtail and small flies, which I have shown to be fairly obvious in early spring, were taken by Wheatears.

Tye found that the only animal parts that passed through the gut were those which were indigestible, such as the elytra or wing cases of beetles or almost complete exoskeletons of ants. Since march flies were rare in the sample but were abundant on the heath, they seem to have been avoided, as apparently happened on Skokholm. Other remains found in faeces consisted of Lepidoptera forewings, spiders' legs, caterpillar head capsules, fragments of snails and possibly grasshopper legs.

While on the subject of indigestible contents, one faecal pellet which I examined contained five small pieces of Old Red Sandstone of which the largest was 3 mm × 2 mm and the smallest 0.5 mm × 0.5 mm. I found it rather difficult to explain the presence of these pieces of stone. Although many bird species pick up grit to aid digestion, the ones best known to use this habit are granivorous; in any case, this Wheatear was excreting them.

Periodically, Wheatears cast pellets of the harder indigestible material: an individual will hold its head slightly forward and downward and then flick it slightly until the small pellet is ejected. Unfortunately my small collection of these was lost before they had been examined. Tye found that the hard remains of bees, which were often caught in the Brecks but which did not appear in the faeces, were ejected in pellets.

In the following paragraphs I comment on such food items recorded on Skokholm and to a lesser extent on Alderney, but obviously this is not a comprehensive list of Wheatear's foods.

Earthworms

Wheatears are chiefly insectivorous but rarely, when females are feeding nestlings, they bring small earthworms to the nest. Female W53 brought three in quick succession within four minutes on a day of heavy rain when the worms might well have come close to the soil surface. Similarly, on another

FIGURE 8.3 *Female with long earthworm*

wet day, female W90, whose second-brood young were five days out of the nest or about three weeks old, pulled out a worm which was so large that she could hardly fly with it: she pecked at it, pulled it along the ground for about 5 cm, then tried to fly but could hardly lift it; finally, she held it in two places, looped, as a Blackbird holds long worms, and flew with it out of sight. Only once did I see a fully-grown Wheatear eating an earthworm, and this was also on a wet day in April.

It is perhaps surprising that earthworms figure at all in Wheatear's diet: as I mentioned in Chapter 2, Wheatears on Skokholm tended to avoid nesting in damp ground on pockets of boulder clay, possibly because neither rabbits nor Manx Shearwaters excavated their burrows in such regularly water-logged soil. Tye (1982) recorded that, after heavy rain, a few earthworms and a number of beetle larvae which had been dug out of the damp sand of the Brecks were taken into the nest. It is likely that earthworms did not often come to the surface in the drier, sandy-soiled habitats where most Wheatears nested — unless rain had dampened the soil. In view of the difference in climate between eastern England, with relatively low rainfall, and west Wales, with half as much rainfall again, earthworms were more likely to figure in a Skokholm Wheatear's diet than in one from Breckland. In any case, Russell (1957) points out that earthworm numbers are generally low both in summer and on dry soils.

Hymenoptera

(1) Ants: Formicidae Most accounts of the Wheatear's diet that I have looked at show worker ants as one of the two commonest items. Most of the identifiable animal remains in Wheatear faeces that I examined were of ants, and the faeces themselves were slightly acid with pH values of 5.6 to 5.8. I dissected one pellet and examined it under a binocular microscope. Most of the parts were from *Lasius flavus*, but there were also parts of a worker and parts of a female *Myrmica scabrinoides*. Table 8.2 lists the complete contents of that one faecal pellet.

The table indicates that this Wheatear had eaten mainly worker ants: the species which featured most commonly as food on Skokholm was *Lasius flavus*, followed by the slightly larger *Myrmica scabrinoides* and by *L. niger*. I have no records of Wheatears feeding on the fourth species of ant known on the island, *L. alienus*. While ants are small and difficult to see, an observer can easily judge from the bird's behaviour when it is eating them at an anthill, although not so easily when it is picking up single insects. Tye also considered that ants were one of the most numerous prey in the nestling's diet, in spite of the distasteful formic acid.

(2) Bees: Apidae Bumble bees are conspicuous and tempting targets for the aerial chases of Wheatears. They are large and dark and, on calm days, fly across the sky in, apparently, a slow flight, but in fact they are not easy to

TABLE 8.2: *Contents of one Wheatear's faeces*

Lasius flavus		
SCAPES	Left 7	Right 6
ANTENNAE	One complete	
TARSUS	1st segment 3	1st segment 9
TIBIA	Left 4	Right 7
JAWS WORKERS	Left 11	Right 7
WINGED FEMALE		
Tibia	*Left*	*Right*
	Front 5	Front 3
	Middle 2	Middle 4
	Hind 5	Hind 6
Femora		
	Front 3	Front 3
	Middle 9	Middle 2
	Hind 2	Hind 5
	One pair jaws	
Ant? sp.	1	1
Myrmica scabrinoides		
Remains of 2 winged females and one worker		

catch and many of the towering aerial chases were unsuccessful. A few were taken into the nestlings. Tye regards bees and wasps as important items of diet for nestlings, and points out that adult Wheatears are adept at removing the stings by banging them on the ground and rubbing their abdomens along the ground. King (1968) saw Wheatears catching bumble bees by the head as the landed on flowers, after which they were banged on the ground as if to remove the sting. Apparently, in southwest Iran Cornwallis (1975) did not find bees as important a food as beetles and ants.

The species that I saw Wheatears catching and eating most commonly on Skokholm and Alderney were *Bombus lucorum*, *B. muscorum* and *B. terrestris*. They were caught flying at heights of up to 20 m, while visiting flowers at about 15-20 cm, or when more or less moribund in cold weather. On other occasions Wheatears looked at bumble bees flying near them but did not give chase.

Orthoptera
Only one species of grasshopper, the common green grasshopper *Omocestus viridulus*, had been recorded on Skokholm, but I never saw Wheatears hunting it. Grasshoppers were apparently common in Breckland, and Tye found them to be the commonest prey item brought to nestlings; but they could be diffi-

cult to catch, leading the Wheatears into some extraordinary chases involving great leaps and bounds.

Lepidoptera

Butterflies and occasionally moths often tempted Wheatears into aerial chases, but adult butterflies were not a major part of the Wheatear's diet: in spite of their slow flight, these insects were apparently not an easy prey. Occasionally females and first-winter birds caught large whites and meadow browns, which the females took into the nest after stripping off wings and legs. Other butterflies that I saw Wheatears catching on Skokholm were small whites and small coppers. Adult moths were less commonly taken by Wheatears, but I was surprised when a female was carrying a cinnabar moth in its beak as if to feed nestlings and a juvenile 36 days old was seen pecking at a caterpillar (but was not seen to eat it). E. L. Turner saw an adult Wheatear taking cinnabar caterpillars to feed nestlings (Coward 1920). Tye has told me that he saw a Wheatear catch an adult cinnabar on the Brecks, but only once, even though these moths were very common there. The flesh of both adult and larvae of the cinnabar are said to be distasteful to birds: the moth's scarlet hindwings and blackish-green forewings, and the caterpillar's orange-yellow colouring with each ring banded with purplish-black, are both examples of warning coloration (Ford 1955).

Generally, however, lepidopterous larvae together with tipulid larvae formed a major part of the nestlings' diet after they were a few days old. The species most commonly brought to nests on Skokholm were the larvae of dark arches and antler moths, both of which fed on the lower stems of grasses, particularly *Poa annua*, and were probably more accessible when the grass was heavily grazed. Even so, Wheatears often had to dig quite energetically among the vegetation, pecking, pulling and digging in order to extract the caterpillars; and finally they were banged on the ground. White ermine and other hairy caterpillars were also beaten on the ground and pecked at, the flesh being eaten but the skin discarded. Tye saw hairy caterpillars being run through the bill a number of times to get rid of irritant hairs and spines, distasteful contents and so on.

Diptera

The flies that I saw being eaten most commonly by Wheatears on Skokholm were craneflies (Tipulidae) and particularly their larvae, which were sometimes brought to the nestlings, and the bulkier bluebottles (Calliphoridae). Brook (1981) found that cranefly larvae of about 25 mm were fed preferentially to nestling Wheatears, whereas adults took those of about 14 mm. Adult Wheatears caught bluebottles sometimes by running very fast with wings raised and scooping them when they were a few centimetres off the ground; high-flying bluebottles were also chased. A whole range of small flies was caught within a few centimetres of the ground, and I described earlier in this chapter how juveniles and adults, rarely, jumped up after them without using their wings.

Coleoptera

As I showed in the ground survey, beetles of various kinds appeared in the short grass quite regularly in May and were sometimes chased by Wheatears, some being caught. Click beetles (Elateridae) were brought to the nestlings but not eaten, and their dried exoskeletons were found in the bottom of the nest. It is possible that these beetles had found their way into the burrow and died there, but I think it more likely that they were brought in as food but were too hard for the nestlings to eat. These click beetles were slow and low fliers and quite an easy catch for Wheatears.

A weevil, found in a Wheatear pellet, was identified by A. E. A. Pearson as *Otiorrhynchus sulcatus*.

Plants

Small fragments of grass blades and some moss were sometimes found in Wheatear faeces on Skokholm, but I thought that these fragments were probably picked up by chance when the bird was taking small insects such as springtails and ants.

Most of the records of Wheatears taking fruit or seeds originate in countries with arid conditions, whether very warm or very cold. Cornwallis (1975) records seeds in stomach contents of Wheatears in southwest Iran. In another arid area in the far north of its range, the Kola Peninsula, one out of five stomachs contained bilberries, whereas the others contained only insects, chiefly weevils and small click beetles (Dementiev and Gladkov 1968). In Alaska, grass seeds and fruits of saxifrages were eaten (Gabrielson and Lincoln 1959). In Greenland, birds such as the Horned Lark and Wheatear, which normally prefer insects to seeds, are forced to feed on a mixed diet since the season for insects is too short in the Arctic to provide a diet exclusively of them. Crowberries, bilberries and the fruits of juniper are the berries most eagerly sought (Freuchen and Salomonsen 1960).

Chapter 9

NESTS AND NEST-BUILDING

NEST-SITE SELECTION

The male's role during the selection of the nest site is largely that of lookout or of a stimulator and bringer of confidence, activities all of which allow or encourage his mate to continue her search for a suitable nest site without too much interference. He tends his territory, defending it against intrusions by Wheatears and other birds, including predators, watching the female and occasionally coming to the nest with her. He stimulates her to further activity by singing, by calling excitedly, sometimes by displaying or pouncing on her. While earlier the male had followed the female everywhere, once she began searching systematically he now often perched on a tussock or a rock and watched. It was almost as if, once accustomed to the territory, neither needed the close proximity of the other to give them confidence as they had done when they first arrived in the territory.

I got the impression that one of the pair was satisfied or content when its mate was using the appropriate behaviour pattern for the season or for the circumstances; for instance, if the female was searching for nest sites busily or the male was attacking a male intruder, the mate would appear relaxed. If the female had stopped searching for nest sites and had been loafing for as much as an hour, however, the male was very likely to follow her more closely, standing erect, bobbing, flicking his wings and, finally, perhaps pouncing on her, after which, almost invariably, she would become active again. The same behaviour occurred later when she had taken a long break from nest-building and, again, the male pounced.

The female sometimes attacked her mate when he did not attack male intruders. A male Greenland Wheatear had established its individual territory within the breeding territory of a pair, and fairly close to the burrow in which the female was building. After a day in which he displayed at and flew at the larger intruder without driving it away, the male of the pair broke off his attacks. At this point his own mate began to attack him. All had been well while her mate had been attacking the intruder, but, once he gave up, her anger at the continued presence of the intruder could be satisfied only by attacking the sole object which could then be attacked — her mate. Not until the Greenland Wheatear left two days later to continue his migration did the breeding female begin building again.

In both these examples, it appeared that the failure of one of the pair to behave appropriately left the 'reciprocal drives' of the mate unsatisfied. In the first case, the male expected the female to search for a nest site and have a rest period; but, if the rest period continued too long, the absence of activity by the female became a cause of the male's dissatisfaction — the male's reciprocal drive was unsatisfied. Similarly, the female attacked her mate when he stopped attacking the intruding Greenlander: i.e., the intruder should have been driven out. The Greenlander, being bigger than both, caused her fear. To satisfy her territorial drive she had to attack something of her own size: her mate!

Almost imperceptibly the business of selecting a nest site changed, and the female began to carry nest material into one or more holes. Although ummated males had occasionally taken material into holes while awaiting the arrival of a female, they never took it into one which was eventually used by their mate; they appeared to take material into whichever burrow was handy at the time. Nor did I see unmated but territory-holding females taking material into holes, but females did not remain unmated and in the possession of territories for long.

König (1965) and Panov (1974) suggested that males chose nest sites and indicated their choice to their mates by displaying at the entrance to the hole, but I saw no evidence to support this suggestion. While males on Skokholm often displayed close to the burrow, I concluded that the spot on which they displayed had no particular significance. The males might become more excited, standing in the erect position, wings flicking, and bobbing when the female was about to enter a burrow, but in my experience the selection was made by the female. I pointed out in Chapter 6 that burrows, holes in walls, even depressions in the ground about 1 m across often had an effect on Wheatears, particularly during the sexual displays.

Part of the difficulty in trying to establish the exact point at which females decided on a nest site was that they often built in several holes and only after a day or two did they concentrate on a single burrow; even then they would change if they were badly disturbed. Female X6948, which was ringed as a juvenile on the island in 1947, built in two holes in 1948 before settling on one and in 1950 did exactly the same, so it was not only first-year females that built in more than one hole at the same time. Other females were recorded taking material in and out of as many as four burrows before they selected one. Tye (1982) records that several pairs initiated several nests but completed only one.

Another female which changed nest sites was seen carrying material into a burrow on 7 April 1951, which was probably within 48 hours of her arrival. A week later she was building in a second burrow, but came back to the first and was building there on 17 April. On 5 May she was building 5 m away, but left that, and the young were reared in a fourth nest burrow. The reasons for these changes were unclear; by carrying nest material into a burrow so soon after her first arrival she might have been behaving as unmated males occasionally did and been stimulated by the sight of one hole to pick up material, although the fact that she finally built in that hole might be significant.

In 1978, a female on Skokholm built in three burrows in succession: she deserted the first after gale-force winds had blown directly into the hole in the wall in which she was building; a day later she was building in another hole in the wall, but deserted after a photographer erected a hide too rapidly and too close to the hole; in the third hole in the wall she reared young. All three nests were within 30 m of each other. Incidentally, the third hole in the wall had, I believe, been used for nesting by two pairs of Wheatears 25 years earlier, but in the intervening years my nest marks had disappeared so I could not be certain, although the position of the hole in the base of the wall and the shape of the rocks next to the entrance struck a chord in my memory.

Some nest sites became traditional sites in that they were used over many years by a series of pairs or individuals. Other ornithologists have told how, after many years, they returned to a locality and found birds nesting in sites which they had occupied many years before. Harthan (1958) mentions finding a Merlin's nest in exactly the same site as it had been 30 years before, Yapp (1970) also mentions that a friend showed him a Goldcrest's nest hanging under exactly the same fork in the bough of a cedar under which he had found a Goldcrest's nest about 30 years previously. Such traditional sites are better known for birds of prey such as Golden Eagles and Ospreys, and H. R. H. Vaughan told me that some Red Kite nests had been used by generations of birds. I think it is probably true to say the same about Oystercatchers on Skokholm: although I cannot be certain, I think that some of the Oyster-catcher scrapes in 1978 were in identical locations to — if not they were within 1 m or so of — those in 1953. Of course the site must be resistant to change, as a hole in a stone wall, and old tree, a rocky ledge or the sides of a rock outcrop. Sometimes such sites are chosen repeatedly because there is a shortage of them anyway, and the Skokholm wall nest site was in a part of the island where there was little depth of soil and rabbit burrows were scarce.

Many birds will carry material into several sites at the same time. Simms (1978) quotes A. R. Lucas, who found six Song Thrush nests being built by the same bird between the rungs of a ladder: sites which had an identical appearance repeated many times. Although some parts of Skokholm were riddled with burrows, the chance of Wheatears actually mistaking the holes into which they were taking material was possible for the first day or two of building, but did not, in my experience, happen often. One female, at least, did not remember exactly where her new site was and, having deserted her first and second sites, began to build in a third burrow, but for the first few hours could not remember precisely where the entrance was. When she flew back with nest material, she either overshot or did not fly far enough, and stopped on the high part of a wall to look around before she flew to the new entrance; but after a few hours she flew directly to it.

Few of the burrow entrances were as identical in appearance as the gaps in the rungs of a ladder, or the holes in a pile of drain pipes, nor did the burrows all face the same way. Female W55 took material into holes 40 m apart and the habitats were quite different, one being in flat grassland and the other in tussocky thrift. Female W19 carried material into two holes about 200 m apart.

I concluded that, when a Wheatear carried material into several holes, it was because the drive to build became so strong that it had to be satisfied, even though the female had not decided between two or more holes. Building in several sites happened quite commonly when a pair had been formed early in the season, four or five weeks before the first eggs were laid. Quick decisions could, however, be made, particularly when second-clutch or replacement nests had to be built urgently: the Wheatears must have seen the possible sites in the course of their earlier wanderings around the territory before the female selected the first site.

Throughout the nest-building period, females accompanied by their mates, occasionally looked into burrows as if searching for a site, and even as late as 15 May in 1948, when she had already laid a full clutch, female W19 was looking into burrows.

NEST-BUILDING

Females collected most of the nest material from their own territories, only occasionally going beyond the established territory boundaries into a neighbour's, particularly if the neighbouring female was not very aggressive. Owning males might just watch intruding females collecting material 3-4 m from them but not attack. As I mentioned in Chapter 5, their 'intrusion', in the conventional sense, might not be an intrusion but rather female A going into her home range; or it could be that her territory boundary was not exactly aligned with the territory of her mate because she, being more aggressive than female B, had a different boundary from that of her mate.

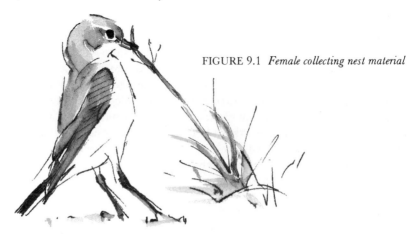

FIGURE 9.1 *Female collecting nest material*

The speed at which females picked up material varied enormously. At times they worked frenziedly, pecking, jabbing at the sward and leaning back as they pulled at a piece of grass still attached to the roots. They bundled together some of the shorter lengths of grass, and it was rather surprising to see them bang this bundle on the ground in the same way as they banged worms, caterpillars, snails and other odd items to stop them moving or to break them open; banging it on the ground was apparently the female's way of dealing with something unwieldy, whether dead or alive. Some of the larger pieces of dead bracken were about 20 cm long and weighed 2 g; these she brought to the nest by half-flying and half-dragging them.

Ten days before laying the first egg, a female's rate of building varied from zero to 35 visits an hour. During the week before the first egg was laid the average rate had dropped to between zero and nine visits an hour, and in the last day or two the females visited burrows once or twice an hour at the most. Of course, the building of replacement nests had to be very much quicker.

After a period of building the females might remain inactive for an hour or more, usually attended by their mates. Once they restarted, their visits were intermittent at first but soon they were bringing material regularly. Sometimes they restarted without any obvious stimulus — I mentioned earlier that males sometimes pounced on their mates if they had been inactive for 30 minutes or so. Otherwise, the sight of a piece of grass, or a feather lying on the ground, released the tendency to build. Of course, it depended on how far the feather was from the nest; a feather lying immediately outside the nest would generally be carried away and not taken into the nest.

Having built a load of material into the nest, females might collect another beakful within 1-2 m of it and dive into the burrow again; or they would hop away foraging, or collecting more material as they went. Chiefly they collected as they moved away from the nest; it reduced the chance of arriving back with nothing. The direction in which they set out from the nest burrow followed no particular pattern; they would fly to the place where they had just been searching or to a new area, but I had no evidence that they set off in a particular direction to find a particular type of nest material.

Occasionally females removed material, such as a piece of egg shell or faeces, from the nest, which they carried off and dropped at distances varying from 1-2 m to about 50 m.

While the females were collecting material males either followed them, standing erect, wing-flicking and bobbing excitedly, or they just stood on a rock or some eminence from which they could see their mates and watch for predators or intruders. Some perches near the nest were regularly used by males, where they preened and sometimes sang quietly some of the peculiarly vibrant 'zweng' notes whose forms and function it is extremely difficult to describe. If a neighbouring male was singing, the owning male would either sing from his perch or fly up to about 10 m in the song flight, and then for a time the two males would sing — sometimes more if their territories were close together.

While the females were down their burrows, the more attentive males sometimes stood at the entrance peering in and singing or calling very quietly, so quietly that until I put a microphone attached to a tape recorder at an nest entrance I could hardly hear the quiet 'purr' call or parts of the warbling subsong. Other males behaved as if the continued disappearance of the female down the burrow caused them anxiety.

If another male entered the territory, the owning male would approach it and begin the erect display, which sometimes resulted in the intruder retreating. If it did not leave, other forms of display such as the advancing display or the flashing display were performed; indeed, the whole repertoire of aggressive displays could be seen during this period of the nesting cycle. Meadow Pipits, Rock Pipits and Skylarks were attacked not only when they were close to the nest: sometimes the male and/or the female Wheatear would fly as much as 200 m to attack a Meadow Pipit, for instance, which might move a few metres but not so far that it left the territory; occasionally a pipit even retaliated. One male I watched on Weeting Heath National Nature Reserve, Norfolk, was very active in defending his territory; I got the impression that this behaviour distracted his mate, who accompanied him whenever he was attacking intruders on the edge of his rather large territory when she should have been on the nest.

Carlson et al. (1985) studied how males of six pairs of Wheatears which formed a dense and asynchronously breeding colony in Sweden watched or guarded their mates. They argued that it was essential for males to guard their female partners closely, particularly during their fertile period. Otherwise, there was a danger that a male from a neighbouring territory would intrude and try to attract and copulate with the female. The authors defined the fertile period as starting three days before the first egg was laid, but they did not say what external signs identify it to an intruding male nor how intruding males know that it is necessary to copulate with these females at this period. The authors also suggest that the intruding male enters the territory to mate with the female when the owner is repelling other intruders. Since they state that most of the intrusions are committed by neighbouring territory-holders, I wondered if, say, male B had been waiting and watching until male A was occupied so that it could intrude and steal a copulation with female A. What, incidentally, had female B been up to while her mate had been sitting and waiting for his chance to head for female A? Had male C been after her? Oddly enough, in their paper, on what they call mate-guarding, there is no suggestion that the male's choice of a high observation point would help him detect the presence of the intruders.

The problem is that this particular part of Carlson et al.'s study was based on a small but dense group of breeding Wheatears, and for one season only; neither the size of the whole area nor territory size is given, so that it is difficult to judge whether the rate of intrusion is likely to have been affected by population density.

On Skokholm, the rate of intrusion or other interaction between neighbouring pairs was much higher when the population was 38 pairs than when it was 11-14 pairs and when territories and nests were further apart. So one wonders whether, when territories are very condensed, every intruding male is bent on transferring his genes.

One male paid so little attention to his first mate that I hardly ever saw him. However, he had time to mate polygynously with an unmated female holding an adjacent territory (each female defended her own part against the other, but this male, when he did appear, roamed over both). He was not particularly attentive to either, and when the young hatched he brought no food to his first family and did not bring material for the second nest. Unfortunately, I had to leave the island at this point, but it seemed as if two females were raising offspring alone and without the benefit of any food the male might have brought them. The question arises: was his behaviour the best way of ensuring the survival of his genes? Unfortunately, I do not know how many of the young survived to the following year to pass those genes to another generation.

Was the close watch males kept during nest-site selection, nest-building and indeed during incubation really concerned only with the prevention of cuckoldry, or was not the male also on the lookout for predators? On Skokholm in the late 1940s and early 1950s, Sparrowhawk, Merlin, Peregrine and Kestrel were not infrequent visitors, causing alarm among Wheatears, even if the Kestrel was not too much of a threat to them.

Carlson *et al.* (1985) do not actually report seeing intruding males sneaking a copulation with the female from another territory. In any case, on Skokholm at least, copulation was not frequently seen since it often occurred in a depression in the ground; and I never had cause to suspect that a female was copulating with a male other than her own mate.

I have two final points to suggest that this proposition is improbable. At exactly the same period, the females also show aggression to intruding female Wheatears and anxiety, although not usually aggression, at the presence of intruding males. They may, however, show even greater anxiety than males in the presence of people and other animals, whether potential predators or sheep, and both parents continued to attack pipits and Skylarks as fiercely.

Occasionally, a male reverted to the behaviour associated with a previous stage: while the female was building, he would suddenly show great energy and excitement and begin visiting other burrows within about 10 m of the nest (one male looked into ten burrows, wing-flicking and bobbing). I have mentioned earlier that individuals often showed traces of a behaviour pattern that was still to come; in this case there was a temporary reversion but it did show that, should an accident befall, the switch to hunting for nest sites would be fairly easily accomplished. Although females made the final choice of nest site, it is possible that the male's behaviour might help in the choice of a second-clutch nest, a choice which has to be made quickly.

A few males brought material to the nest as energetically and efficiently as females, often wing-flicking and showing some excitement. Indeed, some were recorded as building energetically for 15 minutes while the female stood by and watched. Generally, however, males were hesitant about bringing material; they took a long time in actually hopping to the nest; they played about with the material for several seconds; and quite often they discarded it before reaching the nest. In any case, the quantity of material eventually brought could have been of little value to the female. Furthermore, it was quite often unsuitable; once, when the female was bringing rather small fibrous pieces of material for the cup of the nest, the male struggled in with a long piece of dead bracken for which the need had long passed.

Males occasionally removed pieces of grass, feathers and so on which had been lying near the burrow entrance and dropped them some distance away. This behaviour was similar to the habit of removing faeces from the nest or, later, from the burrow entrance. Presumably, if grasses or faeces had been allowed to accumulate there, they could have revealed the presence of a nest to a predator; in this case, the male's actions have survival value. Nest entrances were generally kept clear of all debris throughout the nesting period, except perhaps when the young were actually leaving the nest when, in the general activity of the moment, some could have escaped the attention of the adults.

When the nest was virtually complete and visits rare, and the time for laying the first egg approached, the adults foraged, preened and loafed or lay on their belly in the sun. Unless stimulated by an intruder or some internal sexual tendency, they were fairly relaxed.

THE NESTS

The completed nest of a Wheatear is usually in three parts: the foundation, the cradle and the cup. Whether all three parts are present depends upon the shape of the burrow, and particularly the nursery chamber (if the burrow had originally been excavated by a rabbit for the birth of its young). If the burrow in which a Wheatear was building was too wide for the walls to support the nest, or if there was no natural depression which would give support, then some sort of foundation had to be provided. If, on the other hand, the hole was small, then the cradle and the cup was all that was needed.

In a wide burrow the foundation consisted of a large untidy mass, measuring as much as 25 cm across, of dried stems of bracken, wood sage, ling and other woody plants, as well as flight feathers from Manx Shearwaters, Storm Petrels, Lesser Black-backed, Herring and Great Black-backed Gulls, the whole of which was loosely bound together with dried grasses, particularly Yorkshire fog and common bent-grass. Much of this material was difficult for the female to carry. The foundation of nest W55 contained one heather stem which was 200 mm long and weighed 2.1 g, which is about one-twelfth of the female's own weight, while a second twig 155 mm long, between 7 and 12 mm across and weighing 3.2 g was an eighth of her weight; other pieces were 145 × 6 mm (weighing 1.9 g), 140 × 4 mm (1.4 g) and 180 × 3 mm (0.7 g). In fact, these sticks accounted for exactly a tenth of the total nest weight.

The second part of the nest I called the cradle because, fitting into the foundation, it held the cup and was usually composed of smaller materials; even so, I occasionally found small pieces of dead bracken stems, heather and thrift among the Yorkshire fog and common bent-grass, and these bound the materials rather more tightly than in the foundation. The cradle of a Wheatear's nest was similar to the outer shell of nests which are usually placed in brambles, or among twigs of bushes or the branches of trees.

The third part of the nest, in which the eggs are laid, is the cup, the material of which was even more tightly bound together and, because of its texture, could be quite easily separated from the cradle. The walls were about 25 mm across, and the inside of the cup was about 70 mm across and about 50 mm deep. The fibrous materials were finer and consisted chiefly of common bent-grass and red fescue, with some Yorkshire fog. Other plant materials were mosses, including *Hypnum cupressiforme* var. *resupinatum*, and lichens, including *Ramalina siliquosa* and *R. siliquosa* forma *gracillescens*, all of which were identified for me by Dr. Mary Gillham.

Animal material included wool from rabbits and Soay sheep, and contour feathers from Manx Shearwaters and Storm Petrels. I also found long hairs from our Welsh pony, Sugarback, which lived on the island at that time, as well as a number of short hairs which could have come from one of his rubbing posts and also hairs from our domestic goats.

The cup had no distinct lining of wool or feathers as there is in the nests of finches, for example; feathers were used as another piece of material and played no specialised role in making the interior of the nest softer, although they must have helped with other material in insulating the nest. Feathers do, however, seem important to Wheatears, and a number of authors describing the materials built into Wheatear nests have particularly commented on their presence. The cup had one feature that interested me; when cups were extracted from their cradles and the bottoms of them were examined, they had a rather square appearance, which contrasted with the roundness of the interior of the cup.

Saxby (1874), writing about Wheatears in Shetland, gave a perfect description of a typical Wheatear nest with its large foundation, but what I called the cradle he called the cup and what I called the cup he called the lining. This lining, he says, is in three parts, with twine and worsted carpet, then a thick mass of cows' hair and, third, a great quantity of small feathers on the inside. I think that he must have been describing one particular nest, because the likelihood of a Wheatear being able to find this type of material generally, seems remote.

The shapes and sizes of the cavities chosen by Wheatears in which to build their nests were of such variety that they demanded of a Wheatear quite a skill in adapting an inherent design to meet the circumstances. Table 9.1 gives the weights of various used nests immediately after the nestlings had left. Some nests which did not require any foundation weighed as little as 36 g, whereas others were three times heavier. Where it was necessary for the female to build

TABLE 9.1: *Weights of some Wheatear nests*

	Nest No.	Wt.	Age*	
1949	W49/1	52 g	2°	Max. diameter of nest 210 mm. Seen building for 12 days
	W55/1	93 g	2⁻	Weighed after the young had left the nest
	W57/2	50 g	1ˣ	Weighed after the 5 young had left the nest. Built in 4 or 5 days
	W65/1	45 g	1*	Weighed after the nest and its 7 eggs were deserted
1950	W75/1	121 g	3°	Weighed after young had fledged. With foundation
	W76/1	36 g	2*	Deserted with 2 eggs. No foundation owing to burrow shape
	W80/1	99 g	2ˣ	With foundation after young had left
	W81/1	116 g	3⁻	With foundation after young had left
	W89/1	95 g	1	With foundation after young had left
	W91/1	68 g	1	After young had left. No foundation owing to burrow shape
1952	W136/R	53 g	2	No foundation. Nest pulled out but all material collected

*Notes: The symbols °, ⁻, ˣ and * indicate the same females in both years; * was an unringed female mated with 001443 and believed to be the same in both 1949 and 1950.

a foundation in which to cradle the nest, she had to carry in an extra 50-60 g of material, some of which was awkward to handle. It so happens that most of the nests whose weights were recorded here were first-clutch nests; one was a replacement and only one was a second-clutch nest.

Since it was so difficult to say when a female began to build, it was almost impossible, in most instances, to say exactly how long she took to build a nest. Generally, however, first-clutch nests took between two and three weeks, and, although replacement and second-clutch nests took less time, none of the second-clutch nests was obviously inferior to first-clutch nests. Replacement nest W1 in 1948 was built in only four days between the desertion of one clutch and the laying of the first egg of the replacement clutch. W57's second-brood nest, which weighed only 50 g, was probably completed in four or five days because the first brood left the nest on 31 May and the calculated date for the first egg of the second clutch laid in this burrow was 4 June. Table 9.1 also shows the ages of the females where they are known; while there is too little information to draw any firm conclusions, it would seem that the heaviest nests were built by older females. It should also be remembered that older females tended to return earlier and had more time before it was time to lay eggs.

Nest W75, which weighed 121 g, was built by the pair which were more than two years old, and both had returned to the island in 1950 on 26 March. Their first egg was calculated to have been laid on 24 April, so that they had a month or so from the time they had re-formed their alliance until their first egg was laid in which to select a site and build the nest.

On Skokholm, Wheatears built their nests in holes in the ground, in walls or occasionally under rocks. Of the 260 nests which I had known on Skokholm during 1948-53 and 1978-84, 224 (86%) were in burrows and only 36 (14%) in holes in walls or under rocks. Many of the burrows had been dug by rabbits, but others might well have been dug or modified by Manx Shearwaters and Puffins. At least one burrow, W49 first brood, was occupied by a Manx Shearwater 14 days after the Wheatear nestlings had vacated it. These earth burrows had entrances which measured about 10-13 cm across, and in spring, because they had not been used for some time, grass had grown across the entrance and partly hid them. Later in the season, with the regular passage of adults, the entrance became more obvious. None of the burrows had the long, earth 'doormats' which result from rabbits scraping earth from their warrens, and on Skokholm Wheatears did not nest in rabbit warrens or burrows with wide entrances. In his study, Tye (1982) also found that nest entrances measured 10-15 cm across.

I found no evidence that a Wheatear hatched in a burrow always selected a burrow for its own nest. 001341 and 001451 were both hatched in burrows in 1948 and, of their six nests built between 1949 and 1951, four were in walls (one hole used in successive years) and two were in rabbit stops.

Some of the older writers claim that it is easy to find a Wheatear's nest by the debris left at the entrance. Yarrell (1871) quotes a Mr Salmon, who told him that a Wheatear's nest 'is easily detected by a little observation, for in such situations the old birds amass a considerable number of small pieces of withered stalks of bracken, *Pteris aquilina*, on the outside, at the entrance of the burrow; by noticing these circumstances, its nest is sure to be discovered'. Ten years after Yarrell first wrote this, Stevenson (1866) repeated the statement word for word without mentioning Salmon or Yarrell. In 1891, Borrer, writing about the birds of Sussex, seems to be quoting the same source, although he does not copy it word for word but says that the Wheatear leaves pieces of bracken outside its nest. Walpole-Bond (1938), also writing about the birds of Sussex, casts some doubt on Borrer's statement, pointing out that in Borrer's day bracken was uncommon in those parts of Sussex where the Wheatear nested.

Wheatears on Skokholm removed discarded nest material, including white feathers if they were lying close to the nest, and removed these unwanted pieces in the same way that they did faeces. It is worth remembering the experiment carried out by that excellent naturalist Saxby, who, by accident, left a feather lying near a Wheatear's nest and it was carried away by one of the pair. He repeated the experiment four times, always with the same result.

I have concluded that Salmon's observation, repeated by three other authors, was exceptional, and, at more than 300 nests seen on Skokholm and elsewhere, I have never found material left outside the burrow. Nevertheless, Harrison (1975) wrote in his introduction to the eggs and nests of the Wheatear that '*many* wheatear species accumulate small pebbles or rock fragments as a small low rampart across the nest hole entrances which may substantially reduce the size of the hole'. Axell (1954) found a clutch of Wheatears' eggs in a hole on Dungeness, Kent, which was unusually easy of access; two days later, the hole had been partly blocked with a piece of earth matted with grass roots. At his earlier inspection this piece of light turf, roughly a cube of about 5 cm, had been lying about 10 cm from the entrance, which it now obscured except for a shallow depression the birds had made at the top in order that they might just squeeze by. Alan Tye tells me that many species of wheatear, especially the White-crowned Black Wheatear, accumulate small pebbles across the entrance to their nest burrow.

Many of the nest burrows were rabbit stops or nurseries in which young rabbits had been born. They were single burrows with no branches and only one entrance. Of the 119 nests for which I have sufficient information, 72 (61%) were between 25 cm and 50 cm down the burrow. Campbell and Ferguson-Lees (1972) quote A. Whittaker's nest records as showing that the average distance down the burrow of 13 nests was about 50 cm, and Tye (1982) stated that most of his Breckland nest were about 30 cm from the entrance. The greatest distance I recorded was about 1 m. On the other hand, one or two nests were so close to the entrance that they could easily be seen from the outside, even though they had an earth roof over them.

In the literature, there are several records of open nests. Cox (1921) records that, on 9 May 1903, a nest on Pevensey Level, Sussex, which like neighbouring Dungeness was short of natural Wheatear nest sites, had been built in the open under a small tuft of grass like a Skylark's nest, and Walpole-Bond (1938) wrote that he had heard of similar examples on the shingle of that area. It would seem that, if a Wheatear has found a habitat which suits its gait, which has a food supply of some sort and which has wide vistas, it may well use a somewhat atypical nest site. I qualify the state of the food supply because, as I shall show later when comparing nesting success at Dungeness with that on Skokholm, the average brood size on Dungeness was significantly smaller than on Skokholm.

Another very odd open nest was in the lowest fork about 170 cm up a fruit tree in an orchard; the nest had no covering, but the branches and the foliage of the tree reached the ground on all sides and therefore provided some cover (Sorago 1962). Others, however, may be quite open: Ludwig (1965) found an open nest in sand sedge which was visible from all sides, and John Mitchell told me of a Wheatear's nest and eggs which he found in a clump of bog myrtle on 26 June 1976 in an exceptionally wet site. Tye (1982), in his Breck-

FIGURE 9.2 *Male watching over female*

land study, found five Wheatear nests under heather tussocks; one had a tunnel, and was domed and constructed from dry grass.

Most of the Skokholm nests were built about 10 cm from the end of the burrow, which left a very important space into which the female or the nestlings, when they were old enough to hop, could retreat if a predator entered the burrow: the eggs or the immobile nestlings might be eaten, but the female or the mobile young would have a better chance of being over-looked and thus escape. In the early days of my study, when I had heard that it was possible to catch Wheatears on the nest, and before I learned that they deserted their nests quite readily under these circumstances, I had been somewhat surprised not to find the female on the nest although I had seen her enter the burrow, and I thought that she must have escaped by some other exit. Later, however, when I was able to observe inside the nest, I saw that, if she was incubating and heard a disturbance at the entrance, she hopped off the nest and went further down the passage and flattened herself against the back wall. The nestlings during their last few days in the burrow did the same, flattening themselves against the back wall and occasionally, if the wall was rough enough, scrambling up it. Presumably the dun-coloured plumage of the female and the nestlings helped to make them less obvious. This is, no doubt, why Walpole-Bond (1938) said that very seldom had he succeeded in surprising one, 'even on eggs highly incubated'.

Thirty-two per cent of Wheatear nests on Skokholm were built in holes in earth and stone walls and 4 per cent under rocks and piles of stones. The walls had obviously been solidly built in their time, and some had been maintained until the last farmer had left the island at the beginning of this century. Over the years, however, the rabbits have dug their way into them where, perhaps, a stone had been worked loose by the climate. Some of the smaller gaps and crevices, too small for rabbits, are used by Storm Petrels and Wheatears and the slightly larger ones by Little Owls. The number of Wheatears nesting in the walls was too small to bring them into competition with Storm Petrels, but the presence of Little Owls proved a final hazard to many newly-fledged Wheatears.

Table 9.2 based on the years 1948-53 and 1979 and 1980, shows the direction in which nest entrances faced. It can be seen that the majority faced from northeasterly to southeasterly quadrants, which are the directions opposite the prevailing winds. There was considerable variation from year to year, which was sometimes almost certainly due to the direction from which strong winds were blowing at the time of the final selection of nest site. Table 2.1b showed that most nest sites were sheltered from south to northwesterly winds.

Wind was almost certainly a factor causing pair W11 (1979) to desert a replacement nest on 29 April. The nest burrow was on the west side of a wall, and throughout the day the wind blew from the west at 48-64 kph (30-40 mph) straight into the burrow. The following day the pair was building in a south-facing wall which was not so exposed.

There is another important climatic hazard to choosing a hole in the ground in which to build a nest, and that is the danger of flooding after very heavy rain. Of course, in the wetter areas of the Skokholm marsh, no rabbits or shearwaters dug holes and, if Wheatears had a territory in those areas (which happened only in years when the Wheatear population was particularly high), they nested in walls. In fact, none of the Wheatear nests with young that I found was ever flooded, but, earlier in the season, a few with eggs were lost in this way.

In other parts of its range, the Wheatear is recorded as nesting in a multiplicity of holes of varying shapes and sizes, in holes under stones, rocks or in walls, in drain pipes, old kettles, tins, under driftwood and other jetsam, and so on. Saxby (1874) records nests in peat stacks, walls, quarries and in heaps of large stones; he even found a nest in the crevice in the sun-dried surface of a peat moss, a site which the late Eddie Balfour showed me being used in Orkney by Starlings and Merlins. On the Presely Hills of Pembrokeshire, I found a Wheatear's nest in a crack in the soil where the earth had fallen away from a slab of rock, in an area where rabbits and their burrows were scarce.

Walpole-Bond (1938) records some strange sites on the old military ranges along the shingle of the Sussex coast, such as in the fragments of burst shells — even in a shell, of which the main charge had not exploded — which the adults entered through the holes left by the explosion of the nose cap, and one nest was in an old cannon. Herbert Axell, who for many years lived in the same area, tells that when, as a boy, he was walking along the shingle, kicking at junk, he kicked a tin high into the air when from it fell a clutch of six pale

TABLE 9.2: *Directions in which nest-burrow entrances faced*

	1948	1949	1950	1951	1952	1953	Total %	1979	1980
N	3 (21%)	4 (14%)	5 (19%)	14 (28%)	6 (16%)	4 (17%)	36 (20%)	—	—
NE	2 (14%)	3 (10%)	2 (7%)	6 (12%)	6 (16%)	1 (4%)	20 (11%)	1	
E	1 (7%)	6 (21%)	4 (15%)	8 (16%)	4 (10%)	1 (4%)	24 (13%)	2	
SE	2 (14%)	2 (7%)	2 (7%)	3 (6%)	7 (18%)	3 (12%)	19 (10%)	—	
S	4 (29%)	4 (14%)	6 (22%)	7 (14%)	11 (29%)	4 (17%)	36 (20%)	4	12
SW	1 (7%)	3 (10%)	—	4 (8%)	1 (3%)	1 (4%)	10 (6%)	—	2
W	1 (7%)	6 (21%)	2 (7%)	4 (8%)	3 (8%)	5 (21%)	21 (12%)	—	1
NW	—	1 (3%)	5 (19%)	3 (6%)	—	5 (21%)	14 (8%)	—	1
Vertical		—	1 (4%)	1 (2%)	—	—	2 (1%)		
Total				182					

blue eggs. Later on, he adds to the list of strange Wheatear nest sites, including boots, hats and, on the old-weapon theme, parts of a German flying bomb.

To a very large extent, therefore, Wheatears use whatever suitable sites the breeding habitat supplies; and, indeed, as I have shown earlier, they may also use sites which might otherwise have been considered unsuitable, where the terrain supplied some food and good vistas but nest sites were scarce. For a variety of reasons, I considered that rabbit nurseries were really preferred on Skokholm, and perhaps over most of the British Isles: this site is mentioned first by most authors when describing nest sites in Britain. Of 185 nest burrows that I found on Skokholm in the first part of my study and for which I have sufficient evidence, ten had been used more than once in the six years. One burrow had been used for four different nests by three different pairs (one pair built two nests in it). Two burrows had been used for three nests each: in one of the two burrows all three nests had been built by the same pair (after I had removed the nest material of the previous nest), and in the other the three nests had been built by two pairs (one of which had built two nests in it). The seven remaining burrows had each been used twice, three of the burrows by the same pair and the remaining four by different pairs. There was some evidence to show that pairs preferred to return to the nest site they had used before, and would do so, even for a second brood, if I had removed the first nest after the young had fledged. The hole which had been used for four nests between 1948 and 1953 was in a wall and was, I think used again in 1980, but my nest marks from 1950, which would have clinched the matter, had disappeared in the intervening years. This site was possibly popular since burrows were scarce in this part of the island. Wynne-Edwards (1952), studying Wheatears in Greenland, found a nest site which had apparently been used for eight nests, one being built on top of the previous one; again the scarcity of other potential sites was probably the cause.

Several people have tried to induce Wheatears to breed in areas which seemed to provide a suitable feeding habitat but which lacked nest sites. The most systematic attempt was made by Axell (1954) when he was Warden of the 499-ha (1,233-acre) RSPB Reserve at Dungeness, Kent. He placed a number of old .303 ammunition cans measuring 36 mm × 18 mm × 9 mm around the reserve and half-buried them in the shingle; with a piece of wood he reduced the size of the entrance. These artificial sites were very popular with Wheatears, and he told me that the number of breeding pairs grew from 26 in 1952, nesting mostly in or under bits of junk on the shingle, to 70 pairs in 1954 — more than doubling the population in two years. It is interesting to remember that these were years of high Wheatear populations on Skokholm, and I wonder if western Africa had been a profitable wintering area at that time. Axell kept nest records which showed that the mean clutch size of Wheatears on Dungeness was significantly smaller than that on Skokholm (the probable reasons for this are explained in Chapter 10). Axell also found that the same nests were used, when relined, for second clutches, a behaviour which contrasts markedly with that of Skokholm Wheatears and is an indication of the shortage of nest sites on Dungeness.

Koshelev (1971) has described his attempts to encourage Wheatears to use artificial nest sites. He was studying Wheatears in a part of Turkmenia where natural sites were scarce and where Wheatears had been nesting in piles of wood, building stones, turf stacks and so on, but the nests were usually destroyed when these materials were removed for building purposes. He also mentions that Wheatears had nested in cracks of earth cliffs along water-courses which were often eroded by floods, as a result of which the nests were destroyed. Koshelev rammed metal pipes, which had an internal diameter of 4.7 cm and a length of 80 cm, into banks about 40-50 cm below the top. These pipes were placed in pairs about 10-20 m apart so that, if the first nest failed, a second site was available. Of the 50 pipes prepared for Wheatears, 16 were actually occupied and nest material was found in another seven. Of the 16 sites, nesting went well at only nine and the nests were destroyed in the other pipes. Of the three natural sites in the area, two failed, Koshelev commented that they did badly but he did not say why or how they fared in later years.

Szlivka (1962), in Yugoslavia, also placed artificial nestboxes for Wheatears in piles of stones and left cavities of a suitable shape and size in a brick wall.

I know of only one attempt to move a nest which was in danger of destruction. Generally, Wheatears tend to nest in such remote places that the safety of their nests is not often threatened. A Polish ornithologist, Szczudlowski (1964), tells how a Wheatear had nested in a pile of bricks and rubble which was being progressively cleared away; over six days, he moved the nest in four stages a matter of 10 m to a safer location. The parents were agitated at first, but continued to feed the nestlings for another six days after the nest had reached its new location and after the young had fledged.

Chapter 10

EGGS AND THE BREEDING SEASON

FEMALE ARRIVAL AND THE PRE-EGG-LAYING INTERVAL

For those females which arrived on Skokholm in the last half of March and the first half of April in 1950-52, there was an interval of four or five weeks before they laid their first egg (mean 31.1 days: S.D. ± 4.9; range 18-40 days; n = 42). The shortest interval, of 18 days, came from a female which arrived near the end of that period, and I got the impression that the pre-laying interval of females arriving after 15 April was shorter still.

Brooke (1979), who studied Wheatears on Skokholm from 1973 to 1976, suggested that the interval between the arrival of the female and the laying of her first egg was rather long and was 'occasioned by the time required for the transition from a migratory to a reproductive condition' and, in particular, by the time required for the reproductive organs to develop. From Figure 2 in his paper, I have calculated that the mean interval between the arrival of 13 females and the laying of their first egg was 36.9 days, ranging from 30 to 42 days, which is longer than I recorded between 1950 and 1952, although the difference is not statistically significant.

If physiological change was the only reason for the long interval, then I would have expected that Wheatears with longer distances to travel to their breeding areas from their winter quarters would have even longer pre-laying intervals. The data available from Greenland are scarce, as few ornithologists 'over-winter' in Greenland: Freuchen and Salomonsen (1960) and Meltofte (1976) state that the first Wheatears arrive in early May, although there are a very few records from the last day or two of April, and that the first eggs are laid two or three weeks later, in the last week in May.

I have calculated (assuming that the nesting period is of the same duration as on Skokholm) the laying dates of first eggs from the dates of juveniles leaving the nest burrow given by Asbirk and Franzmann (1979), Nicholson (1930), Sutton and Parmelee (1954) and Wynne-Edwards (1952). The earliest date for a first egg was calculated to have been 14 May at Godthaab, and the latest of the calculated first laying was 20 June at Mestervig (apparently the most northerly known breeding site in Greenland).

169

FIGURE 10.1 *Female flying to nest with tail closed*

Arriving too early in Greenland is full of hazards. Bird and Bird (1941) tell how, in spring 1937, a great number of Wheatears died on account of arriving too early and being caught in hard weather, a danger also mentioned by Snow (1953) and Freuchen and Salomonsen (1960), who describe how in these conditions Wheatears search for scraps around the houses. Snow reminds us that it is not easy for Wheatears to retreat from Greenland in spring if they are caught out by a return of bad weather.

In Alaska, at the extreme northwestern part of the species' breeding range, Gabrielson and Lincoln (1959) recorded the earliest arrival dates of Wheatears in the Point Barrow area of Alaska as 16-23 May. They also recorded parents with juveniles on 10, 13, 14 and 15 July. Assuming that the average date for leaving the nest was 10 July, I calculated that first eggs would have been laid in the first week of June.

It appears, therefore, that the pre-laying interval for Wheatears breeding in Greenland and Alaska, which have a much longer journey to their breeding quarters than those nesting on Skokholm, is only two or three weeks, or a week or two shorter than on Skokholm. It may be significant that the 18-day interval that I recorded on Skokholm was from a late-arriving female. It would seem, therefore, that Brooke's reason for the great length, if it is great, of the pre-laying period, is not entirely correct, and that it is possible for females to change from a migratory to a reproductive condition within less than four to five weeks.

I had also wondered why Wheatears were among the earliest migrants to arrive in Britain. I sought an explanation in the climatic conditions of their winter quarters in Africa, where increasing heat and perhaps the drying-up of the ground, which forced invertebrates to seek shade during the day, combined to drive them northwards. It is clear that Skokholm Wheatears are affected by what happens in their winter quarters: the low population figures on Skokholm since the 1970s suggest that the Wheatear may well have been affected by the Sahel drought, although not so badly as the Whitethroat and other species.

Moreau (1972) points out how harsh the situation in Africa becomes under normal conditions when the temperature starts rising from its January minimum: temperatures 25 mm above the ground in the kind of open habitat which the Wheatear inhabits may be 5°C hotter than in a screen at a standard 1.2 m above ground or 3°C hotter at 75 mm above ground. He also quotes extreme figures for the Sudan, where the screened temperature may be 40°C and the ground temperature 80°C. Moreau quotes Morel as saying that it is little wonder that Wheatears, when not actually feeding, perch on even so slight an elevation as a small stone, implying that the Wheatear perches there because it is cooler; but, even in temperate climates, Wheatears tend to perch on almost anything that gives them a better vista, whether stones, anthills or horse droppings. Indeed, I would suggest that the wider vista gained by perching on small stones and so on would be of greater survival value to Wheatears. It would be interesting to know what effect this increasing heat has upon the ground-level invertebrates that are the main prey of Wheatears.

Whatever may have been the factors that initiated the departure of Wheatears from their winter quarters, other factors controlled the start of egg-laying once they reached their breeding place. Lack (1954) suggested that the 'breeding season of birds is so timed that the young are being raised when their food is most plentiful' and that the most productive clutch size is the one giving rise to the greatest number of surviving young. Tye (1982) showed that, in both years of his study, the peak in abundance of ground-dwelling invertebrates in Breckland, Norfolk, occurred in late May, a date which coincided with the nestling stage of most first broods. Precisely what factors control the start of the Wheatear's breeding season are not understood.

EGG-LAYING BEHAVIOUR

When egg-laying begins, the behaviour of the pair hardly changes. Most females occasionally brought material to the nest in quantities so small that they could have added little to the bulk of the nest, although female W9 visited her nest with material 15 times in 50 minutes on the day she laid her third egg.

For a greater part of the day the pair foraged, preened or loafed 20-100 m from the nest. Presumably because the Wheatears had more time than when they were nest-building, territorial activity and song were more frequent. Males still showed their attachment to their mates and either followed or watched them wherever they went; when the female entered the nest burrow, the male remained close and often sang from a perch or in song flight. Females were also more aggressive to intruding females.

Generally, the eggs were laid daily before 07.30 GMT, but both on Skokholm and at Dungeness there were occasional delays in laying. Exceptions sometimes occurred when the overnight minimum temperature was low (5°C). Delays in laying the second eggs were recorded on three occasions in 1951: it was an exceptionally cold spring in which, as I record later, a number of eggs failed to hatch although they contained fully-developed embryos, I recorded an unusual case on Skokholm in 1948. The first egg of a clutch was laid on 30 April. When I visited the nest the following day, there was still only one egg. I concluded that the nest had been deserted and did not visit it again until 8 May, when there were four eggs. On 12 May there were six eggs, which all hatched on 23 and 24 May, and the young left the nest on 11 June. Bearing in mind that the mean incubation period is 13.1 days, it would appear that the clutch was completed on 10 May, which would fit in with the record of four eggs on 8 May. Therefore, between the laying of the first and fourth eggs there were five days when no eggs were laid.

Another extraordinary case was recorded at Dungeness. A nest with four eggs was found on 19 May. On 30 May, the four eggs were found cold outside the nest but inside the nesting tin. The eggs were replaced in the nest. On 7 June, Axell revisited the nest and found five warm eggs; these were hatching on 22 June. On 27 June five young were in the nest, but on 30 June the nest had been preyed on and only two remained. These left the nest in the first week in July. There are two possible explanations. One is that the female laid a fifth egg at least eleven days after the laying of the fourth egg, that the incubation period for the first four eggs was nearly a month and that, in spite of considerable exposure, the embryos successfully developed. The other, more probable, explanation is that the four eggs replaced in the nest were later preyed on, and that the five eggs found on 7 June were a replacement clutch which had just been completed. As recorded elsewhere, Wheatears on Dungeness re-used previous nests after a shorter interval than on Skokholm, probably owing to the shortage of nesting holes. Experience on Skokholm showed that the interval between the failure of predation of one nest and the

FIGURE 10.2 *Male in song flight*

laying of the first egg of a replacement clutch was as little as 48 hours.

It was in this period — the female's fertile period (i.e. a few days before the eggs were laid and up to the laying of the second or third eggs) — that Carlson *et al.* (1985) suggested that intruding males were most likely to try to steal a copulation and implant their genes in females other than their own mates. They argued that during the fertile period males watched over, or guarded their mates against the copulatory attempts of intruding and neighbouring males. They do not explain why male Wheatears intruded at other times.

It is clear that male and female Wheatears both become more aggressive in defence of their territories in that period, but I think that the evidence suggests that Skokholm males rarely ever succeeded in intruding far into a neighbouring territory without being driven off. Furthermore, there is no evidence that intruding males succeeded in stealing a copulation, although as I describe elsewhere, copulation usually occurred in a depression in the ground and was not easily observable.

We have to remember that, generally, if a male intruded at this period, it was likely to be approached by the owning male accompanied by his mate, and, if a female intruded then, the owning female would generally take action accompanied by her male. Admittedly, some males were not 100% attentive to their mates at this period, but most were and had little time to look for other females.

I regret that I am unconvinced by Carlson *et al.*'s arguments and still think that the male's role is largely to warn the female against predators, which is his constant function throughout the nesting period whether or not the female is in her fertile period, to guard against intruders taking food from their territory, or to give the female 'confidence' by behaving in a way that the female expects.

Nor do Carlson *et al.* provide an explanation for the increased aggressiveness of the female at this stage. Furthermore, they give no hard evidence that they have seen male Wheatears stealing copulations in the field and that a defence against this habit is needed.

EGG SIZE

The mean length of 264 Skokholm eggs laid in 1951 and 1952 was 21.0 mm, ranging from 18.7 to 24.2 mm (S.D. ± 1.0 mm). First-clutch eggs were slightly longer than second-clutch eggs: 221 first-clutch eggs averaged 21.1 mm, ranging from 19.3 to 24.2 mm (S.D. ± 1.0 mm); and 43 second-clutch eggs had a mean length of 20.6 mm, ranging from 18.7 to 22.7 mm (S.D. ± 0.8 mm). These differences were not significant. (See Table 10.1.)

The mean breadth of the 264 eggs was 15.8 mm, ranging from 14.9 to 16.8 mm (S.D. ± 0.4 mm). While the second-clutch eggs were slightly shorter than first-clutch eggs, they were slightly broader: 221 first-clutch eggs averaged 15.8 mm and ranged between 14.9 and 16.8 mm (S.D. ± 0.4 mm); and the mean breadth of 43 second-clutch eggs was 16.0 mm, ranging from 15.4 to 16.0 mm (S.D. ± 0.3 mm). This is a fairly usual result, it being assumed that

TABLE 10.1: *Frequency distribution of Wheatear egg lengths in 1951 and 1952*

| mm | FIRST CLUTCH | | | SECOND CLUTCH | | | Grand Total |
	1951	1952	Total	1951	1952	Total	
18.7-19.0	2	—	2	1	—	1	3
19.1-19.4	4	2	6	2	—	2	8
19.5-19.8	6	5	11	3	1	4	15
19.9-20.2	12	12	24	6	3	9	33
20.3-20.6	18	23	41	6	3	9	50
20.7-21.0	11	14	25	5	3	8	33
21.1-21.4	22	15	37	0	4	4	41
21.5-21.8	21	15	36	2	1	3	39
21.9-22.2	10	7	17	1	0	1	18
22.3-22.6	11	4	15	1	0	1	16
22.7-23.0	3	0	3	1	0	1	4
23.1-23.4	2	0	2	0	0	0	2
23.5-23.8	0	0	0	0	0	0	0
23.9-24.2	1	1	2	0	0	0	2
Total eggs	123	98	221	28	15	43	264

Average length of 1st-clutch eggs = 21.1 mm (S.D. 2 ± 1.0 mm)
Average length of 2nd-clutch eggs = 20.6 mm (S.D. ± 0.82 mm)
Average length of all eggs = 21.0 mm (S.D. ± 1.0 mm)

the female's oviduct was widened with the laying of the first clutch. The difference is not significant in statistical terms. Virtually identical results were achieved when the widths of the eggs of the first and second clutches of the same female were compared. Here, the probability of second-brood eggs being wider was 0.1 with 94 degrees of freedom. (See Table 10.2.)

Witherby *et al.* (1938) give the average size of 100 British Wheatear eggs as 21.3 × 15.9 mm, with the maxima of 24.8 × 15.4 mm and 21.8 × 17.0 mm; the minimum sizes were 19.0 × 15.3 mm and 19.4 × 14.6 mm. Thus, it would seem that the Skokholm mean is fractionally smaller than the Witherby *et al.* mean, but not significantly. As one would expect 34 eggs of Greenland Wheatears are larger still, with a mean of 21.8 × 16.0 mm. Ten eggs of the North African race *seebohmi* from Algeria and Morocco also average larger than the Skokholm eggs, at 21.6 × 15.9 mm. Further east, Panov (1974) gives the measurements of 22 eggs from east Siberia as 20.66 × 15.68 mm and eleven eggs from the Kasachischen SSR as 20.16 × 15.58 mm.

TABLE 10.2: *Frequency distribution of Wheatear egg breadths in 1951 and 1952*

mm	FIRST CLUTCH			SECOND CLUTCH			Grand Total
	1951	1952	Total	1951	1952	Total	
14.9	2	1	3	0	0	0	3
15.0	3	3	6	0	0	0	6
15.1	2	4	6	0	0	0	6
15.2	1	7	8	0	0	0	8
15.3	9	5	14	0	0	0	14
15.4	6	5	11	2	0	2	13
15.5	6	11	17	4	1	5	22
15.6	11	9	20	2	0	2	22
15.7	16	9	25	3	1	4	29
15.8	11	11	22	1	1	2	24
15.9	8	7	15	4	3	7	22
16.0	6	5	11	0	0	0	11
16.1	7	4	11	2	0	2	13
16.2	8	3	11	1	2	3	14
16.3	11	8	19	2	1	3	22
16.4	7	4	11	4	3	7	18
16.5	3	0	3	2	2	4	7
16.6	4	0	4	1	1	2	6
16.7	2	1	3	0	0	0	3
16.8	0	1	1	0	0	0	1
Total eggs	123	98	221	28	15	43	264

Average breadth of 1st-clutch eggs = 15.8 (S.D. ± 0.41 mm)
Average breadth of 2nd-clutch eggs = 16.0 (S.D. ± 0.33 mm)
Average breadth of all eggs = 15.8 (S.D. ± 0.42 mm)

EGG WEIGHTS

In 1951, I weighed 92 eggs from 16 first-clutch nests, one second-clutch nest and one replacement clutch. In the field I used a portable balance in a glass-fronted case, so that it was possible to weigh eggs or nestlings in any part of the island without the balance being disturbed by wind. I usually weighed the eggs within 24 hours of laying and marked each egg with a numeral, using quick-drying stove blacking. I also weighed complete clutches when I found them, but, as incubation had already started, the weights of these clutches are not included in the tables unless otherwise stated. Generally speaking, the numerals on the eggs had little effect on the females, but some apparently turned the eggs so that the numeral was hidden, as Gibb (1950) had found some Great Tits doing in his nestboxes. I could see this only in the few nests over which I had made a trapdoor through the roof of the burrow, or where the next was very close to the burrow entrance.

Later in the study, I used a thermistor electrical resistance thermometer placed among the eggs to record incubation temperature. Once, I discovered that the female had covered it with feathers, which was presumably her reaction to a foreign object in the nest which she could not remove.

Table 10.3 shows that the mean weight of all Wheatears' eggs in 1951 was 2.83 g (S.D. ± 0.2 g; range 2.3-3.21 g; n = 92). Female W103, which was at least two years old, and which laid the two lightest eggs (2.3 g and 2.37 g), also laid a fairly heavy egg (3.09 g) in the same clutch: thus giving a range of 0.79 g for a single clutch, compared with the mean range for 92 eggs of 0.91 g. The two heaviest eggs (3.21 g and 3.2 g) were laid by different females.

I also examined the weights of clutches of five and six eggs. The mean weight of five eggs was 2.87 g (S.D. ± 0.21 g; n = 39.) and for six eggs the mean was 2.83 g (S.D. ± 0.18 g; n = 42), but the difference was not significant.

From the figures that I had available, there was no indication that successive eggs in the same clutch were heavier. In 1951, the mean weight of eleven

TABLE 10.3: *Frequency distribution of weights of freshly laid Wheatear eggs, 1951*

Weight (g)	No.
2.30-2.39	2
2.40-2.29	4
2.50-2.59	9
2.60-2.69	8
2.70-2.79	16
2.80-2.89	15
2.90-2.99	17
3.00-3.09	14
3.10-3.19	5
3.20-3.29	2
Total	92

Mean weight = 2.83 g (S.D. ± 0.2 g)

TABLE 10.4: *Seasonal differences in mean egg weights*

Week	Mean weight	S.D.	Number
30 Apr — 6 May	2.98 g	± 0.99 g	17
7 May — 13 May	2.77 g	± 0.24 g	17
14 May — 20 May	2.80 g	± 0.36 g	28

first-laid eggs was 2.82 g; of eleven second eggs 2.88 g; of 14 third eggs 2.77 g; of 16 fourth eggs 2.82 g; of 15 fifth eggs 2.86 g; and of seven sixth-laid eggs 2.78 g. In one clutch only were successive eggs heavier, and in another clutch the first egg was the heaviest.

I also looked at the seasonal differences in egg weights (see Table 10.4). The mean for the first week — 30 April to 6 May — is significantly greater than that for the second week (t = 3.460; p = 0.01), but the difference in mean weights between eggs laid in the second week and those laid in the third week is not significant. The fact that the eggs laid in the first week of laying were heavier was, so far as I was concerned, rather unexpected and probably due to the fact that most of the females laying in this period were two years old or more rather than to any environmental factor such as differences in ambient temperature.

Unfortunately, I have too few weights of clutches laid by year-old females before the clutch was complete to allow a proper comparison between the egg weights of first- and second-year females. Neither does there seem to be any significant difference between those females known to be two years old or more and the remainder, possibly because some of the females of unknown age were themselves at least two years old.

The mean range in weights of eggs from the same clutch was 0.26 g (S.D. ± 0.17 g), and I have already mentioned the very wide range of weights from nest W103. Some females were very consistent in the weight of the eggs they laid: only 0.1 g separated W113's lightest egg from her heaviest. The range of weights of W102's first clutch was 0.26 g and of her second 0.25 g, and for all her eleven eggs the range was 0.32 g.

I was unable to relate these egg weights to the weights of the newly-hatched nestlings, but as the achievement of this aim would have meant more than one visit to the nest each day, and since Wheatears desert so easily, I decided not to attempt this.

BREEDING SEASON

Figure 10.3 shows the Wheatear's breeding season, based on the laying of first eggs in each clutch, on Skokholm and on the RSPB reserve at Dungeness, where Herbert Axell had been studying Wheatears. It can be seen that laying

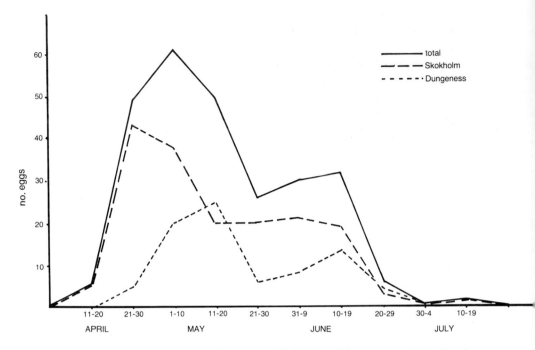

FIGURE 10.3 *Dates of laying of first eggs on Skokholm and Dungeness: unbroken line shows total; long dashes indicate Skokholm, and short dashes Dungeness. Vertical axis shows number of eggs laid, and horizontal shows dates.*

began in mid April and reached a peak in the first ten days of May. Thereafter, the number of eggs laid declined until there was a second but smaller peak in mid June caused by the laying of second clutches. On Skokholm, I knew the history of each pair and whether a clutch was a replacement necessitated by a disaster to the first or a genuine second-clutch nest. Between 1948 and 1952, 47% of Wheatear pairs on Skokholm produced second-clutch nests. Information from Dungeness shows that some pairs produced second clutches, but from the nest records it is not possible to calculate the percentage.

When comparing Skokholm data with those from Dungeness, we see that the peak of egg-laying at Dungeness was about ten days later than on Skokholm but that the peak of second-clutch layings was much closer to that of Skokholm. I wondered whether climate had any effect. Billham (1938) showed that, although the mean temperature in April was the same on Skokholm and at Dungeness, the minimum temperature at Dungeness in March (2.8°C/37°F) was 1.6°C (3°F) lower than on Skokholm. In April and May, the mean minimum Dungeness temperatures of 5°C and 5.5°C (41°F and 41.9°F) respectively were each 0.5°C lower than those on Skokholm. In June, the mean minima were identical at 10.6°C (51°F). Therefore, it is possible that the lower minimum temperatures delayed laying at Dungeness. I have shown earlier that delays in laying occasionally occurred on cold nights on Skokholm, where grass-minimum thermometers recorded temperatures as low as 2.2°C on 27 April 1951 and 1.1°C on 12 May 1951. The lower minimum temperatures at Dungeness may, therefore, account for the later onset of laying.

Lack suggested that birds laid their eggs at a time of the year which resulted in young being in the nest when food to feed them was most abundant. While no such study has been made on Skokholm, Tye's (1982) work on Breckland showed that easily caught insects, which he called 'sitting ducks'; were most abundant when young Wheatears hatched. Tye, incidentally, used hatching date as the central guide to the earliness or lateness of the breeding season because it was easier to determine. There is a greater chance of a nest being found by the time the eggs hatch than by the date of laying of the first egg, but there is also a greater chance that some nests will be preyed on before hatching!

The mean dates for the completion of clutches for first and second broods on Skokholm between the years 1948 and 1952 are shown in Table 10.5. The figures emphasise the annual variation in laying dates. It is obvious that 1948 was a very early year for the laying of the first eggs, with the mean of laying dates being 26 April (S.D. ± 5.28 days; n = 12); and that 1951 was very late, with the mean of laying dates being 9 May (S.D. ± 6.34 days; n = 17), confirming other evidence given elsewhere of the lateness of the 1951 spring. The mean hatching date for 1951 was also late but closer to the overall mean, so that the birds had caught up a little. The mean date for clutch completion at Dungeness was 14 May (S.D. ± 6.45 days; n = 47). Tye's (1980) mean hatching dates for Breckland were 18 May (S.D. ± 4.0 days; n = 6) in 1976 and 20 May (S.D. ± 1.0 day; n = 47) in 1977; his mean second-clutch hatching dates were 20 June (S.D. ± 4.0 days; n = 3) in 1976 and 28 June

TABLE 10.5: *The number of Wheatear clutches completed weekly on Skokholm*

Year	APRIL 24	1	MAY 8	15	22	29	5	JUNE 12	19	26	Total
1948	1	6	—	3	—	—	—	1	2	1	14
1949	—	2	13	3	—	—	1	3	1	—	23
1950	—	3	2	—	2	—	1	1	5	—	14
1951	—	—	5	8	9	1	1	3	6	—	33
1952	—	3	13	3	1	—	—	2	1	—	23
Totals	1	14	33	17	12	1	3	10	15	1	107
%	1	13	31	16	11	1	3	9	14	1	

(S.D. $\pm$ 2 days; n = 6) in 1977. It appears, therefore, from the evidence available that the breeding seasons of Skokholm and Breckland Wheatears more or less overlapped, each place being apparently affected by the local climate.

Laying Dates for Second Clutches

On Skokholm between 1948 and 1953, 47% of the breeding pairs laid a second clutch. What surprised me was the shortness of the average interval between the young of the first brood evacuating the nest burrow and the laying of the first egg of the second clutch in a new nest and, almost invariably, in a new burrow. In most cases, the date of laying of the first egg was calculated with reasonable certainty, either from hatching or, in some cases, from the evacuation of the burrow.

For 28 pairs, the average interval between leaving the nest burrow and laying the first egg of the second clutch was 5.7 days (S.D. $\pm$ 3.5 days), but, in addition, in 1951 two females laid the first egg of the second clutch the day before the nestlings left the burrow. In Breckland, Tye (1980) discovered a female which laid its first egg three days before the nestlings left the burrow.

The dates for leaving the burrow in cases where second broods followed ranged between 27 May and 10 June, but with the majority between 28 May and 4 June (Table 10.6): the latest was 10 June. The table shows two peaks; the first on 28 and 29 May, relates to an early evacuation of the burrow in 1950; and the second, on 3 and 4 June, relates to a slightly later one in 1952. The second part of the table shows the fledging dates of those pairs which did not produce second clutches — at least on Skokholm; as might be expected, the complete range of dates extends much later in the season.

The intervals between leaving the nest burrow and the laying of the first egg would appear short, but it will be remembered that nestling Wheatears begin to leave the nest itself when about eleven or twelve days old and it is perhaps from that point that some psychological change takes place in the female which allows her to lay a first egg of another clutch when the young of the first brood are still in the nest burrow. The whole business of displays, nest-site selection and so on is apparently extraordinarily reduced.

TABLE 10.6: *First-brood fledging dates: (a) those followed by second clutches, and (b) no second clutches*

	MAY					JUNE																
	27	28	29	30	31	1	2	3	4	5	6	7	8	9	10	11	12	13	14	15	16	17
(a)	1	5	5	2	2	5	—	7	5	1	1	1	—	—	1							
(b)			1	1	2	2	1	2	2	1	6	—	—	1	4		2	2	1	1	2	2+

(+24 and 29 June)

TABLE 10.7: *Clutch size on Skokholm and Dungeness*

					NO. EGGS					
	2	3	4	5	6	7	8	9	Mean
Skokholm (S.D. ± 0.8)	1	—	5	48	57	19	1	—	5.7
Dungeness (S.D. ± 0.9)	—	3	13	33	46	1	—	1	5.3
Total	1	3	18	81 (S.D. ± 0.9; n = 228)	103	20	1	1	5.5
%	0.4	1.3	7.9	35.5	45.2	8.8	0.4	0.4	

CLUTCH SIZE

Table 10.7 shows that the average clutch size for Wheatears on Skokholm and Dungeness was 5.5 (S.D. ± 0.9; n = 228), with a range of two to nine eggs. The clutch of nine eggs was laid at Dungeness at the peak laying time. It can also be seen that there is quite an appreciable difference between the clutch sizes of Dungeness and Skokholm: for the former the mean clutch size is 5.3 (S.D. ± 0.9), and for the latter 5.7 (S.D. ± 0.8). The possible reason for this difference is given in the concluding section of this chapter. In Breckland, the mean clutch size in 1976 was 5.8 (S.D. ± 0.3; n = 6) and in 1977 5.7 (S.D. ± 0.1; n = 52) (Tye 1980).

Seasonal Variation in Clutch Size
Figure 10.4 shows how clutch size, based on the date of laying of the first egg, varied with the seasons on Skokholm and at Dungeness. The average clutch size for both localities was 6.1 during the period 11-20 April, rising to 6.2 between 21 and 30 April. Thereafter, the average dropped away fairly steadily, although with a marked rise in the first week of June when the earliest of the second clutches were laid. Between 20 and 29 June, the average clutch size was only 3.9

THE BEGINNING AND END OF THE BREEDING SEASON

The earliest eggs I found on Skokholm were laid in 1948 on 19 April, but I calculated from the hatching date of 3 May 1948 that the first egg of a clutch of six had probably been laid on 15 April that year. The latest broods to fledge left their nests on 23 July in 1949 and on the same date in 1951. Both latter dates occurred in years with late springs and were ten and seven days respectively behind other last fledging dates, which were 18 July 1948, 13 July 1949, 13 July 1950, 16 July 1951 and 18 July 1952.

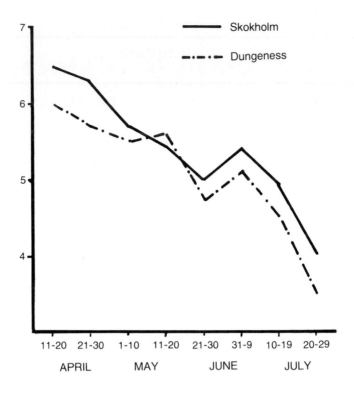

FIGURE 10.4 *Seasonal variation in clutch size: unbroken line Skokholm and broken line Dungeness. Vertical axis shows clutch size, and horizontal shows dates.*

Second Broods

On Skokholm, some 47% of the Wheatear pairs laid second clutches. It was obvious from the Dungeness nest record cards that a number of second clutches were laid there, but it is not possible to calculate the exact proportion. Tye (1982) found that in 1976, the first year of his study on the Brecklands of Norfolk, three of the seven pairs he was observing raised second broods, and less systematic observations on other pairs in his study area suggested that about half of these raised second broods; in 1977 the proportion was lower, with only eight true second broods out of 53 pairs, but in addition there had been five replacement clutches of first clutches that had been preyed on.

Chapter 11

FROM INCUBATION TO HATCHING

INCUBATION

The information for this section comes chiefly from Skokholm, with some from the RSPB's Reserve at Dungeness (Axell, pers. comm.) and some from Breckland (Tye 1980, and pers. comm.). For the purpose of this chapter, I am calculating that the incubation periods of the Wheatear start from the date of laying of the last egg in each clutch, although other aspects of the study suggest that incubation begins with the laying of the last egg but one: the time spent incubating the eggs usually reached the average for the incubation period on the day the last egg but one was laid. The one marked exception was female W194[1], among whose eggs I had placed an electrical resistance thermometer: her incubation temperatures were lower than average during the first five days, probably resulting in a rather long incubation period (14 ± 1 days).

Table 11.1 shows that the average incubation period for Skokholm Wheatears was 13.1 days, with a range of 16-10 days. Its length appears to decrease from May to July and with increasing clutch size. There was also an indication that those females which incubated for longer spells had shorter incubation periods, but the sample is too small to show any significant correlation.

FIGURE 11.1 *Sun-basking*

183

TABLE 11.1: *Incubation periods of Wheatears on Skokholm*

(a) By months

| | DAYS | |
	Mean	Range
April	—	—
May	13.1	16-10
June	12.8	15-11
July	[12.5	12-13]
Mean	13.1 days	

(b) By clutch size

| | DAYS | |
	Mean	Range
4 eggs	13.5	16-12
5 eggs	13.4	16-11
6 eggs	12.5	15-10
7 eggs	[11.0	12-10]

As the incubation period progressed, five females whose behaviour I watched personally spent slightly more time incubating (Table 11.2). They incubated for 71.2% of the time that I had them under observation, ranging from 57% the first day to 77% when the eggs were close to hatching.

A few records from the automatic recorders were not incorporated in these figures. If they had been included, they would have shown — covering as they did all the hours of daylight — a slightly higher incubation percentage, since some incubation periods at dawn and dusk were longer those in the middle of the day. I did, however, include figures from pair W194[1], whose periods on and off the nest were recorded with the aid of the electrical resistance thermometer.

Since female Wheatears are not fed on the nest by males, they have to leave the nest at regular intervals in order to feed, preen, defecate and so on. Compared with the Goldfinch, which I studied when a prisoner of war in Germany (Conder 1948a) and in which the male fed the female on the nest so that she had to leave it only in order to defecate, the female Wheatear left and returned to the nest at comparatively short intervals.

TABLE 11.2: *Incubation percentages of five pairs of Wheatears on Skokholm*

	PERCENTAGE OF TIME SPENT INCUBATING					
Day of incubation	W2	W53	W77	W92	W194[1]	Mean
1	56	65	52	73	39	57.0
2	63	60	67	69	48	61.4
3	68	?	68	60	60	64.0
4	70	60	71	88	36	65.0
5	68	69	64	68	65	66.8
6	68	69	72	64	69	68.4
7	66	75	74	84	77	75.2
8	70	76	?	84	78	77.0
9	62	77	?	76	76	72.7
10	62	69	72	88	69	72.0
11	71	70	74	72	100	77.4
12	79	67	66	70	87	73.8
13					[94	94]
14					[72	72]
Mean	66.9	68.9	68.0	74.6	69.3	71.2

The actual times spent incubating and then off duty by females at eight nests are shown in the three tables 11.3, 11.4, and 11.5. Females at seven nests behaved in a similar way, but W145[1] spent longer periods both on and off the nest so that I have shown her off-duty periods separately in Table 11.5. To the records of the five females mentioned above, I have added records from two nests, W102[1] and W102[2], which had automatic recorders attached to them for the last five days of their incubation periods which gave me more detailed information on their movements in and out of the nest. These eight nests together provided 474 records of spells of incubation, or on-duty periods (Table 11.3), but I have grouped eleven periods (2.3%) in a special category because they were exceptionally long, ranging between 56 and 110 minutes (eight of them were from W145[1]). The remaining 463 records tell us that the average spell on the nest lasted 16.3 minutes, and that 66% of the on-duty spells lasted between six and 26 minutes. The exception was female W145[1], which consistently incubated for longer spells than usual, most frequently between 26 and 45 minutes but also being responsible for eight of the eleven spells that lasted between 56 and 110 minutes. As I shall show later, her off-

TABLE 11.3: *Frequency distribution of spells of incubation (on-duty) for eight pairs of Wheatears*

Mins	W2	W53	W77	W92	W102¹	W102²	W145¹	W149	Total	(%)
					WHEATEAR PAIRS					
1-5	1		1		4	12	—	2	20	(4.2)
6-10	9	7	1	1	53	79	1	3	154	(32.5)
11-15	13	21	5	3	18	48	3	5	116	(24.5)
16-20	8	7	8	5	11	22	5	5	71	(15.0)
21-25	1	4	—	4	3	11	7	3	33	(6.9)
26-30	2	4	1	1	2	4	9	2	25	(5.3)
31-35	—	1	—	—	1	1	6	—	9	(1.9)
36-40	—	1	—	1	—	3	9	1	15	(3.2)
41-45	—	—	—	1	—	—	8	1	10	(2.1)
46-50	—	—	—	—	—	3	5	—	8	(1.7)
51-55	—	—	—	—	—	—	2	—	2	(0.4)
		Mean 16.3 minutes (S.D. ± 10.17; n = 463)								
56+	—	—	—	—	1	1	8	1		
mins	—	—	—	—	88	58	56, 57 58, 59 63, 86 98, 110	57	11	(2.3)

duty spells were even more out of line with the general pattern of Wheatear incubation behaviour.

The mean off-duty spell, when the female foraged, preened and so on, averaged 7.5 minutes, and two-thirds of those spells lasted from four-and-a-half to ten-and-a-half minutes (Table 11.4). These figures, however, exclude female W145¹, whose off-duty spells are described in the next paragraph. Some females were extraordinarily regular in returning to the nest at the end of their off-duty period or in leaving it after incubating: one female regularly returned after six minutes; another remained on for five eight-minute periods in a row; and a third female for four ten-minute periods in a row. There seemed to be some internal rhythm affecting the length of time on and off nests.

Female W145¹, whose movements in and out of the nest were recorded automatically for the last five days before hatching, presented an extraordinarily different picture; and, indeed, looking at the figures, one might even

TABLE 11.4: *Frequency distribution of periods off duty during incubation, excluding W145[1]*

Mins	W2	W53	W77	W92	W102[1]	W102[2]	W149	Total	(%)
				WHEATEAR PAIRS					
1						2	2	4	(0.9)
2								—	—
3	1				5	2		8	(1.8)
4	4	1			3	12	1	21	(4.7)
5	4	11	1		8	29	3	56	(12.6)
6	4	17	4	4	23	42	6	100	(22.5)
7	6	7	9	4	19	38	1	84	(18.9)
8	11	8	2	6	10	23	2	62	(13.9)
9	4	4	3	2	6	11	—	30	(6.7)
10	2	1	—	—	3	6	3	15	(3.4)
11	—	2	—	3	2	5	1	13	(2.9)
12	—	—	—	—	2	7	1	10	(2.3)
13	—	—	—	—	3	1	1	5	(1.1)
14	—	1	—	—	2	3	—	6	(1.3)
15	—	1	—	1	3	4	—	9	(2.0)

Mean 7.5 minutes (S.D. $\pm$ 3.1; n $=$ 423)

Mins	W2	W53	W77	W92	W102[1]	W102[2]	W149	Total	(%)
16+	—	—	1	1	5	12	3	22	(4.9)
mins	—	—	18	17	17, 20 23, 24 26	16(2) 17(3) 19(4) 21, 26 28	28, 34 64		

TABLE 11.5: *Frequency distribution of W145[1]'s periods off duty*

					MINUTES						
1-5	6-10	11-15	16-20	21-25	26-30	31-35	36-40	41-45	46-50	51-55	56+
1	1	1	1	14 ·	9	5	5	5	5	3	5*

Mean 30.3 minutes (S.D. $\pm$ 11.3; S.E. 1.524; n $=$ 55)

*56+ $=$ 57, 61, 71, 71, 96

think that the individual which produced them was a species other than a Wheatear (Table 11.5). From 55 records, it can be seen that the average spell off the nest was 30.3 minutes, ranging from four to 96 minutes; but I excluded from these calculations five off-duty spells lasting from 57 to 96 minutes, although they are included in Table 11.3. It can also be calculated from Table 11.5 that two-thirds of her off-duty spells fell between 19 and 41 minutes, compared with four-and-a-half to ten-and-a-half minutes recorded from the other seven females (Table 11.4).

One might expect that, in view of W145[1]'s long off-duty spells, her incubation percentages would be lower, and indeed, in spite of the fact that the on-duty spells were longer, the mean incubation percentage for the last five days before hatching was 52.45%, against the overall mean of 71.2% and a last-day incubation percentage of about 73.8%. Unfortunately, I do not know the length of her incubation period. Her seven eggs took a maximum of 72 hours to hatch, and the runt died on its fifth day having gained only 2 g. A second died at average weight when 7.5 days old, again for no obvious reason. The average weight of the brood at leaving the nest was 1.4 g below the average weight for all Wheatear nestlings. (It would seem that this was an idle female.) After leaving the nest, none of the young was recorded again.

Incubation Percentage and Weather

Nice (1937) has shown that there is a consistent relationship between the average temperature during the period of observation and the average incubation percentage. From the potentiometer records of five pairs of Song Sparrows taken by Baldwin and Kendeigh (1932), she deduced that the 'cooler the weather, the shorter were the periods on and off the nest, while the warmer the weather, the longer the periods were'. She suggested, as a reason for this, that, the cooler the weather, the more often the birds felt hungry and therefore the shorter the 'on-duty' periods.

Prior to the establishment of Skokholm as a Climatological Station for the Meteorological Office in 1950, such thermometers etc that the Observatory possessed were inadequate for such comparisons, but in 1950 it became possible to compare average temperatures, relative humidity and incubation percentages of pairs W77 and W92. I found no obvious relationship between average temperatures and incubation percentages, but Figure 11.2 shows that there were some signs of a negative correlation between relative humidity and incubation percentages: the higher the humidity, the lower the incubation percentages.

INCUBATION BEHAVIOUR

At the end of a spell of incubating, females emerged from their nest burrows and stood at the entrance, where they would, if undisturbed, preen, scratch, and stretch their wings and legs before they began to hunt. They usually left the vicinity of the nest with a rather undulating flight which Wheatears used

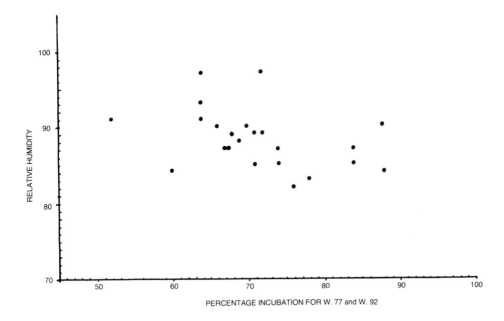

FIGURE 11.2 *Relationship between average incubation percentage and relative humidity (per cent) in pairs W77 and W92 in 1950. Vertical axis shows relative humidity and horizontal shows incubation percentage.*

when they were examining the area for possible intruders or enemies. The distances they flew varied from a few metres to about 50 m, although one female occasionally flew as much as about 200 m (much depended upon territory size), and from there they would begin hunting fairly energetically.

This behaviour was characteristic of this stage of the breeding season, and it enabled me to determine whether a female had been incubating rather than egg-laying. Occasionally, she would fly close to the waiting male and begin feeding near him. Sometimes he followed her, as he had when she was selecting the nest site; sometimes he perched on an anthill and watched; and sometimes he just did not appear. Female W2, which had an observation post within 3 m of the burrow, often flew to it; there she would preen, wipe her beak and then drop to the ground and begin feeding, gradually working away from the nest. At this stage there was no sense of urgency in the rate at which females hunted and they did not necessarily return to the place at which they had been feeding during their previous off-duty period.

When females returned to their nest burrows at the end of an off-duty period, they almost always used a fast direct flight which was a complete contrast to the undulating outward flight. The difference probably reflected the internal situation of the bird in some way: on the outward journey females were perhaps slightly apprehensive about what they would find, but on their inward journey they were responding to the stimulus to get the eggs against their hot brood patches. In spite of that urgency, however, they would divert their flight in order to attack an intruder or a Meadow Pipit foraging close to the nest.

During the first four or five days, males would perch on a tussock of ling or thrift close to the nest site, preening and occasionally hunting rather haphazardly. While this behaviour might be regarded as a continuation of mate-guarding, I considered it, as I had when watching male Goldfinches conducting themselves similarly in Germany, as behaviour consequent upon the female becoming stationary in the nest and thus leaving the male alone, whereas up to that point the pair had usually been close companions. Their way of life had changed — the female became rather sedentary in a hole and the male led a more solitary life — and this took a bit of getting used to.

During these first few days when females were incubating, males would go to the nest entrance and there sing rather quietly: not quite the warbling subsong that was used before the females arrived, but a rather soft and sweet full song without any of the harsh notes to be heard in some versions of songs which indicated aggression. Each male tended to have his habitual way of reacting to this change. Male 001314 was one of the most attentive males and continually visited the nest during periods of observation, both in 1949 and in 1950; he was the most attentive male I encountered during the course of this study.

It was possible that females answered the males, because the males peered into the burrows and turned their heads from side to side as if listening. In 1978, I placed a microphone at the burrow entrance and heard the very quiet purring call, which I heard parents W61 call to their nestlings when I was watching from a so-called coffin hide, but I had not heard a male call like this to a female as I think happened in this instance.

Occasionally, males would go into the holes when females were on the nest and remain down for 10-20 seconds, rarely as long as 30, but I have no idea what they were doing. Some males would also go down when the female was

FIGURE 11.3 *Male singing to incubating female*

away feeding, and one male was seen to bring in some nesting material. I have never recorded males staying in the holes long enough to suggest that they were incubating. Nest-visiting occurred during most parts of the incubation period and presumably helped to maintain the bond between the pair.

While males often sang at the entrance to burrows, they might also perch on a tussock close by, to which they frequently returned, or fly up in a short song flight directly over the nest. At times this song and anxiety 'weet' calls brought the female off the nest, and a visit down the burrow by the male could have the same effect.

Once males had become accustomed to the temporary absences of their mates, they generally loafed about the territory, sometimes foraging in a desultory fashion but then suddenly showing, for no obvious external reason, a sudden burst of activity. Their main function, during first broods, seemed to be to maintain the integrity of the territory and warn of predators. While male W2 was assiduous in these tasks, W53 was continually recorded as ignoring intruders although he joined in song contests with neighbouring males. Male W2 had a Greenland Wheatear male holding a small individual territory within his breeding territory. He attacked the bigger bird during the first two days, but later, except for an occasional battle or flashing display, the Greenland Wheatear was ignored; this is different from female W61's complete cessation of building activities and her ferocious attacks upon her mate when a Greenland mate established its individual territory close to the nest which pair W61 had been building. Perhaps the Greenland male in W2's territory was sufficiently far away from his nest not to cause pair W2 too much anxiety.

NEST TEMPERATURES

Introduction

Because of the failure of eggs possibly owing to cold weather and the low success rate during the cold spring of 1951, I tried to learn something about the temperatures at which the eggs were being incubated and the nestlings were being brooded. A friend, whose name I have regretfully lost, constructed for me an apparatus based on a thermistor (an electrical resistance thermometer) and a potentiometer. The thermistor, contained in a glass cylinder about 20 mm long and about 3 mm in diameter, was inserted between the eggs and was attached by a cable 100 m long to the potentiometer, which had been calibrated against a tested thermometer. Thus, with a nest 100 m from the Observatory buildings, I was able from within the buildings to read off the temperature between the eggs or nestlings when the female was on or off duty. In practice, I used this equipment normally for hour-long periods in the mornings and afternoons and took spot readings at various intervals throughout the night.

During observation periods I operated the potentiometer continuously and recorded the actual temperature once a minute, although I was able to detect changes within a few seconds since the electrical resistance thermometer

191

reacted virtually immediately to changes in temperatures and was sensitive enough to detect differences as small as 0.1°C. In this way, on- and off-duty spells could easily be measured, as well as short periods when the female was off the eggs either poking about between them or changing their alignment. When the female left the eggs, the temperature dropped rapidly at first as the eggs cooled; but even so, as they were cooling, the temperature sometimes rose 0.1°C or 0.2°C or fell rather more quickly than usual. In these latter circumstances, I presumed that the changes in ambient temperature were caused by slight movements in the air.

Nest W194[1], among whose five eggs I inserted the thermistor, was in a rabbit burrow through the roof of which it was easy and safe to construct a trapdoor to allow the insertion of the instrument. Every few days I examined the position of the thermistor as it lay among the eggs in order to ensure that it was in the best position to record temperatures as reliably as possible. Initially the female partially hid the thermistor with nesting material, which I cleared. At first I thought that this might have caused the low temperatures of the first five or six days of incubation, but an examination of the weather records showed another possibility that I discuss later in this chapter. I emphasise that the readings measured the temperature between the eggs, rather than the temperature of the egg itself: to obtain the latter figure the operative part of the thermistor should have been inserted in an egg; for a number of reasons I did not do this.

Unfortunately, 1953 turned out to be my last complete breeding season as Warden of Skokholm, so that I had no opportunity to repeat or improve on attempts to record Wheatear incubation temperatures throughout the nesting period. Nor did I have the facilities to measure the 'deep body' temperature of the incubating female, which Farner and King (1961) suggest is the most reliable way of establishing temperature. From comparing the body temperatures of a number of birds with similar body size (summarised in Farner and King), however, I should expect the Wheatear's body temperature to be between 38.5°C and 43°C.

Length of W194's Incubation Period

Table 11.2 shows that W194's incubation period of 14 ± 1 days was longer than average and also that the percentage of time spent on the nest for the first five days of the incubation period was lower than average. Skutch (1976) points out that the temperature threshold below which embryonic development stagnates varies from species to species: in the Northern House Wren it fluctuates between 16.3°C and 19.3°C.

As the temperature of the egg rises above the threshold, the rate of development increases. According to O'Connor (1985), embryonic development does not begin to any significant extent until the egg is incubated to above 35°C; and, for development to continue, the optimum temperature at which a fowl's egg must be incubated is 38.5°C (Bellairs 1960). Bellairs also says that some development will take place at a temperature as low as 25°C. Blastoderms incubated at 25°C fail to develop beyond the primitive streak stage. Therefore it seems possible that W194's rather long incubation period may well have been due to the low temperatures at which the eggs were being incubated initially, as well as low incubation percentages which were probably associated with strong winds and rain at that time. From the sixth day to hatching incubation percentages remained fairly high, as did the temperature between the eggs.

INCUBATION AND BROODING TEMPERATURES

The temperature fell or rose sharply when the female either left or returned to the nest, and fluctuated slightly when she changed her alignment on the eggs or poked down among them. Table 11.6 and Figure 11.4 show the mean daily maximum and minimum temperatures recorded between the eggs of nest W194 and the nestlings, which hatched between 24 and 25 May. The maximum figures usually occurred just before the female left the nest at the

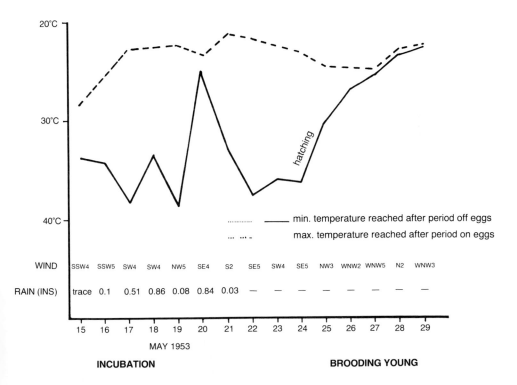

FIGURE 11.4 *Ranges of temperature among eggs and nestlings when female on and off duty. Mean daily minimum shown by broken line and minimum by unbroken line. 15 May 1952 was sixth day of incubation.*

TABLE 11.6: *Mean temperature range (°C) at beginning and end of incubation spells of female W194. (By 29 May, the nestlings were no longer being brooded)*

Date	Maximum	Minimum	Range
15 May	31.9	26.1	5.8
16 May	34.5	25.7	8.8
17 May	37.3	21.7	15.6
18 May	37.5	26.5	11.0
19 May	37.6	21.6	16.0
20 May	36.7	35.3	1.4
21 May	38.7	27.1	11.6
22 May	38.3	22.6	15.7
23 May	37.8	24.1	13.7
24 May	37.0	23.7	13.3
Hatching			
25 May	35.6	29.9	5.7
26 May	35.3	33.0	2.3
27 May	35.1	34.5	0.6
28 May	37.1	36.5	0.6
29 May	37.6	37.1	0.5*

Mean of temperature ranges during incubation (15-24 May) = 10.3°C (S.D. ± 5.4); if the figures for 20 May are excluded the mean temperature range was 12.3°C (S.D. ± 3.7).
The mean of all the daily maxima was 36.6°C (S.D. ± 2.41; n = 43). The mean of all the daily minima was 26.1°C (S.D. ± 5.1; n = 44).
*By this date the nestlings were no longer being brooded.

end of a spell of incubation, and the minimum figures were those to which the temperature had fallen by the time she returned to incubate after an off-duty spell.

Generally speaking, it can be seen that maximum temperatures tended to increase slightly until towards the end of the incubation period, whereas the minimum fluctuated fairly markedly. The exceptionally high minimum on 20th May, as high a temperature as I found under females roosting at night, occurred on a day of heavy rainfall when the female did not leave the nest during the observation period and when the temperature was recorded as falling only 1.4°C.

Once the eggs had hatched, the differences between the maxima and minima decreased quite sharply. At the same time the female was brooding less, and not at all during the observation periods of the sixth day, so that any changes in temperature occurred when upheavals among the nestings exposed the thermistor to ambient temperatures.

Between 15 May (sixth day of incubation) and 24 May, when the eggs began to hatch, the mean range in temperature was 10.3°C, although if the figures for 20 May are excluded the mean was 12.3°C.

The average of the daily maxima reached at the end of all spells of incubation from 15 to 24 May was 36.6°C, ranging between 39.7°C and 30.5°C; while the mean daily minimum reached at the end of the female's off-duty spells was 26.1°C, ranging between 36.6°C and 17.2°C. These figures seem to be similar to incubating temperatures for other species: O'Connor (1985) quotes Baldwin and Kendeigh (1932) as giving temperatures of 37 species of several different orders which averaged 34°C. Drent (1973) states, however, that for most species average temperatures vary between 34°C and 38°C.

Roosting Temperatures

Once or twice a night between 21.45 and 02.33 GMT in the incubation period and when the female was roosting, I made spot checks on temperatures. The mean of 13 observations was 35.7°C (S.D. ± 2.0°C), ranging from 38.0°C to 34.9°C. There seemed to be a slight tendency for the roosting temperatures in the early part of the incubation period to be a little lower than those recorded nearer hatching.

After hatching, spot checks over five nights showed the mean to have risen to 37.6°C (S.D. ± 1.0°C), ranging from 37°C up to 38.9°C as the nestlings grew older.

TEMPERATURE CHANGES IN THE NEST

Rises in Temperature on Female Returning to the Nest

Immediately the female returned to incubate, the temperature between the eggs rose sharply. Where it had fallen as low as 26.1°C (Table 11.7a) while she had been away, the rise during the first minute or two of her return was fairly sharp and about two-thirds of the temperature lost was regained. Some rises in the first and second minutes were so sharp that the temperature originally lost was entirely regained: for instance, in the first minute the greatest gain was 16.9°C, which occurred after the temperature during the female's absence had dropped to 17.2°C, the lowest that I recorded. The slow rise in the third minute was usually caused by the female rising from the eggs to probe among them or to change her position. From the fifth minute and later, the rise remained fairly constant until she left the nest again, although the temperature sometimes fell when she probed among the eggs or changed her position.

Mean Fall in Temperature on Female Leaving the Nest

Once again, the greatest drops in temperature occurred in the first two minutes after the female had left the nest, and particularly in the second. It is possible that the falls recorded in the first minute are rather smaller than they

TABLE 11.7: *Changes in temperature per minute after female left or returned to the nest*

(a) Mean rise in temperature in °C after female returned to the nest from the minimum reached when she was off duty

| | Minimum temperature | MINUTES | | | | | | |
		1	2	3	4	5	6	7
Mean	26.1	6.1	3.1	0.7	0.7	0.3	0.4	0.4
S.D.±	5.1	4.9	2.9	0.6	1.1	0.3	0.3	0.3
Range	36.6	0.1	0	0	0	0	0	0
	to	to	to	to	to	to	to	to
	17.2	16.9	11.9	2.5	4.8	1.7	1.2	1.1
n =		44	43	41	41	32	30	22	12

(b) Mean fall in temperature in °C after female left the nest from the maximum reached during brooding

| | Maximum temperature | MINUTES | | | | | | | | |
		1	2	3	4	5	6	7	8	9+
Mean	36.6	3.0	4.7	2.1	1.1	0.6	0.5	0.4	0.5	0.2
S.D.±	2.4	3.0	3.3	1.4	0.5	0.3	0.2	0.2	0.2	—
Range	30.5	0	0.3	0	0.4	0	0.1	0	0.3	0.1
	to	to	to	to	to	to	to	to	to	to
	39.7	11.8	11.9	6.5	2.6	1.2	0.9	0.8	0.9	0.3
n =		43	43	39	33	29	25	16	11	5	2

should be: the method of temperature-recording that I used in this exercise meant that, during part of that first minute, the female might be on the nest and therefore keeping the temperature high. Nevertheless, there were enough unbiased observations to show that the temperature did not fall as fast in the first minute as it did in the second. Presumably, some heat is retained by the nest material and by the eggs themselves that prevents the ambient temperature cooling the eggs immediately.

From the third minute the fall was fairly steady; in fact the biggest fall in mean temperature (2.6°C) was between the second and third minutes. I was surprised that the rate of fall was not constant, although the variation was very small (0.1°C or 0.2°C), and I can only assume that the ambient temperatures in the burrow changed by that amount; the changes could not have been generated by the eggs or the nest materials.

There seem to be rather few studies of the temperatures at which passerine eggs are incubated. Gibb (1950), studying Great Tits, used an ordinary mercury thermometer inserted through one side of the nestbox so that the bulb was in the centre of the clutch; the mercury column was exposed and could be read without disturbing the sitting bird. There was a steady rise in the daily maximum temperature recorded between the eggs during the course of incubation, from about 34°C to 36°C. Gibb also noted the temperatures when the female returned to the nest after an off-duty spell; he found that the temperature range on the last day of incubation was 5.5°C, when the average time off the nest was about ten minutes. These figures are similar to those for Wheatears.

HATCHING

Duration of Hatching

After 13 days the eggs hatched, and the first indication of the event was often the female flying away from the nest carrying a piece of egg shell, which she deposited at varying distances from the nest. She flew in a rather laboured way, indeed in the same type of flight used later on when she or her mate removed faeces from the burrow.

Hatching lasted one day in 35% of the clutches, two days in 47%, and three

FIGURE 11.5 *(a) Scratching and (b) bobbing*

TABLE 11.8: *Duration of hatching according to clutch size and season*

Hatching prolonged over	No. Eggs in Clutch					Total	(%)
	2-4	5	6	7	8		
One day	2	11	12	1	1	27	(35)
Two days	2	11	19	5	—	37	(47)
Three days	—	9	4	1	—	14	(18)
Mean (days)	1.5	1.9	1.8	2	[1]		

Hatching prolonged over	MONTH		
	May	June	July
One day	20	7	
Two days	27	9	1
Three days	6	6	2
Mean (days)	1.7	2.0	[2.6]

days in the remaining 18% (Table 11.8). There seemed to be a slight tendency for the duration of hatching to increase with clutch size and as the breeding season progressed. While the first result might be expected, I was rather surprised at the second. In any case, the results are not significant.

Number of Young Hatched

Once again, I was able to compare the data I collected on Skokholm with Axell's data from Dungeness and Tye's from Breckland. The mean number of eggs which hatched on Skokholm and at Dungeness was 4.8; for Skokholm the mean was 4.9 young, and for Dungeness it was slightly smaller at 4.7 (Table 11.9) gives details, including standard deviations); but on the small sample available the difference was not significant. The most frequently recorded first-brood size was five young. In Breckland, the mean number of first-brood young hatched in 1977 was 4.9 (S.E. ± 0.2; n = 46) and in second broods 4.2 (S.E. ± 0.2; n = 6) (Tye 1980).

Nestling Period

When the nestlings were about eleven to twelve days old, they scrambled out of the nest and began to wander about in the burrow, although they frequently returned to the nest cup to be fed, to rest and to hide when alarmed. (What is usually called the nestling period should, for Wheatears, more properly be called the nest-burrow period, but since it is possible to see

TABLE 11.9: *Wheatear brood sizes on Skokholm and Dungeness*

	1	2	3	No. of nestlings 4	5	6	7	8	9	Mean	n
Skokholm	2	6	16	33	69	47	11	—	—	4.9	184
%	1.2	3.3	8.7	17.9	37.5	25.5	6.0	—		(S.D. ± 1.2)	
Dungeness	1	7	7	14	31	21	1	—	1	4.7	83
%	1.2	8.4	8.4	16.9	37.3	25.3	1.2	—	1.2	(S.D. ± 1.4)	
Total	3	13	23	47	100	68	12	0	1	4.8	267
%	1.1	4.9	8.6	17.6	37.4	25.5	4.5	0	0.4	(S.D. ± 1.2)	

when the young leave the nest only through some form of observation hole I have conformed with usual practice.) Their legs were weak at first, and their ability to hop poor. The fourteenth and fifteenth days they spent chiefly at the entrance to the burrow, awaiting the return of the parents with food.

On average, the nestlings on Skokholm remained in the burrow for 15.4 days (S.D. ± 2.3), ranging between 21 days and ten days; there is insufficient information from Dungeness and Breckland to make a comparison. There was no seasonal variation in these figures, but there was some indication that larger broods spent rather longer in the nest burrow, although the difference is not significant.

When the nestlings finally left the nest burrow, they sheltered in other holes and burrows, usually alone but sometimes with other siblings. This was the time when they began to isolate themselves and establish their individual distance. When they were about three weeks old, they spent more time above ground. It was at this point, too, that their escape behaviour changed: when alarmed, instead of running into a burrow they flew off.

NUMBER OF YOUNG FLEDGED

The mean number of young which fledged from 221 broods on Skokholm was 4.0, or 1.7 less than the mean clutch size. The number fledged decreased from 4.9 in the last half of May to 4.2 between 1 and 15 June, 3.9 between 16 and 30 June, 3.7 between 1 and 15 July and 2.9 between 16 and 31 July (Table 11.10). Comparison with the Breckland figures is interesting but not simple, since Tye has removed from his figures for nestling success the effects of predation (which was heavy on the Brecks). Even so, for 46 first broods in 1977, the mean number of young fledged was 4.6 (S.E. ± 0.2), compared with 4.9 on Skokholm for the first half of May with predation included. For four second broods in 1977 the Breckland mean was 4.3, compared with 3.9 on Skokholm for the last half of June.

TABLE 11.10: *Mean number of young fledged according to season*

Clutch size	No. of clutches	May 15-31	June 1-15	June 16-30	July 1-15	July 16-31	Mean
2	2	—	2	—	—	—	1.0
3	4	—	—	—	3	1.5	2.2
4	23	—	2.7	2.6	2.5	2.5	2.6
5	75	4.3	3.6	3.3	3.9	4.4	3.7
6	95	4.9	3.8	4.3	4.7	—	4.6
7	20	5.3	4.5	6.5	—	—	5.0
8	1	—	7.0	—	—	—	[7.0]
9	1	—	—	9.0	—	—	[9.0]
Total	221						
Mean		4.9	4.2	3.9	3.7	2.9	4.0

SUCCESS RATE FROM EGGS TO FLEDGING

Of the 1,519 eggs, 82.7% hatched; and, from 73.1% of those which hatched, young successfully left the nest (Table 11.11). That Skokholm had a rather better hatching percentage than Dungeness was due largely to the fact that more desertions were recorded at Dungeness. On the other hand, the losses in the young were rather higher on Skokholm than at Dungeness, probably owing to the rather higher predation rate. I have added comparable figures for Breckland from Tye's (1980) paper. In spite of the different hazards which affected the breeding success of Wheatears in three very different habitats, the final fledging success rates are very similar.

TABLE 11.11: *Egg to fledging success on Skokholm, Dungeness and Breckland*

	Eggs laid No.	Eggs hatched No.	Eggs hatched %	Young fledged No.	Young fledged %
Skokholm	622	526	85	446	72
Dungeness	452	354	78	334	74
Breckland	445	378	85	329	74
Total	1,519	1,258	83	1,109	73

Causes of Egg Losses

There can be little doubt that the figures from Dungeness and Skokholm are biased samples: the most frequent cause of egg losses was desertion. Unfortunately, Wheatears desert very easily if a hand is placed down the burrow when the female is on the nest. As a result, seven clutches of six eggs, one of five, three of four, and one each of two and one egg were deserted. In addition to the desertions, between one and five eggs failed to hatch, presumably because they were infertile, but some might have been chilled during egg-laying during the very cold spring of 1951. As I pointed out elsewhere, 1951 was a cold spring, and on three nights, one in April and two in May, when grass-minimum thermometers showed 2.2°C (38°F), 1.1°C (34°F) and 6.1°C (43°F) respectively, three females failed to lay; in the end, all the eggs in these clutches hatched (Table 11.12).

In 1951, the percentage of first-clutch eggs hatched of those laid was much

TABLE 11 12: *Egg and nestling success for first and second broods on Skokholm*

		No. eggs laid	Eggs hatched		Nestlings fledged	
			No.	%	No.	%
1948	1st brood	71	58	83	55	77
	2nd brood	30	26	87	20	66
	Total	101	84	83	75	74
1949	1st brood	102	96	94	85	83
	2nd brood	14	14	100	14	100
	Total	116	110	95	99	85
1950	1st brood	64	53	83	50	78
	2nd brood	29	26	90	17	59
	Total	93	79	85	67	72
1951	1st brood	138	93	67	71	56
	2nd brood	31	28	90	26	84
	Total	169	121	72	97	57
1952	1st brood	112	102	91	78	70
	2nd brood	31	30	95	30	95
	Total	143	132	92	108	75
Total	1st broods	487	402	83	339	70
	2nd broods	135	124	92	107	79
Grand total		622	526	85	446	72

smaller than usual (67%), while the hatching success of second clutches was about average (Table 11.12). On 30 May, after the other nestlings had left the nest, I collected eleven unhatched eggs and they were examined by A. Vince of St. Bartholomew's Hospital. Of the eleven eggs, only one might have been infertile; the remainder showed some sign of development from a fertile blastodisc, embryos about one-tenth the size of the egg with eyes, to fully developed embryos which had started to chip the egg but which for some reason — perhaps low temperatures — had failed to finish. In other years, the largest number of eggs from a single clutch containing fully-developed embryos that failed to hatch was three (out of a clutch of six). Unfortunately, not all contents of unhatched eggs were checked.

Only four clutches were preyed upon on Skokholm and the predator was thought to be a Carrion Crow or, more likely, Great Black-backed Gulls, which often hunted for noisy Manx Shearwaters. One nest with eggs was destroyed by a rabbit, which cleaned out the burrow. At Dungeness two nest-boxes were blown over in a gale and the eggs destroyed, and a Common Gull built its nest on top of the nestbox, the nest eventually blocking the entrance.

In Breckland, Tye found that, for first broods, almost one egg per clutch, or about 11.1% of the eggs, failed to hatch.

Causes of Loss of Nestlings

Predation was the most frequently recorded factor causing loss of nestlings: altogether, 17 of the 30 nests to which a cause for the loss of nestlings was assigned were preyed on. Sixteen of the nests on Skokholm were pulled out by crows or gulls and the young eaten, and three nests were destroyed by rabbits, Manx Shearwaters or Puffins. In five broods, individuals which had failed to compete successfully for food died. At one nest on Skokholm which I had under close observation, I had noticed that the faeces brought from the nest were of an unusual colour, and two days later one of the young died. One unringed nestling was caught up in the grass at the entrance to the nest and died. At Dungeness, one nest-tin was run over and crushed by a lorry, and one brood was thought to have been taken by a rat.

The predation noted on Skokholm and at Dungeness is not perhaps typical of the factors likely to be encountered in other parts of the Wheatear's range. In Breckland, for instance, Tye (1982) has shown that the major predators on Wheatears were stoats, although rats and possibly hedgehogs may have caused some of the losses attributed to stoats. Foxes and dogs also excavated and destroyed a few broods, although some well-developed nestlings managed to escape by retreating down the burrow beyond the nest when foxes were digging. Of the 98 completed nests that Tye studied in 1976 and 1977, 27 (28%) were preyed on when containing eggs (eight) or young (14) or on the day of hatching. Predation seemed more of a hazard in Breckland than on Skokholm.

While Cuckoos were not recorded laying eggs in Wheatears' nests on Skokholm, at Dungeness or in Breckland, they have been recorded as doing so in other parts of Europe (Cramp 1988).

202

DIFFERENCES IN BREEDING SEASON BETWEEN SKOKHOLM AND DUNGENESS

There are several remarkable differences between the records from Skokholm and Dungeness. In the first place, it was obvious that breeding was earlier in the west than in the east, and possibly lower minimum temperatures at Dungeness might have been responsible. Second broods appeared at approximately the same time in both places.

Another striking difference was in the clutch sizes, the mean for Skokholm being about half an egg larger than for Dungeness, and this difference is statistically significant. This may be a good example of a clutch size being adapted to the amount of food available to feed the parents or the young. On Skokholm, the habitat is chiefly rabbit-grazed grassland, and there was a very dense population of rabbits as well as Manx Shearwaters and Puffins, all of which excavated burrows. Dungeness, on the other hand, is a vast shingle area with vegetation only on parts of each ridge and with a locally distributed rabbit population. Here the Wheatear population had been increased when Axell put out nestboxes, thus providing artificial nest sites in an area where natural sites were scarce. Knowing Dungeness a little, I came to the conclusion that the vegetation was not extensive enough to provide sufficient food — which, on Skokholm, was normally grass-eating moth larvae and cranefly larvae at this season — for the same sort of population of Wheatears as was found on Skokholm. The provision of these nest sites had induced Wheatears to breed in a habitat which might otherwise have been unsuitable. It was also obvious that the size of the nestboxes provided was not the cause of the smaller clutch sizes.

The shortage of nest sites at Dungeness led to behaviour recorded both there and at other places where nest sites are scarce (e.g. Baffin Island: Wynne-Edwards 1952), but not on Skokholm. At Dungeness, first-brood nests were frequently relined for second broods in the same season, something which never happened on Skokholm. The only occasions when a second-brood nest on Skokholm was constructed in a first-brood burrow in the same season was after I had removed the first nest for parasite examination, and in those circumstances, obviously, a new nest had to be built; but, if the first-brood nest was not removed, then the burrow would not be used again until the old material decayed, which usually took about two years.

Chapter 12

CARE OF NESTLINGS

BEHAVIOUR OF ADULTS

On the day the eggs began to hatch the female spent almost as long covering her newly-hatched young as she had spent incubating the clutch — often a few eggs had yet to hatch. On the first day the brooding percentage ranged between 77% and 29%, averaging 49%, and thereafter the amount of time spent brooding declined (Table 12.1). Of the nine females for which I have sufficient information, one stopped brooding after the fourth day, five after the fifth day and three after the sixth day. Such a range of brooding percentages might be expected when hatching was spread over up to two days and the nest inspected only once a day. As I shall show in the next section, by the time the female gave up brooding the nestlings had more or less gained temperature control, or were homoiothermic.

The behaviour of the males changed little during the first day or two of the nestling period. Some peered into burrows when the eggs were due to hatch, and automatic recorders also indicated changes in the usual pattern of nest-visiting which could be interpreted only as a visit to the nest by the male when the female was brooding.

FIGURE 12.1 *Male bobbing*

204

TABLE 12 1: *Brooding percentages (percentages of time spent on nest) during the nestling period on Skokholm*

| Female | | | DAY OF NESTLING PERIOD | | | | |
	1	2	3	4	5	6	7
W2	77	40	45	30	6	0	
W19	61	31	32	20	6	0	
W30	31	51	54	51	13	0	
W53	68	50	46	29	?	15	0
W63	41	26	21	11	9	0	
W67	?	52	36	26	0	0	
W92	54	70	52	64	18	10	0
W102[1]	35	29	32	21	21	18	0
W149	29	31	17	10	11	0	
Mean	49	42	37	33	7	5	

A male could discover that the eggs had hatched in several ways, of which the female's changing behaviour was the most obvious: she left the burrow more frequently, possibly hunted more energetically than she would for herself and returned to the burrow more quickly, sometimes re-emerging almost at once. The male might have learnt something while he was peering into the burrow but, when watching from about 20 cm, I did not hear the young calling until they were three days old. It is said, however, that embryos of some species, when still within the shell, can hear each other calling and that this ability helps them to synchronise their hatching; perhaps other Wheatears could hear something, although I could hear nothing. A few males went into the burrows while the eggs were hatching, and something must have stimulated them to do so since, for most of the incubation period, they hardly approached it.

Males still defended the territory against other Wheatears, although less energetically than during egg-laying and incubation periods. The less frequent use of song and the less energetic defence of territory at a time when abundant and easily accessible food was so necessary seem to throw doubt on the suggestion that one of the primary functions of territory is to safeguard the food supply. Surely one would expect defence against other Wheatears to be fiercer when food was most urgently needed; but perhaps the territory boundaries had become so well recognised by this time that attempts to intrude were fewer. Of course, intrusion by other Wheatears was also reduced at this period, since they, too, were concentrating on feeding their offspring. When second broods were being fed, first-brood juveniles usually left the vicinity of a nest if they were threatened by the owners, although, as I have pointed out elsewhere, juveniles of different age groups in Greenland were sometimes tolerated by adults feeding nestlings (and Alan Tye tells me that this happens in some other wheatear species).

If defence activities against other Wheatears were reduced, their anxiety at the presence of potential predators or creatures that they feared, such as humans, sheep, gulls and raptors, became more intense as the time for leaving the nest approached. This again suggests that fear of attack on self and offspring is of greater importance than fear that another Wheatear will take their food supply.

THE RATE OF ATTAINING HOMOIOTHERMY

Immediately the eggs hatch, the range of temperature between the maximum and minimum is reduced. The female broods the nestlings less and less, so that the maximum temperature does not build up so high as it did just before the hatch, nor does the minimum fall so low when the female leaves to find food (Figure 11.4). Presumably, while the nestlings cannot yet maintain their body temperatures, they lose heat less rapidly than the eggs and every day there must be an element of improved temperature control.

Wheatear nestlings apparently gain temperature control relatively quickly. By the sixth day female W194 had given up brooding them, and the changes in temperature recorded by the thermistor were usually less than 1°C and could usually be accounted for by the nestlings' movements in the nest exposing the thermistor to the ambient temperatures for short periods. It is, however, possible that it was only in the nest that they could maintain that control. Visiting one second-brood nest when the nestlings were six days old, I found that the nest had been pulled out of its burrow, probably by a Great Black-backed Gull, and only one chick was lying on the earth of the burrow, cold and comatose. It was replaced in a makeshift nest but was dead on the following day. Both parents were still present, but I rather think that the chick was too cold to solicit food and the female had given up brooding.

The age at which the young apparently gained control seemed to be similar to that of nestlings of a similar size, but possibly rather short if one thinks of Wheatears as fledging or leaving the nest burrow at about 15 days; it is, however, similar to those species whose fledging period is about ten days. I pointed out earlier that at about eleven to twelve days young Wheatears leave the nest and begin wandering up and down the burrow, so that in some respects to say that the nestling period is about 15 days is misleading. Unfortunately, to determine more precisely the length of time that Wheatear young spent wholly in the nest (rather than using the nest for increasingly short periods as a resting place) would have caused considerable disturbance.

FIGURE 12.2 *Watching*

Orientation of Burrow Entrances

Ricklefs and Hainsworth (1969), in discussing the nest cavities of Cactus Wrens, point out that the orientation of the entrances to the prevailing wind is important in moderating the ambient temperature within the cavity. In the case of Cactus Wrens, which live in deserts where temperatures can rise so high as to be lethal to nestlings, it is often important that the cavity entrance should face prevailing winds to keep the interior cool. In temperate zones, even in the cool climate in which Wheatears nest, shelter from the prevailing winds, however, is more important and, as I pointed out in Chapter 9, not only did the majority of burrow entrances face away from the prevailing winds but the actual sites were often in some form of wind shadow. Nest W194, situated in the northeast part of Skokholm, was sheltered in this way from the west, but it was exposed to easterly winds.

Between 9 May, when the fourth egg was laid, and 14 May, when incubation temperatures were low (see Figure 11.4), the wind blew fairly strongly from easterly quarters: on four of the five days the strength was force four or five (21-39 kph/13-24 mph), which, as I have commented earlier, was the strength at which the progress of birds began to be impeded. Of course, temperatures are low during the first days of incubation proper, but I concluded that easterly winds were probably responsible for these low figures, and might well have been responsible, either directly or indirectly, for the low incubation temperatures shown in Table 11.6 and Figure 11.4.

I also mentioned earlier that heavy rainfall and a southeast force 4 wind on 20 May was probably the reason why the female remained on the nest throughout the period of observation. We shall see later in this chapter that rain affected nestling weights.

FEEDING NESTLINGS

In the early days of this study and, later, when I returned to Skokholm between 1978 and 1983, I studied the rates at which adult Wheatears brought food to the nest by watching the nest-burrow entrance. This I did either from a hide within a few metres of the nest or by sitting in the open 50-100 m from the nest with my back against a rock so that I was less obvious. Then, from either of these positions, I counted the number of times males and females went in and out of the burrow, how long they remained inside, in which directions they set off to find food, and how they reacted to the events around them.

In early June 1949, I used the coffin hide, which I described in the preface to this book and watched nestlings W67 from a distance of about 25 cm. It then became apparent that parents often brought more than one food item. Table 12.2, based on that study, shows the number of times that adults visited the nest per hour and the mean number of beakfuls fed to the nestlings. Because of the nature of the nest burrow, I still had difficulty in determining which parent was visiting. The factors that seemed to affect the number of food items brought by the parents to the nest were strong westerly winds (see day eight). The nest site was sheltered from southeasterly winds by a range of rock outcrops known as South Crags, and from northerly directions by the West Knoll range of outcrops.

It was also clear that on warm sunny days with little wind the parents

TABLE 12 2: *Number of food items given to nestlings W67*

Age in days	No. visits per hour	No. food items	Mean	Weather		Max °C(°F)
1	?	?	?	r.o,	SW6-7	11.7 (53)
2	13	21	1.6	bc,	SW3-4	18.0 (57)
3	18	23	1.3	bc,	SW5-6	13.3 (56)
4	19	21	1.1	bc,	SW5-6	18.0 (57)
5	25	29	1.2	b,	SW2-3	18.0 (57)
6	25	34	1.4	b.o,	S1-6	15.0 (59)
7	34	48	1.4	o.r,	S4	15.0 (59)
8	30	30	1.0	c.bc,	W5-7	16.1 (61)
9	40	50	1.3	b.bc,	SW1-2	16.1 (61)
10	37	49	1.3	o.d,	SW3	15.6 (60)
11	29	39	1.3	o.d,	SW5-2	16.1 (61)
12	14	22	1.6	bc.b,	NE1-2	17.2 (63)
13	44	69	1.6	b,	E1	17.2 (63)
14	30	34	1.1	bc,o,	Var.1	18.3 (65)
15	38	45	1.2	o.r,	W4	?

b = blue sky; bc = bright intervals; c = cloudy; d = drizzle; o = overcast; r = rain.

produced more items on each visit. For instance, when the nestlings were 12 and 13 days old, the mean number of food items per visit was 1.6; the weather on those days was recorded as mostly sunny with light easterly winds and a maximum temperature of 63°F (17.2°C). By contrast, on the eighth day, with cloudy weather and strong south to southwest winds, the mean number of food items brought at each visit was low. Unfortunately, Skokholm had not been appointed a climatological station by the Meteorological Office in 1949, and consequently the instruments we were using were untested or unchecked.

When watching the parents bringing food to the nest, it was obvious that, as the young grew, the adults were bringing in larger prey items. This meant that, not only were the young receiving more visits from their parents with caterpillars, other insect larvae, worms etc, but these items were increasing in bulk.

The time I could spend watching nests like this was limited by other aspects of the Wheatear study and also by the daily chores required of a bird observatory warden. In 1951 I began using automatic recorders, based on a design by John Gibb, who had been using them on tit nestboxes in Wytham Woods, Oxfordshire (Gibb 1950), although I had to redesign the trigger mechanism to fit rabbit-burrow entrances. A small platform was placed in the entrance to the nest burrow in such a way that a bird entering could not avoid treading on it. The platform was made of a flexible sheet of copper, which, when trodden on by Wheatears, bent and made contact with another sheet of copper mounted on wood, which completed an electrical circuit.

In the damp atmosphere of a Celtic Sea island, however, corrosion regularly impeded the circuit, so I made another contact. A piece of copper wire was soldered to the bottom of the copper platform and, when the platform was pressed down as a Wheatear stepped on it, the copper wire dipped into mercury contained in a glass tube — a practice which would probably be frowned on today for health reasons. When the circuit was thus completed, it operated a solenoid and caused a pin to pierce a length of telegraph tape which was being drawn on to a reel mounted on the hour hand of a clock. At the end of the 24 hours I marked the time on the tape and, before removing it from the reel, I made a cut across the accumulated tapes, since the reel revolved once an hour, these incisions marked the hourly intervals.

By measuring the length of tape between the incisions, I was able to calculate the number of millimetres that had passed the solenoid each hour. In practice, about 8.5 m of tape passed the solenoid every 24 hours, or about 6 mm a minute. This method of recording was obviously very advantageous and, in spite of the fact that humidity caused corrosion which often impeded the electrical circuits, I gained a lot of useful information.

I used these automatic recorders in 1951 and 1952 at a number of nests. When all was working well, they gave me a 24-hour record of the number of visits to the nest by one or other of the parents (but not which of them). I also had to assume that the adults brought food at each visit.

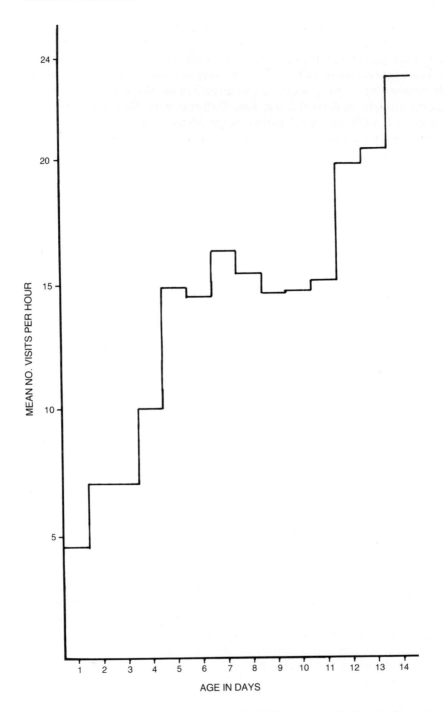

FIGURE 12.3 *Mean number of visits per hour in nestling period, including data from automatic recorders and from personal observations.*

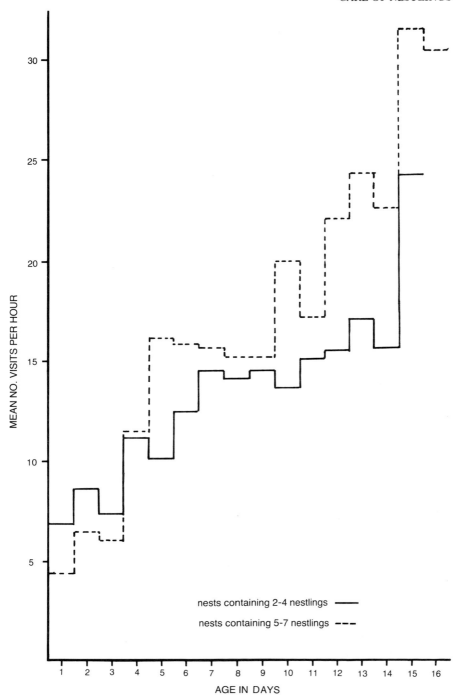

FIGURE 12.4 *Mean number of visits per hour by adults to nests containing 2-4 nestlings (unbroken line) and 5-7 nestlings (broken line).*

211

Feeding Rates

Figure 12.3 shows the average rate at which the nest was visited every hour, based on records collected by visual observation at nests by myself or by my helpers, and also on records from automatic recorders. The graph shows a fairly steep rise in the rate of feeding during the first three or four days, then a rather slower increase until the nestlings reached the age of eleven or twelve days old, when they began wandering up and down the burrow; at this time the feeding rate increased again, reaching a peak when they were about 13 days old, after which the rate decreased slightly. This final statement is based on visual observations only, since the nestlings, in hopping up and down the burrow, also hopped up and down on the platform of the automatic recorder, invalidating the record if not actually stopping the tape.

For the first 15 days, the mean number of hours spent recording feeding rates daily for all nests combined was 44.2 (S.D. ± 8.9). Because nestlings interfered with the trigger mechanism on days 14 and 15, I had adequate records for only 23 and 24 hours respectively.

This curve differs, unexpectedly, from the curve showing the average daily weight increases of the nestlings. Their weight generally increased slowly for the first two or three days of their lives, steeply for the next seven, and slowly again from the eleventh day until they left the nest; in the last phase it was subject to great daily fluctuations (Figure 13.4). In the case of Figure 12.3, the graph curves upwards fairly steeply during the first two or three days, then the average number of visits to nests with food slackens between days 8 and 11, and the rate increases for the last few days. I shall be returning to this point later when discussing weight increases.

I tried to discover whether visiting rates with food depended upon the number of nestlings in each brood, but I had sufficient data to show differences only in the feeding rates for small broods of two to four and for larger broods with five to seven nestlings (Figure 12.4). As might be expected, the feeding rates for broods of five to seven were generally higher than for the smaller broods. Again, the figures from 15 days and older have been omitted, since it was obvious that the nestlings were hopping on and off the trigger platform. Tye (1982) examined the rate at which parents brought food to broods of different sizes: in general, members of large broods received fewer visits than members of smaller broods. He points out that, because members of larger broods could keep warmer in the nest than smaller broods, the energy required for thermoregulation would be less.

The Role of the Sexes in Feeding the Nestlings

Taking the nestling period as a whole, the female averaged 12.4 visits an hour, compared with 5.2 visits by males (Table 12.3). In only one of the six pairs (W19) for which I have sufficient information did the feeding rate of a male equal that of a female. A very few males brought food from the time of hatching, but others were either slow to appreciate that the chicks had hatched or took two to four days before they brought food. In 1980, I watched

TABLE 12.3: *Average number of visits per hour with food by females and males in the nestling period*

Pair	Female	Male	No. Hours
W 2	13.3	6.4	22
W19	9.2	9.2	29
W30	13.5	6.7	13
W52	7.7	6.7	27
W63	10.3	2.0	32
W67	20.5	0.25	8
Mean	12.4	5.2	—
Total hours	—	—	131

a male which was mated to two females, of which one was feeding young that had just left the nest burrow and the other was still building a nest. During the week that I watched, the male helped neither, brought no food for the juveniles which had already left the nest burrow, and ignored the second female building a nest.

Brooke (1981) records that in six broods he watched on Skokholm most feeding of nestlings was undertaken by the female. Tye also found that in the early stages females spent 10-30% more time foraging than males, and again during the nestling stages of second broods females spent more time foraging.

Some males visited first-brood nests with food more often than they did second broods; at this latter stage some males were already beginning to moult, and were less energetic than they had been. If a male came to the nest with food and the female was sitting, he would sometimes leave the burrow still carrying the food and wait for the female to come off. As they had during incubation, males often listened at the burrow entrance when the female was on the nest, and when they could well have been calling quietly to each other. Sometimes males peered into their nest burrows when females were off. Once, when female W19 was away and her nestlings were about two days old, the male spent about 30 seconds peering down the burrow and warbling the quiet subsong; finally, he flew off without entering the burrow. He behaved as he would when calling the female off the nest.

Female Taking Food from Male

Once or twice, a female was seen to take food from the male during the greeting display when they met at the entrance to the burrow. It is tempting to think of this as 'courtship feeding', but I never saw the female swallow the food: she took it into the burrow. One female was already near the nest when the male flew up to her, spread his wings and quivered them. She hopped to him and took the food, flew with him to the burrow entrance and stood there for a minute. Eventually the male went in, and after he had left, having presumably fed the nestlings with another particle of food, she entered. She did not seek more food.

On another occasion, a male flew to the nest with a caterpillar. The female emerged and greeted him. The male moved away, but the female came after him, still greeting. Again he moved away, but still she followed. Finally, he dropped the caterpillar, which she retrieved and took into the nest.

In each case the basic soliciting posture was used, first by the male and second by the female, in circumstances which were typical of the greeting display except that, because food was passed, one might be tempted to call it courtship feeding. Since the females did not eat the food themselves but passed it to the nestlings, the circumstances were different. Furthermore, the female took the food from the male: he did not offer it to her. I never saw this behaviour in the pre-egg-laying period, when courtship feeding normally occurs. This behaviour may have been mistaken for courtship feeding by some ornithologists in the past.

Behaviour of Adults when Searching for Food

When an adult emerged from the nest burrow after it had fed the nestlings, it usually looked around for a second or two and then flew off in search of more food. Sometimes it flew as little as 5 m or up to about 250 m before it began to search. Sometimes one would hop away from the nest, hunting as it went, or it returned to the area in which it had been foraging successfully earlier. While some adults would sometimes return to the nest immediately if they found a food item less than 5 m from the burrow, others would pick it up and continue searching as they hopped or flew further from the nest, hunting as they went. Usually they hunted within the territory or, if the population density was low, as much as 250 m from the nest.

Brooke (1981) studied how an adult Wheatear used its territory when feeding nestlings. He found that, the further Wheatears travelled from the nest, the more food items they brought back: that they exploited areas close to the nest more intensively but met similar feeding conditions at all distances. They also tended to return to those parts of the territory which they had recently visited, particularly after gathering a meal for the young rapidly.

Both parents showed anxiety when predators or large animals such as Soay sheep were near a nest containing young, and this anxiety became rather frenzied when the nestlings were about to leave or were actually leaving the nest burrow. In the last few days, both parents were apt to call the low-intensity anxiety notes 'tuc-tuc' almost continuously when in the vicinity of the nest, but when they were anxious this changed to 'weet-tuc-tuc'. A human close to the nest or the appearance of a falcon could stimulate them, and one or both of the pair might hover to get a better look at the cause of alarm. It was at moments of stress such as this that, on Skokholm, the parents might attack any juveniles — their own or strangers — that were near the nest and which they might have ignored under normal circumstances: juveniles were an attackable substitute for whatever it was that caused the fear. I am reminded of female W61, which attacked her own mate fiercely when he gave up attacking a male Greenland Wheatear which had established its individual territory within a few metres of their nest.

Anxiety is a state which intensifies throughout the breeding season: it starts becoming obvious when the time for egg-laying draws near, builds up to the time of the hatch, and reaches a peak when the young leave the nest. Carlson *et al.* (1985) argue that increased anxiety and aggressiveness shown by the male during the female's fertile period — just before and after the laying of the first egg — is linked with a possible danger that a strange male might 'steal a copulation' with his mate and thus pass on the stranger's genes rather than his own. These writers tend to ignore the increasing anxiety shown for the safety of the eggs and young which is manifested throughout the nesting period, until it reaches a climax when the nestlings leave the nest; this anxiety affects males as well as females.

Presumably, if male A was trying to steal a copulation with the female of pair B, then one might expect that male B was trying to steal a copulation with female A. If this was generally advantageous to males within a population, I wonder why Nature 'permits' Wheatears to indulge in such an energy-consuming territorial system. If it is to prevent other males doing it to which other males are we referring?

In England and Wales at least, some Wheatears have second broods, but the authors make no suggestion as to whether Wheaters mate-guard in that situation. From my observations, territory defence by the male is rather more relaxed at that stage.

Visiting Rates Throughout the Day

Automatic records showed that the parents visited the nestlings at frequent intervals for the first three or four hours of daylight. The rate slackened appreciably after midday, and then, depending on the pair, the visiting rate would increase again towards nightfall. Some adults were very idle just after midday; others continued visiting, but rather more slowly. The visiting pattern at the end of the working day was also varied. With some pairs, the intervals between visits gradually became longer and longer until night intervened. On the other hand, one pair visited its nest almost frenziedly during the last 30 minutes or so of the day, and then stopped abruptly for the night.

A number of factors reduced the visiting rate sometimes so seriously that the nestlings had lost weight by the following day. Unfortunately, one of the disadvantages of automatic recorders was that they could not often tell me what those factors were.

Some pairs just seemed inefficient. W116 produced two broods in 1951: the first had six young, all of which fledged, and of the five second-brood young only four fledged. Throughout the nestling periods of both broods the recorder showed that the visiting rate was lower than that of some other pairs, and the weight of the six nestlings of the first brood was below average by 1-2 g

(22.5 g, compared with 24 g). Only two of the nestlings from W116[1] achieved average weight and none exceeded it. It was probable that the weather affected the visiting rate: the first three days after hatching were wet and, later, strong winds blew from the north and northeast. I have no explanation for the drop in weight of brood W116[2] on 5 June 1951, when they were 13 days old, a date on which the weights of brood W103[1] also dropped. The meteorological records show a perfect summer day.

Pair W149 was also erratic in the way the adults came to or left the nest: for instance, towards the end of the incubation period, the female left the nest for 13 minutes at 02.05 GMT on 13 May, about two-and-a-half hours before sunrise and before she normally left the nest for the first time. I have no explanation for that behaviour. There were days when the visiting rate of adults W149 was low, but on 22 May, when the nestlings were five days old, no visits were made to the nest for 235 minutes before noon, and it is not surprising that on 23 May the brood had lost weight when they should have gained about 3 g. Again, I have no idea why this pair failed to visit their brood at a time when they were normally visiting at a fairly high rate. It was interesting to see, however, that on the following day the feeding rate was 76% higher, and when I weighed the nestlings 24 hours later they were gaining weight once again. Thereafter this pair had no major problems, although between 14.00 and 15.00 GMT they regularly took breaks lasting between 20 and 43 minutes when they did not visit the nest.

NEST SANITATION

For the first three days after the eggs had hatched, the majority of faeces were eaten by the adults. Female W67, which was watched from about 25 cm, was seen eating something on nine occasions during an hour's observation on the second day of the nestling period. The early faeces were usually black in colour, and white ones were brought out of the nest from the third day onwards. Occasionally black faeces re-appeared later, possibly when the nestlings were unwell: female W63 was seen carrying out black faeces when her nestlings were seven days old, and it was at about that period that two died.

Both parents removed faeces, but females, since they usually brought more food more often, also tended to take more faeces away. On average two or three faeces an hour were removed, but towards the twelfth day the figure had risen to five or six. After this, the number seen to be removed dropped, perhaps because the nestlings were spending more time in the burrow rather than the nest.

If the faeces were taken in transit, the adult generally held the faecal sac by the end which first emerged from the cloaca. Later in the nestling period, when the faeces had been deposited on the rim of the nest, the adult held the sac across the widest part. When the adult emerged from the nest burrow, it did not stop to look around as it did normally, but flew off directly in a rather rigid and purposeful type of flight with no undulations and wings quivering rapidly, rather as in the moth flight. The head was usually held higher than in the usual flight position, the tail depressed, and the bird usually flew within 30 cm of the ground; the flight was easily recognisable for what it was. The faeces were usually carried 30-100 m from the nest and often placed carefully on the ground, after which the adult wiped its beak and began searching for food.

During the last two days that the nestlings were in the nest burrow and when they were old enough to come to the mouth of the burrow, faeces tended to accumulate on the ground outside the entrance. The parents allowed the majority of these to lie, although some were removed. I got the impression that, if the adults saw a juvenile dropping faeces outside the nest, they removed them.

When the juveniles had left the nest burrow and taken up temporary residence in other burrows, the faeces they dropped outside were ignored. By that time the faeces had lost their tough covering and would have been difficult to pick up anyway. Since only one, or perhaps two, juveniles were sheltering in a burrow, there would be less chance of collections of faeces drawing the attention of predators.

THE WORKING DAY

One of the benefits of automatic recorders was that, during the incubation period, the time that the female first left the nest and the time she retired to roost on the nest could be ascertained reasonably accurately. Similarly, during the nestling period, the time that the nestlings received their first and last feeds and therefore the length of the Wheaters' working day could be determined, However, because automatic recorders were not used much during the incubation period, and not at all immediately after the clutch was complete (because of the danger of desertion), I have rather few records for the length of the working day during incubation.

In west Wales, the average interval between sunrise and sunset for a week in mid May, which is the period for which I have ten records of a Wheatear's working day during incubation, was 940 minutes (15 hours 40 minutes); and the average time that Wheatears were active was 879 minutes (S.D. $\pm$ 83.0 minutes), or 14 hours 39 minutes (Table 12.4a). In the last half of May, the average interval between sunrise and sunset had increased to 996 minutes (16 hours 36 minutes) and the average length of five days of incubation in W102's second clutch had increased to a mean of 960 minutes (16 hours). The differences in the length of the working day are not statistically significant.

The intervals between sunrise and the females' first emergence during incubation also varied enormously, and for no obvious reasons. Two females emerged from their burrows an hour or two after sunrise: W102[1] delayed her appearance for as much as two hours after sunrise and W149 for nearly three hours, but others came out for the first time a few minutes before sunrise.

TABLE 12 4: *The working day of Wheatears on Skokholm 1951-52*

(a) Incubation

		MINUTES		
		1-15 May		16-31 May
Mean		878.8		960.2
SD	±	83.0	±	17.0
n =		10		5
Longest		991		976
Shortest		748		941

(b) Feeding nestlings

			MINUTES					
	16-31 May		1-15 June		16-30 June		1-15 July	16-31 July
Mean	1,006.3		1,040.8		1,045.6		1,038.6	1,036.4
SD	± 35.41	±	18.0	±	13.9	±	23.6	± 18.6
n =	3 6		13		7		24	5
Longest	1,077		1,065		1,063		1,081	1,060
Shortest	877		994		1,023		998	1,010

At the other end of the day, the time females went to roost in the burrow also varied inexplicably so far as I was concerned. The intervals ranged from two minutes to 122 minutes before sunset, averaging 24 minutes. After sunset the mean interval was also 24 minutes, but one female would remain out for up to 34 minutes. On the other hand, W102 was, again, a bit of a maverick: she left the nest on four different occasions for two or four minutes between sunset and sunrise — well after dark, twice before midnight and twice after. There was no obvious reason for her to do so (unless she was 'taken short' and had to leave the nest in order to defecate!).

Once the eggs had hatched and the adults had begun to feed the nestlings, the length of the working day increased more or less in relation to sun time (see Table 12.4b). It can be seen that their average working day in the last half of May had increased to 1,006 minutes (16 hours 46 minutes), and to 1,041 minutes (17 hours 21 minutes) in the first half of June. It was five minutes longer (1,046 minutes) in the last half of June, and then decreased again to 1,036 minutes (17 hours 16 minutes) in July. None of these differences is statistically significant.

The longest working day that I recorded during the nestling period was 1,081 minutes, or 18 hours 1 minute, and the shortest was 877 minutes, or 14 hours 37 minutes. At this stage, the majority of Wheatears became active well before sunrise. In May, on average, the adults first fed the nestlings 34 minutes before sunrise (S.D. ± 14.2; n = 32), and in July 25 minutes before sunrise (S.D. ± 13.1; n = 32). Females generally visited more often than males (Table 12.3); the chances were, therefore, that females started the feeding. One of the few pairs that was late in starting to feed was W102, which, on two mornings, began visiting 35 and 36 minutes after sunrise.

In the evening, the last visit in May averaged 23.8 minutes after sunset (S.D. ± 14.1; n = 47); the last in June 27 minutes after sunset (S.D. ± 10.3; n = 24); and in July 37.5 minutes after sunset (S.D. ± 13.5; n = 31). The longest times in July between sunset and the last visit to the nest were 64 and 63 minutes, by pair W102 during their second brood.

In Senegal, with a day length of approximately twelve hours, Wheatears become active about 20 minutes before sunrise and feed until well after sunset — until it gets too dark to feed (Tye, pers. comm.).

FIGURE 12.5 *Snoozing*

Chapter 13

Nestling Weights

Using the portable, glass-fronted weighing machine which I have described in Chapter 10, I weighed the nestlings from 21 first broods and five replacement and second broods in 1951 and from 16 first broods in 1952. All those in 1951 were weighed as individuals, but in 1952 only seven broods were weighed as individuals and the others were weighed in bulk. This does lead to a slight discrepancy, since brood weights on a certain day may include nestlings hatched 24-72 hours apart. Nevertheless, when looking at the relationship between the weather, visiting rates by parents and so on, it is the daily bulk weight of the nestlings which is of importance. Figure 13.2 shows the combined mean weights of all Wheatear nestlings in 1951 and first-brood nestlings in 1952 (the thicker, unbroken line), the average weight of 1951 first-brood nestlings (thinner, broken line), the average weight of 1951 second-brood nestlings (mixture of dashes and spots), and the average weight of the 1952 nestlings (line of dots).

The shape of the curve as shown on the graph — a logistic curve (O'Connor 1985) — is similar to that which is seen in many other passerine birds, e.g. Great and Blue Tits (Gibb 1950). Generally, nestling Wheatear weights increased relatively slowly at first, accelerated between the third and tenth days, and then gradually slowed until many of them reached their maximum weight when they were 15-17 days old.

FIGURE 13.1 *Female removing faeces*

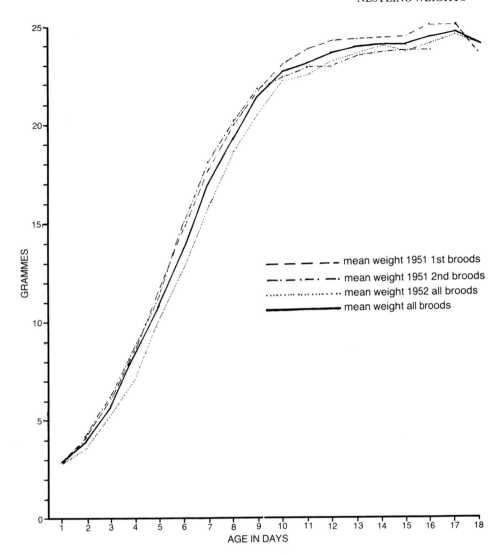

FIGURE 13.2 *Mean weight of Wheatear nestlings, 1951-52. Thicker, unbroken line = all 1951 and first-brood 1952 nestlings; thinner, broken line = 1951 first-brood; dashes and spots = 1951 second-brood; dots = 1952 first-brood.*

Table 13.1 and Figure 13.3 establish that the largest daily weight gains occurred up to the tenth day and that thereafter, at a time when the chicks were very active, they were still generally gaining weight, although some gains were very small. Finally, there was often a drop in weight on the last day, which may well have corresponded with the fact, generally, that the parents' visiting rates dropped on the last day that nestlings were in the nest. The automatic recorders had been rendered valueless at this stage by the young jumping up and down on the trigger mechanism!

221

TABLE 13.1: *Mean daily weight gains of all Wheatear nestlings weighed on Skokholm, 1951-52*

Age in days	Average weight (g)	No.	Average increase (g)
1	2.8	167	—
2	3.9	189	1.1
3	5.7	194	2.2
4	7.9	194	2.2
5	10.8	193	2.9
6	13.8	186	3.0
7	17.0	178	3.2
8	19.3	170	2.3
9	21.4	156	2.1
10	22.7	145	1.3
11	23.1	156	0.4
12	23.6	151	0.5
13	23.9	149	0.3
14	24.0	136	0.1
15	24.0	111	0
16	24.4	64	0.4
17	24.6	31	0.2
18	24.0	11	−0.6

Comparison with the Breckland figures of Tye (1982) apparently shows that, up to five days of age, Skokholm nestlings were heavier than the Breckland birds (10.8 g, compared with 8.5 g). At ten days old, however, Skokholm birds averaged 22.7 g, whereas the Breckland birds averaged 23.5 g, and at fledging they were 24.4 compared with 25.9 g.

O'Connor pointed out that the growth curves of some altricial young birds do not reach their final magnitude until after the young have left the nest, and that for a few days after leaving the nest the young continue to put on weight. Figure 13.2 shows that young Wheatears reach this weight while they are still in the nest burrow, but the young begin to leave the nest and wander up and down the burrow when eleven or twelve days old (Chapter 14). So, in spite of the appearance of the graph, the nestlings had reached a point at, say, eleven days old when their behaviour changed markedly; they left the nest but, instead of hiding in undergrowth as do other members of the Turdidae, they remained hidden in burrows until their escape reaction changed from cowering and hiding in burrows to flight.

Rain, wind and other factors affected the growth rate of the nestlings, particularly in the first three days after hatching and also after they were ten days old. I was, however, impressed by the fact that the amount of variation was slight during the period of rapid growth between four and nine days, even though the apparent feeding rate might fluctuate quite markedly in the same period.

The other problem for which at present I have no explanation is that the curve showing the average visiting or feeding rates (Figure 12.3) for all broods appears to have a different shape from that showing the increases in nestling weight (Figure 13.2). I should have expected the curve showing the rate at which parents brought food to the nest to have run parallel to the curve showing the increase in weight; while the greatest increase in weight was from day 4 to day 10, however, the fastest feeding rates were in the first three and last five days of the nestling period. Perhaps the large food items and the greater number of items the parents brought at each visit compensated for a lower rate of bringing food to the nest. The rise in the feeding rate from about eleven days old is comprehensible, because this is the period when the nestlings are beginning to leave the nest and hopping up and down the burrow.

Tye's figures giving the frequency of feeding visits by adults to individual nestlings show a very similar curve, rising sharply at first, flattening out until day 11, then rising again. The figures are not strictly comparable but show similarities.

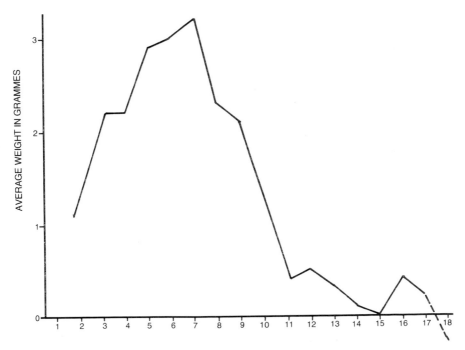

FIGURE 13.3 *Mean daily weight gains of all Wheatear nestlings, 1951-52.*

Newly-hatched Nestling Weights

When weighing hatchlings for the first time, the nestling down of some individuals was still damp and presumably these birds had only recently hatched. The mean weight of 31 hatchlings which had damp nestling down when I weighed them was 2.4 g, ranging between 1.9 and 3.2 g (S.D. ± 0.3 g; S.E. ± 0.053 g), which is about 0.5 g smaller than the mean weight of day-old chicks. Some nestlings whose down had already dried weighed 1.7 g, so presumably their hatching weight was even lower. The nestling weighing 1.7 g and which had hatched within 24 hours of its siblings died three days later having gained no weight, while its siblings had doubled theirs. Tye tells me that the newly-hatched chicks in Breckland averaged 2.5 g.

I compared mean weights of clutches of eggs with the weights of young birds when a day old. Because I was apprehensive of causing the female to desert when the eggs were about to hatch, however, I weighed only one clutch the day before the eggs were due to hatch, which showed that the average weight loss during the incubation period for each egg was about 0.3 g.

FLUCTUATIONS IN WEIGHT

Generally, individual nestling weights increased steeply and in parallel until they were about nine to ten days old, when they tended to fluctuate quite appreciably: a nestling which was the heaviest one day could be the lightest at the time of weighing the next. When averaged out, however, the growth rate of the whole brood tended to show a gentle increase except for the last day or two, when weights generally fell (Figure 13.2).

Possibly one of the reasons why weight was no longer increasing so rapidly towards the end of the nestling period, even though the feeding rate continued to increase, was that when about ten days old nestlings became more active and might have been using much more energy. The variations in weight, sometimes gaining and sometimes losing, were in some pairs quite clearly a reflection of the effect of rainfall on visiting rates, a matter to which I return in a later section.

Another cause of fluctuations was the behaviour of nestlings when the adults brought food to the nest (Chapter 14). One nestling would accept feed after feed until satiated, or until it had to stand on the nest rim in order to defecate and in doing so lost the prime position in the nest. If I had weighed it before and after it had defecated, I suspect that its weight would have been very different.

SOME FACTORS WHICH AFFECTED FEEDING RATES AND WEIGHTS OR CAUSED THE DEATHS OF NESTLINGS

Having looked in general terms at the rate at which the nestlings were fed by adults and also at their growth rates, I shall now describe some of the factors which reduced feeding rates and consequently affected the weight of the nestlings. Although Wheatear nests have been preyed on or destroyed by gulls, shearwaters or rabbits, none of the nests that I was using for this part of the study was destroyed and most disruptions to the smooth development of the young originated with the behaviour of the parents or the weather. Since most of the information used in this section came from automatic recorders I am unable to say which adult was chiefly involved, but I have indicated in Table 12.3 that, for those pairs that I actually watched, females fed the nestlings more often than males.

Parents caused problems to their young, or some of them, in two or three ways. The laying of the eggs could be spread over three days, and I discuss the implication of late hatching in the next section. All I shall say here is that all nestlings which hatched three days after the hatching of the first died within two or three days.

Some adults did not bring food to the nest so often as others: their visiting rates were erratic. W116[1] provides one example. The reason they were slower than average is unclear, since the weather — wind and rain — was about average: in fact, one of the biggest drops in nestling weight that I recorded when weight should have been increasing, occurred in brood W116[1] on 5 June, when the weather was perfect. Only two of the fledglings reached average weight, while the remainder fledged below average. The eggs of W116[2] hatched over three days, and the last chick died two days later at its hatching weight.

For 235 minutes on 22 May, neither of W149[1] parents visited their nestlings and instead of gaining the expected average of 2.9 g the chicks lost an average of 0.4 g (range 0.2-0.7 g). Once again, the reason for this gap is unknown.

I should point out that there are two kinds of runts: those which are hatched several days late and, second, the smallest in a brood all of which were hatched within 24 hours. Those in the first category all died, while those in the second category had a chance of fledging the heaviest in the brood.

Earlier I mentioned that from the ninth or tenth day weights tended to fluctuate fairly markedly; it was a period of weight adjustment in some broods — a nestling which had a low weight at ten days could well be the heaviest on fledging. If one nestling had been more successful than its siblings in taking a bigger share of food, its success usually depressed the weight of the others. If it left the nest before the others, their weight usually increased until they attained a similar fledging weight.

From 1947 to 1953, I had no evidence of any deaths attributable to starvation because of a shortage of food in the environment. When I was revisiting the island, however, the severe spring drought of 1984 reduced Wheatear brood size (E. Gynn, pers. comm.). Tye (1980) found that starvation was a relatively unimportant mortality factor in 1976 and 1977 in Breckland.

While wind and rain might have immediate effects on the rate of feeding, only once in my study was rain a possible indirect cause of a nestling's death. This individual was hatched two days late during a period of three wet days and had no chance of recovering its weight when its siblings were also underweight. In most cases, nestlings were able to regain within a day or two weight lost because of rain and to continue to grow thereafter.

I knew of no instances on Skokholm where heavy rainfall drowned nestlings in burrows. In early spring the situation is different: the soil is already water-logged after the winter, and nests in burrows in areas of shallow soil over solid rock were occasionally flooded and the parents then deserted the eggs.

Rain occasionally slowed the growth of nestlings in the period between three and nine days old, but it could cause considerable variation from about the ninth day until they left the burrow. A general picture of weight increases and the effect of rain is provided by the six nestlings of brood W97[1] (Figure 13.4). One of the nestlings J4705, was hatched about 24 hours after the others and, although lighter than its siblings until 13 days old, it actually left the burrow the heaviest of the brood.

The lines showing individual growth between three and nine days old are not as straight as they usually are and show the effect of rain on 20 May 1951, when the nestlings were six days old: rain began to fall at 10.05 GMT, and I weighed them at 16.05 after rain had been falling for six hours. On 22 May rain began at 04.30 GMT and continued until 13.00, and I weighed the nestlings that day at 15.15 GMT. On the following day rain fell all morning, with showers in the afternoon; when I weighed the chicks at 15.45 GMT, it was obvious (Figure 13.4) that the growth rate had slowed markedly. Rain in varying quantities fell from 24 to 27 May, but at night or in the early morning or evening and rarely during that part of the day when the adults were hunting; and throughout this rainy period weights increased, although with some fluctuations. Except for the last hatched, all chicks were above average weight, as they had been since hatching. On 28 and 29 May, however, rain fell almost continuously, and the nestlings were weighed at 15.00 and 21.00 GMT respectively: all showed a fall of varying amounts, although they all remained heavier than average. Persistent rain presumably made insects and their larvae more difficult to find. The parents of brood W97[1] were experienced, and this might have been important; they had bred on the island before and were better known to me as George and Margaret (see page 303).

In 1952, rain days of 27, 28 and 29 May also slowed up the development of W116[1] and W145[1], where some nestlings lost weight. In brood W147[1], LH013 lost 2 g between 28 and 29 May from which, in the continuing rainy weather, it never recovered and it left the burrow below average weight, having earlier been about 2 g above it. In nest W149[1], J5058, which was hatched 24 hours after the others and was the runt, lost almost the same amount on the same day, where its three siblings lost nothing; it remained in the burrow after the others had left and its weight increased steeply until, on leaving, it weighed as much as the others had when they left.

I think that one can safety conclude that when rain is falling prey items are less easy to find, and that if it persists nestlings put on weight only slowly and in some cases lose weight — as much as 2.9 g in 24 hours. If the rain clears, this weight is sometimes regained.

226

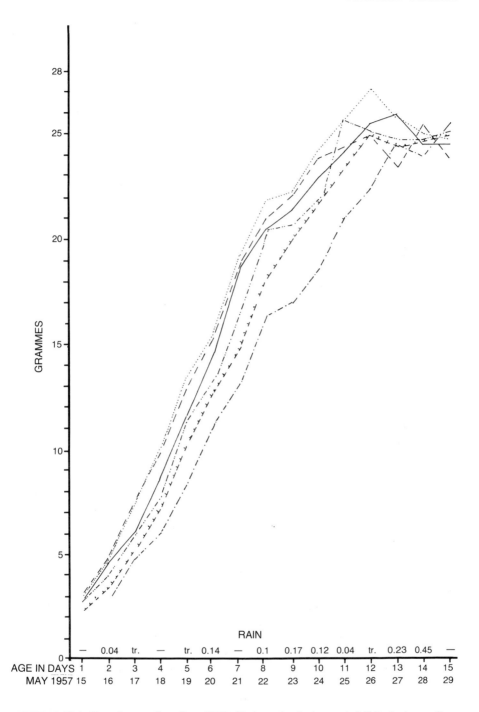

FIGURE 13.4 *Growth rates of nestlings W97. Horizontal axis shows rainfall (inches) as well as age in days, and dates; vertical axis shows weight (grams).*

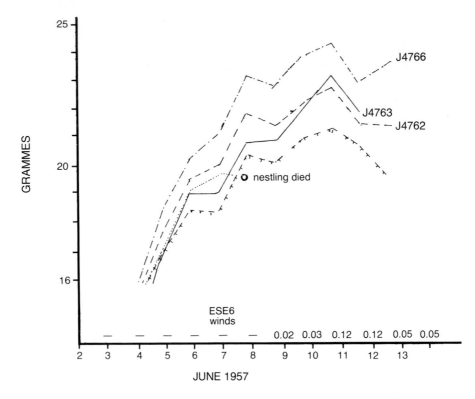

FIGURE 13.5 *Growth of nestlings W100; all nestlings seven days old on 5 June. Horizontal axis shows one entry for wind, rainfall in inches.*

The principle that the weakest nestlings get less food when it is in short supply (as when rain is falling) came out very clearly at times but, given fair weather later, they caught up again within a day or two. The one possible fatality was a nestling hatched three days after the first of the brood at the beginning of a period of three rainy days. When first weighed it was 1.7 g, and it died at the same weight two days later. Of course, the opposite principle also applied: the heaviest nestlings of a few broods when food was in short supply, managed to gain weight when the remainder of the brood lost theirs. Presumably, being heavier and stronger, they got most of the food.

To a lesser extent than rain, wind influenced feeding rates and consequently the weight of nestlings. However, I was not always sure whether the wind disrupted the adult's activities or those of the prey. The whole island was not equally exposed; this meant that some adults might not be affected at all by the wind, or that they could hunt in more sheltered parts of their territory.

On Skokholm, most nests were sheltered from the westerly quadrant but often remained exposed to the east (Chapter 9). A force 6 easterly wind (34-50 kph/25-31 mph) on 7 June was followed by a weight loss in four out of the six of W116[1]'s nestlings and all three of 127[1]'s nestlings, and, since easterly winds of 21-29 kph (13-18 mph) persisted on 8 June, weights did not recover for two days. Brood W119[1] lost weight when the wind was southeast 21-29 kph, but that nest was in a gulley close to the cliff edge up which south-easterly winds were funnelled rather strongly.

The weights of nestlings W100[1] demonstrated clearly (Figure 13.5) the effect of an east-southeast force 6 wind (34-50 kph). Without exception the growth rates of all the nestlings were depressed and one nestling died, although the cause of death is not known. A few days later, most of the nestlings lost weight on a rainy day.

Some weight increases could be disastrously spectacular. The six nestlings of nest W137[1] were all hatched on 12 May 1952. Their weights ranged between 2.4 and 4.8 g on the first day of weighing. They all grew without complications until they were five days old, when the heaviest put on less weight than the others. On the next day it regained the weight, but it died when seven days old (Figure 13.6). On the same day, another nestling from

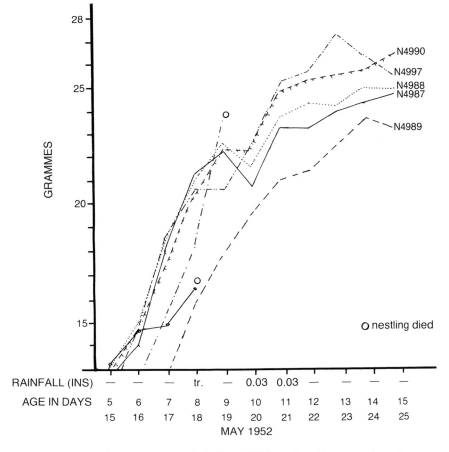

FIGURE 13.6 *Growth of nestlings W137, 1952; on 17 May, all nestlings seven days old.*

the middle of the brood's weight range suddenly gained 5.6 g in 24 hours (rising from 18.3 to 23.9 g) and was dead the next day. This last weight increase is almost double the mean maximum weight increase for that age (Table 13.1). There are no obvious reasons for this individual to have taken all this food. One sibling lost weight that day and the remainder had lost weight by the following day. It seemed as if this one individual had, somehow or other, accepted most of the food brought to the brood and died of overeating!

SURVIVAL OF LATE-HATCHED NESTLINGS

Of the 32 nestlings in 1951 and 1952 which hatched 24 or 48 hours after the remainder of a brood, three died within two or three days of hatching and one after eight days, having failed to gain weight. Of these four hatchlings, three had hatched 48 hours late and the other only 24 hours late.

One nestling which hatched 48 hours late successfully left the nest at 13 days with the remainder of the brood but with a low weight of 18.8 g, which is less than the overall average at 7.5 days. The mean weight on leaving the burrow at 16 days was 24.4 g.

Of those which hatched 24 hours late and survived, 19 eventually fledged at an average weight and eight fledged with rather low weights. It would seem, therefore, that nestlings hatching 24 hours late started with a disadvantage, but that two-thirds overcame this and fledged at approximately average weight while one-third fledged at weights below average.

It should not be forgotten that leaving the nest successfully represents only the first stage in a bird's life and that the first month is, for most bird species, the time when the heaviest mortality among juveniles occurs. We do not know the qualities that allow a male or a female to fly thousands of miles each autumn and spring and then undertake a rigorous breeding programme. Nor do we know the qualities parents pass on to their offspring which enable one, two or (once) four individuals from the same brood to return to breed on Skokholm for between one and five — possibly seven — years, whereas frequently whole broods are never seen again on Skokholm, although some may have found a suitable nesting habitat elsewhere.

Those That Returned

I tried to discover whether any particular factors, weight among others, made some nestlings more successful in surviving the non-breeding season with all its hazards and in returning to breed a year later on Skokholm. I examined the records of those that returned to the island in 1952 and 1953 and which had hatched and been weighed as nestlings on the island in 1951 and 1952: 13 weighed in 1951 returned in 1952, and six from 1952 survived until 1953 or longer. Of the 19, only five nestlings turned out to be males when seen in adult plumage and 14 were females. Sadly, only ten, all females, came from broods that I had weighed: their average weight on fledging had been 24.2 g (S.D. ±

FIGURE 13.7 *Female hovering over weasel*

1.49; range 22.1-25.9 g), which seemed to be about the general average fledging weight on the island (Table 13.1).

Incidentally, two of the lightest birds at fledging — at 21.1 g and 22.6 g — that returned the following spring were siblings, J4775 and J4776 from nest W121.

I was surprised that the number of returning females was so much greater than that of males and so decided to check all the records of Skokholm-bred

TABLE 13.2: *Sexes of nestling Wheatears which returned as adults to Skokholm*

Ringed	Males	Females	Totals
1946-7	5	14	19
1948	6	9	15
1949	8	12	20
1950	14	6	20
1951	2	11	13
1952	3	3	6
Totals	38	55	93

nestlings which had returned to the island either to breed or just to visit. Table 13.2 shows the results.

The majority of these birds returned to nest on the island in the following year. There were three exceptions. One male, WS986, was ringed in 1947 and was not seen again until June 1949, when it was trapped, and, a week later, was identified by its colour rings, but thereafter it was never seen again. Female 001445, ringed as a nestling in 1948, was recorded twice early in 1949 and then disappeared. Another female, 001448, unrelated to 001445, was also ringed as a nestling in 1948, was not recorded in 1949, but returned in 1950 to become the second mate of X6875; together they were known as George and Margaret. These gaps of a year suggest that some Wheatears return to their birthplace not only for their first summer but, if they are unsuccessful in finding a territory or mate in their first breeding season, come back in their second to try again.

It is odd that in some years the majority of nestlings hatched on the island and which returned to it in the following year were females and that only in one year did males outnumber females. The mean number of females was 9.2 and of males 6.3; the difference is not significant (p = 0.2692). It will be remembered that almost throughout spring migration on Skokholm males outnumbered females. I have no explanation for that discrepancy — perhaps the sample is too small.

*C*hapter 14

DEVELOPMENT AND BEHAVIOUR OF NESTLINGS

FEATHER DEVELOPMENT

Hatchling Wheatears are unattractive creatures: a grey natal down emerges from five feather tracts, but otherwise they are naked. The eyes are closed but the blue-grey of a large eyeball is prominent beneath the closed eyelid. The large bill has a lemon-yellow cere, yellow gape and a conspicuous white egg tooth on the tip of the upper mandible. The ears are apparently fused and do not open until the nestlings are three days old.

The hatchlings lie curled up in an embryonic position, with their head over their belly and wings over legs. They move little, but they can — when freed from the nest — stretch out their legs and apparently try to move with alternate leg action. They stretch up their open beaks on quivering necks even when adults are absent, as if, said Skutch (1976), in a silent plea for food. They also move their anal regions a little in order to defecate.

FIGURE 14.1 *Juvenile*

Natal down emerges from some ten to eleven points around the upper part of the eye (the inner supraorbital tract) (Figure 14.2); on each side of the nape there are four points of natal down in two pairs (the occipitalis); on the scapulars there are ten points (the humeralis); down the spine there are two rows of ten points (the spinalis); and there are two points on each of the femoral tracts. The down is smoky-grey in colour and tends to stand erect once the nestling is dry. There is no down on the underparts. I doubt whether this down provides much insulation against air temperatures, but its grey mistiness hides the nestlings when they are sleeping and lie quiet in the nest. Ingram (1966) suggested that the chief function of natal down was to hide the nestlings, but Skutch found this explanation difficult to accept. However, my observations of nestlings W67 within a burrow, which I describe later in this chapter, support Ingram's view. During the first 24 hours, the nestlings hold their heads down in the embryonic position so that the pink of their bodies is then well hidden by the down, which usually persists on the tips of the growing body feathers — on some individuals at least — until after they have left the nest.

On the second day, feather sheaths begin to show beneath the skin as small black specks on various tracts in the nestlings' upper- and underparts, but they do not pierce the skin until the third day. Figure 14.2b shows the tracts where the feather sheaths are visible on a two-day-old nestling. The first sheaths pierce the skin on the tracts around the eye, on the nape, and on the tibio-tarsus. On the fourth day, sheaths appear in most tracts, indeed in all except other parts of the head. The rictal sheaths around the edge of the gape have also made their first appearance. The feather vanes on the tibio-tarsus are already fanning out; on this tract the sheaths break open close to the skin.

On the fifth day, the sheaths continue to grow: feathers on the central spinal tract, the head and down the sides of the body are the most advanced. On the head, sheaths are appearing on the crown. Along that part of the wing from which the secondaries appear, 2 mm of sheath emerge and nearly 1 mm of feather vane have broken out on the secondary coverts. Between the sixth and eighth days, more sheaths are beginning to break open at their tips and the feathers beginning to fan out.

It is not until the ninth day that the primary and secondary feather sheaths begin to break open and the feather vanes to fan out. The rictal bristles protrude 2 mm from their sheaths. Over most of the dorsal portion of the nestlings there is, by now, a moderate covering of feathers, but as the sheaths have broken to their bases they probably do not retain much warmth. On the back there is a small bare area between the lower spinal and femoral tracts

FIGURE 14.2 *Development of feathering on nestlings: (a) showing feather tracts from which nestling down emerges on day-old Wheatear; (b) showing areas where quills are visible beneath the skin (|) on two-day old nestling; (c) seven days old, showing areas where quills have pierced the skin and where feathers are beginning to fan out (⊥); and (d) 12-day old nestling showing where quills are beginning to fan out (⊥) and where feathers are fanned out to base (∧).*

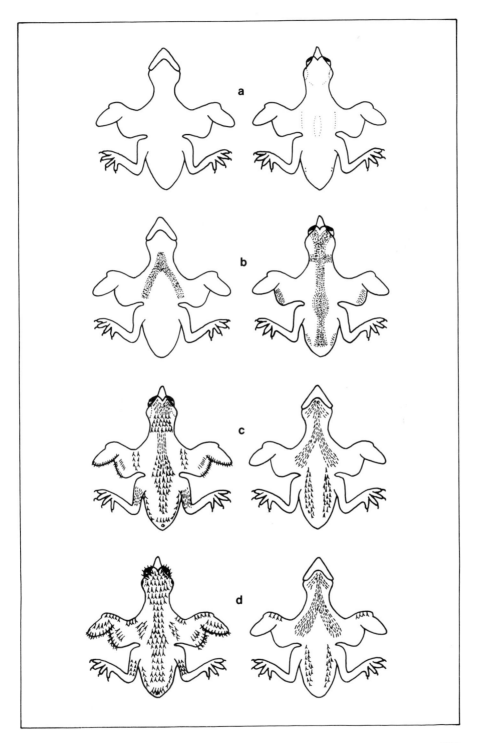

which is not covered by the fanning feathers. On the ventral surface another bare area extends from the lower breast to the cloaca, and another on the side of the breast which is not covered by the folded wing. Of course, these ventral areas are not as a rule exposed to air temperatures since for most of the time the nestlings are still huddled in their nests unless they are being fed by their parents.

Feather development is now fairly straightforward: sheaths continue to grow and the tips are breaking open rapidly, probably assisted by the nestling, which is beginning to preen itself. Most of the contour feathers reach their greatest length between the twelfth and fourteenth days, and the sheaths on all tracts are breaking down rapidly.

The flight feathers continue to grow for some time after the juveniles have left the nest and while they are still being fed by their parents. The shortness of the primaries means that the white rump is continuously exposed, which may help the parents to find them when bringing food. On the other hand, it also makes them more obvious to predators, but at this age juveniles on Skokholm remained close to burrows into which they rushed as soon as alarmed (see Chapter 15).

FIGURE 14.3 *Jumping*

SOFT PARTS

The egg tooth by which the hatchling has cut its way out of its shell is very prominent on the first day, but becomes less obvious as the days pass until it is visible only as a faint white spot on the sixteenth and seventeenth days and possibly for some time afterwards.

Very cautiously — because of their fragility — I measured the rate of growth of the bill, wings and tarsus of some nestlings, and because of my caution the results were not too exact. Owing to the difficulty of finding the base of their skulls, and to the dangers of so doing, I decided to measure the bill from the nostrils to the tip. The modal adult bill length from nostrils to tip was 9 mm, and it seemed that the nestling's bill grew about 0.5 mm a day until it measured about 8 mm on the ninth or tenth day; after that, the growth to full length was slower. The bill hardened slowly, but the pale cream line of the gape and the pale spots at the corners of the bill persisted into late summer, even when the Wheatear had acquired its first-winter plumage.

The tarsus also seems to reach its maximum length of 28 mm on the ninth or tenth day: it was still fat and fleshy, but grew thinner, harder and darker after the tenth day.

I have said earlier that the wing length continued to increase after the young bird left the nest. Adult wings measure between 90 mm and 100 mm, whereas the wing measurements of the two nestlings about to leave nest W78[2] were 66 mm and 64 mm respectively, so that they had approximately 30 mm to grow. At that time the total length of the fourth primary on the 66 mm bird was 43.5 mm, of which the fanned vane occupied 28 mm.

Once the young left the nest at 15 or 16 days, I had only a few chances of catching them in order to measure them before they were nearly adult.

NESTLING BEHAVIOUR

Over nest W67 (whose six eggs hatched on 27 May), which was in a longer burrow than usual, I had set a coffin-shaped hide made of a dark material. As I described in the Preface, it had no observation slits so as to reduce the light to that in which it was just possible to write notes. I then inserted a metal-polish tin with its bottom removed through the soil above the nest, and placed a microscope slide over the opening to prevent the draught blowing straight into my eye. Although the dim light inside the hide allowed me to see the notes I was writing, it was too dark to allow the female or the nestlings to see me. In this way I was able to watch the behaviour of the adults and nestlings from a range of about 25 cm. When the nestlings became very active and moved around the burrow, more soil was removed and a bigger box with a glass top replaced the metal-polish tin so that I had a wider view. Generally speaking, this hide was highly successful for observing what went on in the nest without causing disturbance to the adults. I began observation on 28 May, when the nestlings were a day old.

During the first four days that I used this hide, the female was still brooding the young. When the nestlings were able to wander around in the burrow, they used the observation funnel as another bolt-hole when they were alarmed by unusual noises outside. Eventually they detected some movement and noises above them, which, occasionally, they used to watch for and listen to; this did not, however, prevent them from climbing up into the observation funnel when alarmed.

Parental Attention

Table 12.1 showed that female W67 brooded her nestlings until they were four days old, decreasing her brooding time from 52% of the period of observation on the second day to 26% on the fourth. While she was brooding, she spent most of her time facing the entrance to the nest burrow, and often closed her eyes for 10-20 seconds. Occasionally she stood on the rim, looked down at the nestlings and poked down among them, mandibulated their natal down, or ate any faeces that might have been dropped. After eating faeces she sometimes left, but otherwise she settled back, placing her feet on either side of the cup and gently lowering herself on to her chicks, and finally shuffling herself from side to side. She ate faeces only for the first three days, after which she or the male removed them from the nest burrow.

Table 14.1 shows the activity of the female during the 60-minute observation periods. The number of quiet, 'strangled', food calls she used tended to increase in relation to the number of times that she fed the nestlings; as they

TABLE 14 1: *Female W67's activity on nest during 60-minute periods*

Age of chicks (days)	Calls	Preens self	Looks down	Pokes down	Eating faeces	Carrying faeces	Nibbles	Preens chicks	♀ on nest (mins)
Two	2	1	12	8	11	—	—	—	31
Three	18	—	10	8	2	1	1	8	22
Four	23	—	10	5	—	6	—	—	16
Five	17	—	12	6	—	7	—	—	—
Six	13	—	13	13	—	4	—	—	—
Seven	17	—	11	2	—	4	—	—	—
Eight	1	—	6	—	—	6	—	—	—
Nine	—	—	—	—	—	10	—	—	—
Ten	—	—	4	1	—	6	—	—	—
Eleven	—	—	—	—	—	2	—	—	—
Twelve	1	—	—	—	—	12	—	—	—

developed and as they slept less, so the number of these food calls diminished. When the nestlings began to cower, showing the onset of fear, on the twelfth day, the female used the call again, however.

The female preened the nestlings, sometimes very vigorously, until they were five days old, and from the seventh day they began preening themselves.

Her 'poking down' into the nest not only stimulated the nestlings to defecate, but sometimes she re-arranged them. One which had defecated on the rim of the nest was still lying awkwardly on top of the others when she returned. After she had fed them she began poking among them and, when she finished, the nestling which had been lying on top was back in the nest.

NESTLING DEVELOPMENT

Patterns of Behavioural Development of Wheatears

Nice (1943), in her study of the Song Sparrow, grouped the major features of the development of the juveniles into five stages, which O'Connor (1985) has modified. He calls the first 'Post or late embryonic', in which the behaviour is mainly involved with nutrition. In his second stage, 'Preliminary', comfort movements begin to be seen; in the third, 'Transitional', comfort movements mature or develop further; in the fourth stage, 'Locomotory', the juveniles leave the nest and fledge; and in the fifth stage, 'Socialisation', we see the adult social interactions develop.

This pattern fits the development stages of the Wheatear very well, although, because the Wheatear raises its young in a burrow over a period of 15 or more days compared with the ten days in the open nest of the nestling Song Sparrow, there are some interesting differences with regard to both the age at which various developments occur and the order in which they appear. The shorter time in the nest of the Song Sparrow means that some of its development is, or has to be, telescoped, while growing up in a burrow puts constraints on the development of the nestling Wheatear. Table 14.2, which is based chiefly on nestlings W67 but also on observations on the young of nest W128R, shows the age at which different behaviour patterns appear.

POST-EMBRYONIC STAGE

This stage covers the period from one to four days old and is concerned chiefly with receiving food, defecating, and a developing facility to communicate with parents.

Two Days Old

When watching the nest from the coffin hide, I first knew that the female was approaching the nest when the light flickered as she hopped down the burrow. On reaching the nest, she looked down for a second or two and then called the 'half-strangled' food call, which can be imitated by placing one's teeth against the upper lip and drawing in air in short bursts. At this note, one or more of

TABLE 14.2: *Appearance of activities in the young Wheatear, based chiefly on nest W67*

Stage:	I				II			III				IV			
Age (days):	1	2	3	4	5	6	7	8	9	10	11	12	13	14	15
Gaping	—	—	—	—	—	—	—	—	—	—	—	—	—	—	—
Defecating	—	—	—	—	—	—	—	—	—	—	—	—	—	—	—
Feathers appear			—	—	—	—	—	—	—	—	—	—	—	—	—
Food calls				—	—	—	—	—	—	—	—	—	—	—	—
Yawning				—	—	—	—	—	—	—	—	—	—	—	—
Ears open				—	—	—	—	—	—	—	—	—	—	—	—
Eyes open		+	—	—	—	—	—	—	—	—	—	—	—	—	—
Reacting to shadows				—	—	—	—	—	—	—	—	—	—	—	—
Stretching wings					—	—	—	—	—	—	—	—	—		
Stretching legs						—	—	—	—	—	—	—	—		
Preening						—	—	—	—	—	—	—	—		
All facing entrance							—	—	—	—	—	—	—		
Standing							—	—	—	—	—	—	—		
Climbing on rim							—	—	—	—	—	—	—		
Claws grasping							—	—	—	—	—	—	—		
Exploratory pecking							—	—	—	—	—	—	—		
Shuffling wings							—	—	—	—	—	—	—		
Localising sound								—	—	—	—	—	—		
Location call									—	—	—	—	—		
Out of nest												—	—	—	—
Hopping on tarsus												—	—	—	—
Cowering												—	—	—	—
'tchi' rudimentary alarm call													—	—	—
Full soliciting posture														—	—
Fluttering wings														—	—
Bobbing in anxiety														—	—
Sound experiments															—

the nestlings stretched up their necks and gaped and they were then fed, after which they subsided again. Their reaction to the food call indicated that they could hear from hatching, and, like many other passerines, they might have been able to hear during their last day or so in the egg. So, although the ears appear to be closed for the first day or two of their life, perhaps the protective membrane allows the nestlings to hear.

Generally, the first nestling to gape and not necessarily the one nearest to the female was given the food. When three or four nestlings opened their beaks at the same time and the female had fed two, she looked down at the others before choosing which to feed next. Sometimes, while she was feeding one nestling, the beaks of the others closed and their necks drooped. If she had more than one food item, she touched another chick on the beak and if it did not respond she called two or three times until other nestlings gaped and were fed.

Once she had fed them she poked among them, and she not only took faeces directly from the cloaca as they were being evacuated and swallowed them, but also took any that were lying in the bottom of the nest. The nestlings apparently made an attempt to stretch themselves up to defecate, but were not yet developed enough to reach the rim of the nest. Then the female settled to brood but every few minutes got up and poked down. She usually left the nest after this poking down, and occasionally after the male had sung loudly from just in front of the nest.

While the female was out of the burrow the nestlings kept their heads and beaks down more or less in the embryonic position, but occasionally they stretched up their necks with closed beaks pointing upwards. Movement among the nestlings, although mostly of an unco-ordinated nature, was almost continuous, as if one was uncomfortable and moved, thus upsetting the others.

The colour of the gape and as much of the throat as I could see was yellow, probably shading into red on the lower throat. The yellow gape and the unfeathered neck stood out well against the grey down of the nestlings in the rather dim light, and this patch of colour probably made it easier for the female to see the nestlings.

Three Days Old

There was no marked change in the female's behaviour, and the male did not visit the nest while I was watching. The nestlings, however, were now stretching up their heads more vigorously: at times four or five heads stretched together. As before, the female usually fed the chick which gaped first in answer to her food call, but occasionally she put the food two or three times into a beak, withdrawing it each time because the chick failed to grasp it quickly enough. Once she shifted the food to another gape.

Two further pieces of nestling behaviour became apparent. What looked like yawning or jaw-stretching was quite common: the beak was opened and shut quite swiftly, with the head and neck in the normal position, and sometimes it was accompanied by a nibbling movement of the beak. The action looked like yawning, but Ferns (1985) thinks that there is no exhalation or inhalation of air and the movement just stretches the jaw muscles.

I also heard the nestlings calling for the first time. Just after they had been fed I heard, just once, a very thin call sounding like 'ee-ee-ee-ee'.

Defecating, even though more co-ordinated than on day 2, was still a

241

laborious process, taking about three seconds; the struggles of the nestling as it tried to extract itself from its siblings indicated to the female that an individual was about to relieve itself. Then, when the anal portion of its body was lifted well up, the cloaca was enlarged by extending its inner surfaces outwards rather like the mouthparts of a sea anemone and eventually the faeces were evacuated and taken by the female. The chicks, however, still could not get their cloaca to the rim of the nest.

The beak of the nestling was an odd shape. The nostrils seemed red and swollen; the cere around the gape was swollen and yellow; and the open mouthparts were very conspicuous against the grey down which still covered the upperparts of the body.

Four Days Old

More co-ordination was obvious: one chick was able to get its cloaca on to the rim of the nest in one swift movement, whereas another was still struggling. Heads were more obvious most of the time as the nestlings abandoned the embryonic position. When the female called, they all stretched up for about two seconds before they began to subside again. For about a minute after being fed, a chick might make a swallowing movement with its bill that lasted for two seconds, which was different from the rather jerky movement by which the nestlings accepted food. After they had been fed the chicks all snuggled down; the female watched them, and if the movement continued longer than usual it was often a sign that one of them was about to defecate.

THE PRELIMINARY STAGE

It was in the period from five to seven days old that the nestlings began to put on weight most rapidly. Their eyes were open and they could now see and understand the meaning of the flickering shadow in the burrow, as well as see and hear their parents' food call. I also saw the first of the comfort movements: stretching wings and legs, as well as preening.

Five Days Old

The nestlings were now calling for as long as ten seconds, often without the female calling first, and it became apparent that they could see well enough to detect the flickering shadow in the burrow that heralded the female's approach and that they understood its meaning; even so, when not being fed, their heads still pointed towards the centre of the nest. On one occasion, the female was standing on the nest rim and all the chicks had their heads down, but, when she moved, the movement of her shadow apparently stimulated one chick to gape with neck stretched upwards as it was fed. She still had to call periodically to stimulate further gaping if she had brought more than one food item.

Settling down was not always simple, and shuffling went on for several seconds; one nestling had to struggle to get more into the centre of the nest while another was lying on its side trying to right itself. The chicks were growing so large that the female had to straddle the nest, one foot on either rim, to poke among them. She may have been searching for faeces but she really seemed to be sorting out the young, pushing them about, perhaps preening them a little: I did see her pulling at some down.

The male brought food at least once and was very noisy about it, singing full song and subsong outside the nest and then a rather harsh version of the 'weet' note.

Six Days Old

The most noticeable change was that the chicks almost invariably started gaping when the light flickered in the burrow, and the adults used the food call less often. Another possible stimulus to gaping was the sound of the female's rings jangling together as she hopped down the burrow. I could hear them well, but, since the male was not ringed and the chicks gaped at his shadow, it was clear that the shadow was the more important. There were still occasions, however, when the chicks did not gape at once, perhaps when they were asleep, and then the adult had to call.

The male brought food occasionally but was less practised than the female. Once, he brought a large caterpillar which he dumped in a chick's beak; the chick could not swallow it at one gulp, and when the male left the caterpillar was still dangling out of the chick's beak. At this age the chicks usually swallowed quickly when food was put into their gapes.

When they were stretched up, their necks still quivered as if with strain, and when the chicks finished gaping the necks subsided limply, rather like a deflating balloon. On this day the wings of one or two of the chicks were also quivering as they moved, but they were not being beaten.

Seven Days Old

The nestlings were now so large that they filled the nest, and I found it difficult to discern which parent was visiting; the male sometimes announced his arrival outside the nest by a burst of quiet song, and sometimes I could hear the jingling of the female's rings. The nestlings still slept a lot and sometimes the flickering shadow did not alert them, so the adult used the food call. Provided that the chicks could extricate themselves from their siblings they deposited their faeces on the rim of the nest, and the adults removed them at the next visit. Not being very mobile, the chick then had a problem in getting back into the nest and often lay on top of its siblings for several minutes.

Several new activities were seen: the wing-quivering or twitching that I recorded at six days was seen more regularly; one nestling stretched its leg, lifting up its body as it did so; and at least one other began to use its beak, pecking at the side of the nest and also preening first its wings and later its back for a second or two. The female also had a quick nibble at the down. Preening apparently started earlier than this in hand-reared American Redstarts (Ficken 1962). Oddly enough I did not record head-scratching,

243

although the American Redstart made its first attempts when seven days old. I wonder if this and bill-wiping tend to occur earlier in a hand-reared bird. In a tightly packed Wheatear's nest, such an activity would be difficult to accomplish. O'Connor (1985) points out that it is difficult for nestlings to make sideways movements until they get out of the nest (at twelve days for Wheatears), and until then the young are effectively too constrained for space to practise many of these activities.

THE TRANSITIONAL STAGE

The third stage, when the nestlings were between eight and ten days old, saw a lot of action, which was concerned chiefly with the maturation of activities that had occurred before. Now the chicks are all sitting facing the entrance, making much more active use of their legs, beginning to peck, and, very important, beginning to localise sound.

Eight Days Old

The tendency for the young to align themselves in the nest so that they faced the burrow entrance was more marked, and even those whose bodies were facing in another direction turned their heads to the entrance and gaped when an adult arrived. They climbed on the rim to defecate, sometimes with difficulty, and then they had problems in getting back into the cup; for a time they lay on top of the others and did not gape very vigorously. They seemed temporarily sated.

The quivering of wings and other parts of the body was commoner, and several new activities appeared; at one point one chick shuffled its wings, an action which an adult used to shuffle its own wings into its breast feathers. Heads were moved in a number of different ways: chicks yawned or stretched their gapes; they made nibbling movements with their bills; and they shook or flicked their heads. One preened its underwing, and another stretched out a wing to prevent overbalancing.

Nine Days Old

The heads of the chicks, in two rows, faced the burrow entrance, and I saw more clearly how climbing on the rim of the nest to defecate helped to ensure a more equitable distribution of food. Those in the front of the nest received a greater proportion of the food and when satiated they climbed on to the rim of the nest to defecate; and those in the second row pushed forward into the vacated space, where they began to get a bigger proportion. The chick which had been on the rim of the nest struggled into the space at the back of the nest, where it dozed and occasionally gaped but less energetically than the others. In a few minutes another of the front row got up to defecate, and had then to go to the back of the nest. Something similar was recorded for a number of hole-nesting species ranging from Southern House Wrens to woodpeckers (Skutch 1976).

Otherwise, the most important change was the nestlings' power to localise sound. If I imitated the parents' food call, the chicks switched their heads towards me and some shuffled around in the nest a little; another raised its head to look upwards when I tapped on the glass of the observation box. They turned their heads apparently trying to locate the sound.

The nestlings' reaction to the flickering shadows was instantaneous: as soon as the burrow darkened, all those interested in food simultaneously stretched out their necks to the uttermost, almost as if they had been drilled to it.

Ten Days Old

Few new developments were recorded, just a general maturing of activities that I had seen before. Preening was quite common; chicks which had climbed on to the rim of the nest to defecate still had problems getting back into the nest, as they were still so weak on their legs. A new call was heard which sounded like 'tchi': it had a vibrant twang to it, which is difficult to describe and which is characteristic of one or two of the adult calls.

THE LOCOMOTORY STAGE

This stage covers the period between eleven days and the departure from the burrow (having really left the nest itself when they were twelve days old). Whereas in the previous stage the nestlings had learnt the ability to localise sound, in this stage they started to use the location note, which ultimately developed into a call used by adults in territorial circumstances. In Nice's (1943) arrangement of the various stages of development the Song Sparrow leaves the nest at ten days, whereas the Wheatear actually started leaving its nest when twelve days old (although this is not obvious to an observer outside the burrow). Fear also came to them and they cowered whenever the parents called 'weet' or 'tuc' notes, but it was not until the following day that they began to utter the rudimentary version of the Wheatear's anxiety call.

Eleven Days Old

There were no marked changes in the nestlings' behaviour, but their alertness and their ability to move had further matured: they were less like automatons reacting only to their parents.

Twelve Days Old

Fear came to the nestlings for the first time: when an adult called 'weet-tuc-tuc' alarm notes, the nestlings crawled back and away from the nest and crouched as close as they could against the end of the burrow — the escape chamber. Even there, however, they continued searching for other positions; one tried to climb up the observation funnel. Presumably, they were trying to get away from the light. When an adult arrived to feed them, the nestlings watched but did not solicit, until the adult used the food call. From then on, the nestlings spent some time crouching at the end of the escape chamber.

This was also the first day that the chicks showed that they could scramble out of the nest (Nice, 1937, commented that the age at which nestling Song Sparrows start to crouch in reaction to their parents' alarm calls is broadly the age at which they can first hope to survive outside the nest).

It was now easy to distinguish two calls: first, the call by which the young solicited food — 'see-see-see-see'; and, second, the vibrant, twangy, staccato note — a location note — which sounded like 'tchi' or 'tzee', and which is virtually impossible to describe phonetically (Tye tells me that he wrote it down as 'breezh'). This latter call, or one very similar to it, was used by adults in territorial affairs, and I shall be suggesting later that this call used by the nestlings signified the very beginning of the territorial tendency that became more obvious once the chicks left the nest and began to isolate themselves from their siblings.

At the same time as the young showed fear for the first time, the male also appeared rather more excitable. It seemed as if there might have been some link between the male's increased excitability and the arrival of fear in the young.

This was also the first time that I saw all the chicks out of the nest at the same time; some perched on the rim, and another hopped across the empty cup from rim to rim.

Thirteen Days Old

No major changes in behaviour were seen: the chicks spent much time preening themselves; legs and wings were stretched. A new call was heard, rather hard and sounding rather like an early version of 'tuc', which is the general anxiety note. I described it in my notes as rather like a plunger being drawn out of a rather wet plug-hole.

Fourteen Days Old

All the young were out of the nest and hiding in the escape chamber with their yellow gapes hidden, and the adults had to call to attract their attention. The nestling which was fed bowed and bobbed its body while flicking its wings upwards in a fully developed food-soliciting position of a young Wheatear calling its own food calls. When the adult left the burrow all settled down in the nest, but only temporarily for as soon as an adult returned it faced a battery of open gapes and fluttering wings. Most of the young were venturing up the burrow towards the entrance, where they were fed, although adults occasionally came as far as the nest. In fact, there was a constant movement up and down the burrow.

There was one new activity: one nestling bobbed its body and tail up in the anxiety display, which may have been related to the appearance on day 13 of the elementary version of the 'tuc' call, with which bobbing is related for most of the year.

Many of the activities which had already appeared matured further. The young reacted to the 'tuc' notes from adults by cowering with gapes hidden in the escape chamber. When the male sang they sometimes just listened, but on other occasions they all snuggled down in the nest; this presumably indicated that song caused them some slight anxiety, and they relieved their anxiety by returning to the nest, which, up to a day or two previously, was their place of safety. At other times of mild anxiety they sought their nest. The male's song, however, did not cause as much anxiety as the 'weet' and 'tuc' notes.

In addition to the 'tchuc' ('tuc') call, the food call now sounding like 'tchū', and the 'tzee' vibrant call, another call was heard which I called a sneeze. This I described as starting off like the buzzing of a fly in a web, then continuing like an intake of air through soft and very wet lips; as the young made this noise, its whole body shook.

Fifteen Days Old

It was by now almost impossible to observe all the activities of the young birds through the observation funnel, since for most of the time they were hopping in and out of the nest and up and down the burrow towards the entrance. Some returned to the nest and snuggled down; they might well have been those which were temporarily sated because, when it was clear from the calls at the entrance to the burrow that the adult was feeding chicks, the bird in the nest solicited food by calling but did not move. When one of the adults called alarm notes, all returned and shuffled down with a side-to-side motion that lasted a second or two, almost cowering at times and thus hiding their beaks, which were otherwise a conspicuous feature; and there they remained until an adult came into the burrow and fed the first one that rushed out of the nest to meet it.

Out of the nest but in the burrow, the chicks usually fluttered wings and bobbed and bowed quite deeply while soliciting food. Those in the nest still reacted to a flickering shadow, even when caused by another chick at the entrance to the burrow. Activities observed included preening, stretching half-folded wings, yawning or jaw-stretching.

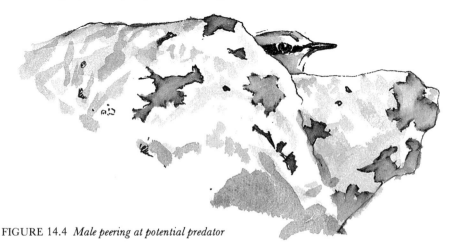

FIGURE 14.4 *Male peering at potential predator*

Perhaps because the chicks were so mobile, the adults seemed less assiduous in removing faeces that had been voided on the rim of the nest or in the burrow near the nest.

In addition to the four recognisable calls described on the fourteenth day, there was a whole range of sounds which seemed to result from the chicks' experiments with making noises. It is impossible to convey them accurately on paper and, although one in particular was similar to one of the notes used by adults in sexual circumstances, it was difficult to link others with calls in an adult's repertoire even though this is very varied.

Unfortunately, at this point I had to leave the island temporarily so was unable to see the chicks leaving the burrow from the inside, although I have watched them doing this from outside a nest. Joan Keighley, the island's assistant warden, told me that the last of the young to leave the burrow was 18 days old when it did so (13 June).

RELATIONSHIP OF WEIGHT INCREASE TO DEVELOPMENT OF BEHAVIOUR

From Table 13.1 or Figure 13.3, which show the mean daily weight increase for all nestling Wheatears in 1951 and 1952, it can be seen that the stages in the development of behaviour outlined above more or less coincide with the different rates at which chicks put on weight during the nestling period. For instance, weight increases during the first four days were slightly slower that in the second stage of development between five and seven days, when the growth curve is very steep; and then, in stage three, the curve slackens again before it fluctuates about an asymptote in stage four.

The locomotory, and fifth, stage in the development of juvenile behaviour, when the young leave the nest burrow, will be discussed in the next chapter.

C*hapter 15*

AFTER NESTING

The young, which had almost lost their nestling down except perhaps for a tuft on the orbital tracts or the tips of the primaries, usually vacated the nest burrow when they were between 15 and 17 days old. The extremes were twelve days old, which was the age at which the nestlings usually scrambled out of the nest into the burrow for the first time, and 21 days old, when they usually gave up using burrows to escape danger when alarmed. Of course, the change that they made was not too extreme because they moved only from their nest burrow to other burrows. Indeed, the move from the nest burrow was really only an extension of hopping about at the entrance of the burrow, which some nestlings began when they were 14 or 15 days old.

The juveniles were growing larger and noisier. Whereas it had been economical and convenient for the adults to have the newly-hatched nestlings in one place for easier feeding and protection, there was now an increasing danger that the nest would be found by a predator, and the quicker the young could disperse into separate burrows the better. Dispersal became a possibility as soon as the juveniles could move and start foraging for themselves and needed less attention from their parents.

Another of the advantages of the move was to escape from the nest, which was giving shelter to an increasing number of fleas, flat flies and feather lice

FIGURE 15.1 *In flight, showing underwing pattern*

which were parasitising the nestlings. I have seen juveniles just after they have left the nest so heavily infested that they were stopped suddenly by an involuntary twitching of skin, wings or tail which necessitated some energetic scratching or preening, pecking or probing into awkward corners of their plumage, as well as shaking all their feathers vigorously — all very different from the more leisurely tempo of the morning preening routine. One had toppled forward on to its head while pecking very energetically at the centre of its belly.

The parents also removed faeces that the juveniles had dropped as they stood about at the entrance of the burrow, so long as the membrane remained tough enough to be picked up. Once that was no longer secreted, however, the dropping remained on the ground and became a signpost to an occupied nest, although the change to soft faeces usually occurred when young were about 16 days old.

When the moment came — suddenly — to leave, the juveniles hopped quickly away from the nest burrow; none had done more than flutter its wings when in the burrow. Sometimes the adults were present, but often they moved off on their own. At nest W19, two of the nestlings had been wandering as much as 3 m from the burrow entrance when, quite suddenly, one hopped a further 6 m and disappeared into another burrow; after a minute or so, it came out and moved yet further away. In this case, I saw no behaviour on the part of the adult which could have been interpreted as trying to entice the juveniles away from the nest. At nest W53, the first nestlings also left without any special behaviour on the part of the adults: the nestlings just hopped energetically — looking very long-legged — away from the nest burrow. While this was going on the male returned and began calling alarm notes, but the nestlings kept moving (they do not 'freeze' in the open) and in a few seconds had disappeared into holes of their own choosing. They did not stop to investigate by listening or looking into the hole to see whether anything else lived there.

Diving down holes was not always diving to safety. On 16 June 1950, I saw male W76 taking food to a juvenile in front of a burrow. When I returned later on to catch and examine the juvenile, I discovered that it was dead, crushed by the other occupant of the burrow — a Manx Shearwater incubating an egg. In 1952, G. V. T. Matthews found two freshly dead Wheatear juveniles in occupied shearwater burrows. At that time Little Owls nested in burrows in walls, and in July in particular Little Owl pellets, which often contained the remains of Storm Petrels, occasionally contained recognisable remains of Wheatears, including their leg rings. The question arises as to whether the owl caught the Wheatears in the open or down the burrow. I should have thought that the foraging times of Wheatears and Little Owls did not overlap very much, so it was less likely that the Little Owls caught them in the open than in a hole in the wall.

At nest W53, one nestling seemed as if about to leave the nest on 31 May 1949. It was standing in the open and looking around, but at that point the female arrived at the burrow entrance with food and the nestling turned around and hopped back to the burrow to be fed. On the following day, both adults were feeding the nestlings at regular intervals when suddenly the male warbled a subsong which was scarcely audible to me. A nestling which had just been fed by the male hopped on to a tussock within 30 cm of the entrance, beat its wings and then hopped away strongly. Three minutes later it was 3 m away and still hopping. A second nestling also left at that time, but two remained in the nest. Finally, the whole brood left and within 24 hours all were settled as a group about 30 m away. In this case, the subsong may have had an effect on the nestlings, although any effect was not immediately apparent. Possibly it was more of an expression of excitement on the part of the male.

If humans or other animals were near the nest when the nestlings were leaving, both parents, but more particularly the female, would go through a whole gamut of displays linked with anxiety: the 'tuc-tuc' notes being the mildest, then 'weet'-ing with increasing intensity. The female would come close to the intruder, bobbing, watching and hovering at about 5 m.

Once or twice I got the impression that the female was enticing the young in a certain direction by flying ahead of them and calling the 'tuc-tuc' notes. The juveniles hopped and fluttered towards her. Once she went into a burrow down which a juvenile had gone and in which it remained for some time. She flew to another juvenile which was leaving the nest burrow and flew ahead of it as she had before. Five minutes later she returned to the burrow entrance and called, but none of the remaining nestlings emerged. When the first excitement had died down a little, the female returned to the first young bird and fed it twice, picking up food items from within 30-60 cm.

Within 24 hours and usually less, a brood of juvenile Wheatears would be dispersed in several burrows between 10 and 70 m from the old burrow, but usually within the original territory. However, I know of two fairly major exceptions. In 1948, W18 had nested in the eastern section of The Neck about 130 m from nest W3, which was still further to the northeast and in a very small territory with sea on three sides and W18 on the fourth side; to the west of W18 was W8. Female W3 had laid her first egg on 26 April, five days before W18 was calculated to have laid her first egg. The nestlings from all three nests were colour-ringed. It was with some surprise that I discovered two W18 juveniles, now 23 days old, and male W18 about 350 m from the nest which they had left a week previously; in moving this distance, the male and the juveniles had had to pass through W8's territory until they found themselves on either side of South Haven, a popular habitat with *Phylloscopus* and *Sylvia* warblers but not particularly frequented by Wheatears.

It would seem that pair W3 with their brood of more advanced young found themselves so contained in their movements by the sea cliffs that they expanded into W18's territory. Family W18 retreated into and through W8 territory into less suitable Wheatear habitat.

Subsequently five of this brood were recorded again quite regularly in their post-dispersal headquarters in the Home Meadow 500 m from their nest, and two were in the same place when nine weeks old. The parents built their second-brood nest in their new territory about 300 m from the first-brood nest.

The second exception on Skokholm was provided by pair W62, whose territory was bounded by Spy Rock on the north side and by the territory of a very aggressive male, 001447, on the other three sides (incidentally, 001447 was a member of the 'super' family whose two sisters and another brother, all from nest W23, survived between two and four years: Figure 5.5). As soon as W62 juveniles were able to move, the female escorted the whole family north, and when the juveniles were 24 days old they were already 240 m from their nest site: to achieve that they had had to cross about 100 m of mature bracken to get to the grass sward on which they foraged for some days.

What interested me about the first incident was that W3 was able to force a neighbour out of its established territory into a less suitable habitat, which it could have occupied when it first arrived. However, it chose to establish itself on the edge of the territories of W3 and W8, both of which were earlier arrivals. The incident lends support to the idea that Wheatears which arrive early and establish a territory, but subsequently lose part of it, still treat the lost part as theirs; this could be why adults W3, when very short of space to feed young, expanded into their instinctive territory, forcing W18, the later arrivals, to move. W8, also an earlier arrival than W18, did not apparently give away any more of its instinctive territory, so W18 had to move still further until they landed up in a habitat on either side of a small valley the banks of which were three-quarters covered with bracken and the remainder with grass and heather, and which was not often used by Wheatears; it did not have a wide vista either. The year 1948 was the first in which I studied the breeding behaviour of Wheatears, so it is also possible that W3 had bred on the island, and in that territory, in the previous year, without my knowing it.

In the second incident, territories tended to be rather small in the Spy Rock area — the mean size of six territories in this area was 0.9 ha (1.7 acres) compared with the mean of 1.5 ha (3.8 acres) for 99 first-brood territories (see Chapter 5). Nevertheless, I would not have considered that 001447 was particularly hemmed in, and I think that it was his excessive aggression that caused him continuously to attack pair W62. Interestingly enough, 001447 established his territory before the arrival of W62, and again it is possible that 001447 continued to treat the land occupied by W62 as his territory.

Whilst these moves were exceptionally long for dependent broods on Skokholm, on Baffin Island in the Canadian Arctic, where the Wheatear population is smaller and more dispersed, Sutton and Parmelee (1954) recorded that a juvenile they had ringed just after it had left the nest had moved with its siblings one-and-a-half miles (2.4 km) in a week and was still being fed by its parents in the new location.

Once the juveniles have left their nest burrows, only rarely do they return. Even so, the parents do not lose interest in the nest burrow at once: I returned to one nest two days after the young had left it to examine the material for parasites; the young were located about 40 m away, but as I put my arm into the burrow the male called the 'weet' anxiety notes. On the same day, the male had also shown anxiety when a Carrion Crow was foraging near the old burrow. Another pair displayed at me, the male in song flight and the female with the tail spread in the threat display, as I passed their empty nest burrow; two days later, the male was still agitated by my presence near the nest and picked up and dropped pieces of dried grass. Of course, parents occasionally used the same nest burrow for a second clutch if I had removed the remains of the old material, and they used the same nest in the following years if the material had rotted away. So, it would appear that Wheatears remember a previous, recently vacated, nest site sufficiently well for several days to be anxious if an enemy is near it, or, possibly for several years as a future nest site.

Having left the nest burrow, the juveniles exchanged one troglodytic existence for another as they localised themselves in new burrows. The change, however, is marked in another way: it is a complete separation from the place in which they had been hatched. In the first two weeks of their life, they had developed confidence in the nest burrow as a shelter and they had demonstrated this confidence by diving back into it whenever they were warned of danger or disturbance. Now they had abandoned it. They had moved on average 40-50 m, but at the first alarm after this change they dived with blind confidence into a new burrow. Although they continued to make use of holes for safety and rest for another five or six days, they did this less and less so that, when they were 21 or 22 days old and had gained skills in flying, they virtually gave up their dependence on burrows for shelter from danger.

By the time the young leave the nest burrow, they have lost most of the nestling down and are adequately covered with feathers which are still growing. The feathers of the upperparts — the forehead, crown, nape, ear-coverts and mantle — are greyish-buff, tipped and edged with darker grey, which gives the juveniles a speckled appearance when seen at close quarters and a generally cryptic colour when viewed at a distance. The slight supercilia are creamy-buff, tipped and edged with darker grey, and the feathers of the lores are a blackish-brown, tipped with creamy-buff. The underparts are similar, creamy-white on the chin with feathers tipped pale grey, and finally the lower belly and the undertail-coverts are a pale creamy-grey with a few darker tips. Their flight and tail feathers will be retained until they moult in the following July, while the contour feathers will be lost almost immediately when feathers of the first-winter plumage begin to grow.

Except, therefore, for the gape and its edges, which are yellow, and the white rump and outer basal two-thirds of the tail (all of which can be very conspicuous), most of the plumage is dun-coloured, which camouflages a motionless juvenile against many backgrounds but particularly those which are earthy or rocky. Even the black feathers of the wings and tail have up to 4 mm of rufous edging, which tends to soften the blackness, while the bill and the legs, which later become black, are grey or greyish-brown.

At this stage, the juveniles look rather dumpy birds: they are not fully grown, and both wings and tail feathers have several millimetres to grow.

Actually, there is a rather nice balancing act going on: as the tail grows, so more white is appearing at the base of the tail feathers, making the juvenile more conspicuous; the wings, however, are growing fast enough for their tips to hide this lengthening white patch and thus prevent it becoming too obvious when the birds are at rest. The tarsus is quite long and, while it makes them look a bit top-heavy, it gives them at this stage in their development, their main means of escape. Juvenile Wheatears also appear to have rather uptilted bills: an impression which is emphasised because they tend to sit in a rather hunched posture and the puffy yellow edges at the corner of the gape, which still indicate to a parent where to put the food, tend to be drawn below the eye.

In their new locality the juveniles would spend the greater part of the first day or two within 2-3 m of the burrow, which now provided them with shelter and an escape route into which they could hop when the alarm was called or when they were frightened by a gull flying overhead. Yet even here I saw variations, and one juvenile might be much more venturesome than its sibling. One family I watched on Skokholm in June 1980 consisted of three juveniles; two were lodged in burrows 0.5 m or so apart and the third was in another about 3 m away, but the latter sometimes wandered away as much as 10 m, although it returned to the same locality.

The burrows remained a key point in the juvenile's lives until they became independent. They might begin to hop 30-40 m away or, a day or two later, fly 70 m across the territory and spend much of the day away from the burrow; but in the later afternoon and evening they would return to the vicinity of the burrows and roost in them overnight. Finally, when they became independent, they sometimes moved even further away, when they roosted in other holes or in the shelter of cover such as the edges of thick clumps of bracken, among heather tussocks and so on.

THE CHANGING BEHAVIOUR OF ADULTS

Once the juveniles had located themselves in their new burrows, the pressure on parents slowly declined. It was at this stage that some males gave up feeding their young altogether, leaving the females with the main burden. They would appear close to the nest from time to time, watch, follow the female but ignore the young. Other males continued to feed the young, although the rate at which they brought food gradually declined still further. Table 12.3 showed that, on average, males had visited nestlings with food less than half as often as females. In the Brecks, Tye also stated that, except for one brood, the females' contribution was greater than that of the males', but he went on to say that the exception showed that a male was capable of raising a brood alone.

In contrast to what went on in Britain, Moreno (1984), watching seven broods of young in central Sweden, said that the parents divided the responsibility for feeding the brood from when the juveniles were about three weeks old until they were independent at about four weeks old. Moreno stated that both parents fed their respective group of young at different sites, and the young of the same family unit appeared to be aggregated in territories.

Neither in the years 1948-53 nor in 1978-83 did I see any sign of division of labour, except perhaps for an hour or so when an individual might be fed exclusively by one parent, but in any case females fed most. As I mentioned above, Tye found that once the young had left the nest some males gave up feeding nestlings altogether. Nor was brood division recorded by Brooke (1979, 1981), who worked on Skokholm between 1972 and 1976. In central Sweden, where second broods are unknown or uncommon, males would possibly have more time to care for fledged young, whereas in Britain some males would already be searching for second nest sites.

This failure on the part of the males to feed fledged young was particularly noticeable with second broods and might be linked with the fact that males seemed to come into moult about ten days earlier than females, at which time they often — but not always — disappeared from the territory altogether. If this happened at the end of the first broods, the female might mate poly-gynously with another male which already had a mate. Armstrong (1955) records that male Wrens lose interest in their young as they fledge or, at the most, give them a morsel of food very rarely.

Females foraged within a range of 30 m or so of the juveniles. Generally, they picked up one food item at a time and brought it back immediately, whereas when the female had been feeding nestlings a day or two earlier she had been bringing as many as two or three food items at a time and feeding up to three nestlings. The juveniles begged food from her in the same noisy manner as they had earlier, making themselves very conspicuous with their wing and body movements, the flashing white rump and yellow gape and the rather vibrant call notes. Only rarely did they follow the female, even when she was hunting quite close to them, although from one brood a juvenile regularly followed the female for 1-2 m and even flew up and displaced her from the top of a 25 cm-high bracken stem on which she had been perched.

Being localised was clearly helpful, since it meant that the female did not have to search too far afield for a juvenile when she had food for it. It was rather remarkable how rapidly the female was able to fix the location of five or six burrows — or as many burrows as there were young — and, once she had visited a juvenile in its new burrow, she usually flew back to the same burrow without hesitation. If the female did not see the young early in her approach, she changed from her fast low flight to the undulating flight or 'bouncy' search flight which had been used earlier in the season to locate intruders, and as soon as she located one of the young she would redirect her flight to it. If they were all hidden, she perched on rocks until she could see one of them. The juveniles, as soon as they saw her, were signalling to her most conspicuously that they were hungry.

As the days went by the juveniles hunted more for themselves, and, although they often solicited the female when they saw her, it was the hungriest which kept the closest watch for her, reacted most quickly to her

255

appearance and most frequently managed to attract her attention. Those which had been more successful in their own hunting were not so much on the alert for her.

I have written earlier in this book about the apparent attraction to Wheatears of holes or depressions in the ground. To adults, holes and depressions were places in or over which they copulated, danced, built nests and raised young. It was interesting to see that depressions had some significance even for the juveniles. They had spent their first days in a burrow, where they were fed and were safe. From a fortnight to three weeks old they were beginning to adapt themselves to a life in the open, and to use burrows to a lesser and lesser extent. When first out of the nest they returned to their new burrow to be fed, and I have recorded juveniles, six days out of the nest burrow (22 days old), running a few feet towards a shallow depression in the ground (not a burrow or anything as deep as that) in which to beg and be fed by the female.

These juveniles seemed to be reconstructing the situation they knew when being fed in the burrow, but we were also seeing the birth of a behaviour pattern which will recur in the breeding season. Now it is without sexual significance (or with very little sexual significance), but next spring that will change. Later in this chapter, we shall see in juvenile Wheatear behaviour the framework of many other behaviour patterns which develop sexual significance at the beginning of the breeding season, whereas at this early stage their sexual valency is low.

JUVENILES ISOLATING THEMSELVES

From the first five days to a week after they left the nest burrow, the juveniles began to isolate themselves from their siblings in separate burrows which could be from as little as 25 cm to 5 m apart. The common contact was the female when she brought food. Apart from that, they began to avoid each other, a phenomenon which was remarked upon by the Heinroths as long ago as 1928, when they stated that the young of wheatears of the genus *Oenanthe*, of chats *Saxicola* and of thrushes *Turdus* tended to isolate themselves when they left the nest. Tinbergen (1939) also recorded this behaviour from Snow Buntings. He further commented: 'Its function is obvious: it decreases the chances of destruction of the whole brood when a predator discovers one of them.'

Unlike Snow Buntings, which, when they became independent of their parents, gathered in flocks, Wheatears persisted in their independence. Whether one remained within its parents' breeding territory or moved a few hundred metres further away, it tended, for a time, to localise itself in an area which it might then defend against other juvenile Wheatears.

I had previously described this behaviour (Conder 1949, 1956), but it was only in 1980 that I realised that what I had observed 30 years earlier was the first stage of a juvenile establishing an individual territory. In 1949 and 1956 I had called this area 'an individual distance'. Now I think that the area around the bird is better called the *individual area* since the area remains mobile, with the term *individual distance* retained for the minimum distance at which an individual will tolerate another of the same species: in fact, the individual distance is the radius of the individual area. Once the Wheatear leaves the nest it maintains the individual area at all times, although actual form of this may be modified by a number of factors. Once the area becomes localised, we see the first stage towards the establishment of a territory, which is usually defended.

In the early days, when they were still being fed and were using their legs rather than their wings to escape, juveniles generally kept close to the ground, only rarely perching on tussocks up which they could clamber or bracken to which they could fly. As they used their wings more to escape from danger, however, they abandoned their close proximity to the ground, which provided a good background for their cryptically coloured plumage, and flew on to rocks and walls, which provided the high look out and the open vista which is one of their essential habitat requirements.

The juveniles were becoming more efficient as hunters. The female would occasionally bring food, and whichever juvenile was approached begged and called while it was being fed, but stopped once the food particle had been eaten; this is in marked contrast to their behaviour when they first left the nest burrow. Apparently, because of their own successful foraging, they were not very hungry.

While in the nest, the young had started exploratory pecking, picking up and mandibulating various objects within reach. They continued as they hopped around at the entrances to their burrows, finding and eating those creatures which were neither too large, nor too hard or distasteful. Part of the experience was in learning what not to eat. Occasionally the young picked up distasteful items and, with a flick of the head, the offending item would be discarded. In some ways, the flick of the head was similar to that used to discard regurgitated pellets of undigested material, although the regurgitation of pellets did not require so much energy as disposing of something which appeared thoroughly distasteful. Tye (pers. comm.) found regurgitated pellets common around burrow entrances, which is something I failed to observe.

At 18 or 19 days old, the juveniles were doing something more than pecking about in an exploratory manner: from their burrows they hopped 1 m or more, rather slowly, and would peer for a second or two into the vegetation, perhaps the grass on which they were standing or into a clump of thrift or wood sage, and then peck at a spider or some insect they saw walking over a leaf or up a stalk. The pecking was all superficial — items from the surfaces of plants — and not digging among the grasses as the adults will do when they are searching for caterpillars to feed the nestlings.

In their exploratory pecking they began to discover the types of food they could eat. To some extent, the food that the female had been bringing since they had left the nest could have given them a visual image of what food was. While they were nestlings, she and the male had brought fairly large caterpillars such as those of the dark arches moth, cranefly larvae, which were quite common among the grass roots, or two or three smaller caterpillars concealed in her gullet. After the juveniles had left the nest burrow, she brought smaller insects and an occasional caterpillar, all of which she sometimes collected within sight of the young. On these occasions, the juveniles would have been able to see how the female hunted and could have acquired an image of the type of food the female was bringing.

More often, however, the female hunted some distance away, and, sharp-eyed though the Wheatears were, I found it difficult to believe that the juveniles learnt very much about hunting methods from their parents. Later, I saw juveniles hunting in ways which I am fairly certain had not been used by adults in sight of the young during the first week out of the burrow. As they developed, the juveniles — possibly quite hungry for much of the time, since males fed them only irregularly once they left the nest — gradually developed the hunting methods of the species and these were gradually being modified by their everyday experience. At the same time, the female was bringing sufficient food to prevent them from starving. The juveniles obviously had some independent drive to hunt because, even when the female was still foraging on their behalf only a few feet away, they continued to hunt; earlier they would have been watching and begging from her.

By their twentieth day, the young attempted to catch insects which were flying 2-3 cm above ground by rushing at them in a rather crouched position with their heads low. Often they would dash quickly after several different insects in quick succession. If the weather conditions were suitable for flying insects, which usually meant little wind, the juveniles used this method almost as commonly as the hopping and pecking-down method.

At about the same age, the juveniles began stretching up and snatching at insects flying past them, or even jumping up a few centimetres without open-ing their wings. It was a method of hunting which over the years I had not often observed. It was about 15 years or so after seeing an etching by Gould that I saw this behaviour in the field; funnily enough, I then saw it performed not only by juveniles, but also by adults in a different locality (see Chapter 8).

By the time that the young were 23-24 days old, they had grown longer tails and had lost the rather tubby look of a juvenile. They were finding food for themselves, using various methods of hunting and flycatching. They were also becoming increasingly competent at preening themselves, scratching them-selves leg-over-wing, or stretching leg-under-wing with tail spread, although even at this stage they were not entirely steady on one leg and occasionally still looked as though they might topple over.

Except at roosting time in the evening, or in heavy rain, they had virtually given up the shelter of the burrow and consequently spend most of the day hopping about in the open. When they wanted to rest, it was often in the lee of a tussock of thrift, grass or heather. Even when it was raining they did not always go into burrows, but, sitting with their heads hunched on their shoulders in whatever shelter they could find, they sometimes got rather wet.

Tye (1982) noticed that Wheatears often used the perch and pounce method of hunting when rain was falling. On the other hand, if the sun was shining and they were resting, the juveniles, like their parents, would sink to the ground and sun-bask for 20 minutes or more unless they were disturbed. After that, they would perhaps preen before beginning to forage again.

FIGURE 15.2 *Song in flight*

Chapter 16

INDEPENDENCE TO MIGRATION

When they were 21 days old, a few juveniles began fluttering 0.5 m or so into the air: the beginning of flycatching techniques used commonly by adults hunting insects which they had spotted 10 m or more away.

It was imperative that the juveniles should gain some efficiency in feeding themselves because the female was leaving them for longer and longer periods. Some females continued to give an occasional feed until the juveniles were about four weeks old, but others were too busy looking for second-clutch nest sites, and a few females had already laid the first egg of a second clutch.

It was also now, before they became independent of their parents and while they still formed groups, that they performed the various displays, which I described in Chapter 6. I have also described, in Chapter 7, the vocalisations used by young birds. Whilst the emergence of behaviour patterns, such as flycatching, which were of immediate value to their self-preservation were to be expected, attempts at displays more usually performed in the breeding season and associated with territory and sex, such as the dancing display, were less expected.

FIGURE 16.1 *First-winter plumage*

By the time the juveniles were four weeks old, they were capable of looking after themselves, of finding sufficient food to maintain and increase their weight and generally continue their development.

Progress towards independence was a mutual affair: as the fledglings aged, the parents were losing interest in their young. Some first-brood parents were beginning to look for suitable holes for second nests, and those which had already produced a second brood were beginning to moult. Even though parents had given up feeding their young, a juvenile would still, in a chance encounter, beg for food, and I have a record of an eight-week-old bird (eight weeks out of the egg) trying to get an easy meal without success. The oldest juvenile successful in begging was five weeks old, but it had an injured wing and could not fly (whether this success was a coincidence or not I do not know). Generally, adults either ignored older juveniles or hopped towards them in the advancing or threat display, and perhaps even chased them a few metres.

While the parents did not chase their offspring out of the territory — indeed some remained in the remoter parts of the original territory until they completed the moult — both parents occasionally threatened the young, and one male attacked its young if they came within 5 m of the second-brood nest site. This made life rather difficult for some juveniles when they were 26 to 28 days old: at one moment the female might feed them, and the next the male might attack them. Within a day or two, however, the adults would be nest-building energetically and the young would be on their own, and, provided they avoided the vicinity of the new nest, they were able to stay on the outskirts of the territory. Some females were less fierce and tolerated their offspring when they were building, but they attacked juveniles from other broods when they came within 20 m of the second nest.

Although as a general rule juveniles on Skokholm became independent of their parents when they were about a month old, it would appear that more northerly Wheatear populations do not behave in the same way. Nicholson (1930) and Wynne-Edwards (1952), who were studying Wheatears of the Greenland race *leucorrhoa*, recorded well-grown young, some even in partial moult, close to adults feeding their nestlings; uncharacteristically, the adults did not attack the older juveniles. Nicholson saw young of two distinct age groups present at a nest where the male was feeding nestlings. He also observed that, occasionally, a moulting juvenile went into the burrow to feed the nestlings. Nicholson and Wynne-Edwards both suggested that Wheatears in Greenland might be double-brooded and that the young from the first brood are not driven away from second broods.

On Baffin Island, Sutton and Parmelee (1954) saw nothing to indicate that juveniles attended adults feeding nestlings of another brood or of second broods. They did, however, record a second male feeding the brood; apart from Nicholson's record mentioned above, the only other record of a second male feeding nestlings of which I am aware is that at Dungeness, Kent, mentioned by Axell (1954).

Non-mated birds are recorded as attending or helping at the nests of breeding members of their species, and carrying out some of the behaviour patterns associated with nesting, including the feeding of nestlings. Gaston (1985) records that species of 32 families or subfamilies of birds are known to indulge in co-operative breeding. Many of these are colonial, and it seems odd to find such a strictly territorial species as the Wheatear behaving in this way. It might be thought that the extra food which is brought to the nestlings could increase the number of young reared and thus the chance of the parents' genes being carried on, but Gaston suggests that often help of this kind does not result in more young being reared.

One of the main problems that Greenland Wheatears would have to address is the post-nuptial moult. On Fair Isle, Williamson (1957a), who examined a number of Wheatears in moult which were either local breeders or migrants from the northwest, claimed that the severity of the wing moult restricted the Wheatear to a single brood on Fair Isle, as well as further north (including Faroes, Iceland and Greenland). I think that many other aspects of their breeding behaviour would have to be telescoped in order to achieve a second brood: it is the shortness of the season even more than the moult which makes second broods in Greenland impossible.

Whether there was a second brood or not, both Nicholson and Wynne-Edwards described how well-grown nestlings, some even in partial moult, were close to adults feeding their nestlings, a behaviour which caused them to think that Greenland Wheatears might be double-brooded. One explanation of this grouping of different aged nestlings could be that sometimes they are polygynous, albeit uncommonly: thus a male, involved with two females, would be able to produce two broods within a shorter period than required to raise two broods with one female. Even so, the normal territorial boundaries between polygynous females and their offspring would have to break down to allow the juveniles of two different age groups to attend a nest. With the available information, it is impossible to be certain of the explanation of this behaviour. Too few ornithologists have had the opportunity to study marked Greenland Wheatears from the moment of their arrival on the breeding ground.

By the middle of the moult from juvenile to first-winter plumage, the appearance of an adult no longer stimulated birds of the year to solicit food from adults. Indeed, the need to distinguish adults or even the sex of another Wheatear had disappeared for the time being.

The failure of a juvenile to return to roost in the area in which it had been localised after leaving the nest was another mark of independence: it had previously wandered 100 m or so from the point at which the brood had first been localised but had usually returned to roost with its siblings. From now on, young Wheatears would lead a nomadic life, often joining other young Wheatears for a time, and later being joined by adults whose ties to their breeding territories were now severed. The groups were in effect a rather loose and open assemblage of individuals which might include siblings but which lacked the tight cohesion of finch or Starling flocks. When alarmed, Wheatears flew from the source of the alarm but they did so individually in different directions, not as a flock; and, when the cause of the alarm had disappeared, they generally returned as individuals to the area in which they had been localised. They would stay together until some other stimulus set an individual off on another stage of its nomadic life and, once again, the composition of the group would change.

In a group, each individual would generally keep about 2-10 m from a neighbour, the distance at which they preferred to see neighbours and which I called the 'social distance', and this resulted in groups forming. Within that 'social distance', of course, each individual maintained its own individual distance and moved away from or attacked another Wheatear if the latter came too close.

At first, groups tended to form on the edge of their parents' territory or between two territories, but as the adults lost their aggressiveness and territorial boundaries disappeared groups formed where food was abundant and easily accessible. As I pointed out in Chapter 4 when discussing the behaviour of migrants on Skokholm, Wheatears in autumn collected where the vegetation was most heavily grazed, which was a different habitat from that selected by Wheatears on spring migration. A wall, rock outcrop or hillock on which they could perch was an added attraction; they preferred positions with wide vistas and often competed with others for the highest points.

While they were with the group they could well remain localised for several days, even weeks, before moving on; but while they were localised, in the sense that a certain area was their headquarters, they might move as far as 100 m — even 200 m — and eventually return either by direct flight or by hopping and flying to the headquarters area. Once an individual became localised in such a place, it would generally attack others which came too close. In this way, individual territory was established.

From the time that juveniles had lost their dependence on burrows, they demonstrated on the one hand a tendency to group with others, but on the other a tendency to act aggressively to an individual that came too close. Some which had strong aggressive tendencies might attack neighbours even though the latter had not moved, and it would seem as if the sight alone of another was sufficient stimulus to set it off.

Clearly some young Wheatears were confused by this changing situation: they flew towards others in an attempt at grouping but discovered that they had encountered a localised bird and were attacked. Often they tried a second approach but were attacked again. Some of the proximity attacks, where an aggressor suddenly flew at a neighbour, might develop into chases lasting as long as 20 seconds, and even show some of the characteristics of display flights such as sexual chases where, normally, the pair flew in circles or figures of eight, or the zigzag flight which was used in territorial circumstances. The aggressor bobbed frequently.

Although chasing was common in the early days of their independence, those juveniles flown at rarely resisted: they flew off. Occasionally the attacked bird held its ground while the attacker flew over its head in the pounce, and, exceptionally, juveniles were recorded in the 'head to head fight'. Juveniles and first-winter birds sometimes flew from one location to another with their tails fanned, which is a self-advertising display usually indicating that the individual is defending some kind of territory.

At the end of 30 days, just after it had become independent, a juvenile was heard singing a five-second phrase of subsong, and throughout the remainder of their stay on the island subsong might be heard from young Wheatears whether they were locally bred or on passage. A burst of subsong could last up to 30 seconds, but it was much quieter and more difficult to hear than the warbling subsong used by males in spring either before the arrival of their mates or during some of the sexual encounters: indeed, one would normally have to be within about 10 m of the singer to hear it. Very rarely, young Wheatears would fly up in the song-flight pattern, but they did not attain the height of the spring flight nor were so many song phrases used.

The act of isolating themselves from their siblings almost as soon as they left the nest, their aggressiveness to siblings and others which came too close while they were part of a group or assembly, and their subsong were all elements in the establishment of a mobile individual area, which developed into an individual territory whenever they localised themselves for two or three days. Once localised, they were less inclined to retreat when another came too close and so they attacked the intruder and defended the area.

Even when about three to four weeks old, some juveniles could be aggressive to other species: they chased Meadow Pipits, Rock Pipits, Skylarks and Linnets, and later Yellow Wagtails and Pied Flycatchers when the latter were resting on the island.

Young Wheatears spent much time foraging, chiefly by hopping 1-2 m and then looking down in the typical Wheatear fashion, but also using other hunting behaviour described in Chapter 8. They were at this stage more efficient hunters than during the first few days out of the nest, but they were still learning and pecked at non-edible objects such as plants — not at the insects which they might hold, but at grass blades and so on. Sometimes the

FIGURE 16.2 *First-winter Wheatear looking up at potential predator*

action of one juvenile stimulated another to copy it, as when a juvenile flew at a meadow brown butterfly and failed to catch it and a second juvenile tried the same thing a few seconds later. Cinnabars, both moths and caterpillars, were very common on the island, and I have frequently seen them in the beaks of Wheatears, young and old, but never recorded Wheatears actually eating them. Cinnabar moths and caterpillars are both distinctively coloured and said to be distasteful (Ford 1955): perhaps these young birds were learning the hard way about the value of warning colours.

In the intervals of foraging, Wheatears often rested by perching on a wall in the sun but sheltered from the wind by one of the stones, or they sank down until they were sun-basking, resting on the belly, sometimes on the grass or a pile of stones. Whether they were still in juvenile plumage or had moulted into first-winter plumage, they were difficult to detect against those backgrounds. There they would doze for a time, then perhaps preen a little, suddenly see some tempting insect, dash out after it and then return to their resting place. Or instead they would explore their habitat, shelter under rocks, look into holes in walls, depressions in the ground and sometimes burrows, although they would rarely enter them outside the breeding season in the way that Wrens do; that some did, however, enter burrows is testified by the corpses found in Manx Shearwater and Little Owl burrows, although those found dead in the owl burrows could have been killed outside and taken in.

When alarmed, the juveniles used many of the behaviour patterns of anxious adults, such as hovering to observe, or standing behind a rock or stone on a wall and just putting the top of the head and the eye high enough to look over the top. On other occasions they would stand on a wall looking around, turning their heads upwards and sideways to watch gulls as they passed over: if a gull looked as though it was getting very close, the young Wheatear would fly off 20 m or so in the undulating flight with tail fanned.

DISPERSAL OF JUVENILES FROM NESTING TERRITORY

I kept track in a rather casual way of the directions and distances in which colour-ringed young Wheatears dispersed from the territories in which they had been hatched. Generally, the identification of these birds had not been a major part of my Wheatear study, but when I saw a colour-ringed individual I identified it and recorded its location. When Jeffery Boswall was Assistant Warden at the Observatory for seven weeks between 13 August and 30 September 1951, he concentrated on plotting the positions of colour-ringed individuals. Several young Wheatears were also retrapped. As a result of Boswall's and my own efforts, I had 549 records of movements of independent individuals in the year of their birth.

Skokholm was not an ideal site for this aspect of Wheatear behaviour since the sea probably deterred the birds, initially at least, from travelling the distances they might have done if they had been on the mainland. On Baffin Island, for instance, where there was no such barrier close to the nest site,

Sutton and Parmelee (1954) shot one of a family of Wheatears one-and-a-half miles (2.4 km) from where it had been caught and ringed a week previously just after it had left the nest. They also state that these juveniles were still being fed by their parents; they were therefore less than a month old.

Table 16.1 shows the average distances the young were found from their nest sites in relation to their age. From the time they lost their dependence on burrows (at about three weeks old) or their dependence on their parents (between three and four weeks old), most dispersed in various directions. Between four and five weeks old, 68% were identified within 300 m of their nest sites, although four individuals (two from one family) had already travelled as much as 800 m. By six weeks, 75% were identified within 500 m, although three (including two of the four mentioned above, but not the siblings) had moved between 1 km and 1.25 km from the nest site.

Table 16.1 also shows that at six weeks the outward spread had slowed a little and that the average distance from the nest remained about the same until the ninth week. This more or less stationary period coincides with the moult from juvenile to first-winter plumage. Throughout the remainder of the period in which some colour-ringed young Wheatears were recorded on the island (the oldest was 14½ weeks), the average distance gradually increased again.

There seemed to be two tendencies: one part of the population appeared to remain within about 600 m of the nest site, whereas others, at as early as eight weeks of age, had dispersed more widely over longer distances. Table 16.1

TABLE 16.1 *Average distances travelled by ringed juveniles after leaving the nest*

| | Age in weeks | | | | | | | | | |
	3-4	4-5	5-6	6-7	7-8	8-9	9-10	10-11	11-12	12-13+
(a) Average distances (in metres)										
1st brood	215	273	393	604	585	705	774	661	482	900
Replacements & 2nd broods	118	280	436	237	424	215	907	860		
Mean	193	275	398	521	505	534	833	748	644	855
	← Dispersal →			← Moult →						

(b) Ranges (minimum to maximum distances)

	3-4	4-5	5-6	6-7	7-8	8-9	9-10	10-11	11-12	12-13+
	0	0	0	0	0	100m	100m	150m	200m	250m
	to	to	to	to	to	to	to	to	to	to
	600 m	800 m	1.25 km	1.3 km	1.7 km	3.4 km	1.55 km	3.8 km	2.1 km	1.7 km
						(4 km)		(4 km)		

TABLE 16.2: *Dispersal directions of juveniles after independence*

Compass directions	No.	%
O	26	10.3
N	25	9.9
NE	46	18.2
E	34	13.5
SE	28	11.1
S	23	9.1
SW	34	13.5
W	23	9.1
NW	13	5.1
	n = 252	
Away from sea	164	65.1
To short grass	168	66.6
Family together: Yes	54	65
No	29	35

shows that maximum recorded distances from the nest had increased steadily, and that some individuals had probably crossed the sea barrier. In writing about these distances I must emphasise that they were in fact the minimum distances moved, because several days, even weeks, might elapse between sightings during which the birds could have moved in a variety of directions and over some distance.

The problem of determining the direction in which the young dispersed from their nesting territories was complicated by the fact that most nests were close to the sea so that the birds could not travel more than 1 km, but chiefly much less, without arriving at the cliffs and the sea barrier they represented. It was therefore inevitable that four-week-old juveniles either remained where they were or moved away from the sea or along the coast; in fact, 65% of the juveniles took directions initially which were away from the nearest cliffs (see Table 16.2). Bearing that point in mind, I have analysed the compass directions from the nest in which the juveniles were first recorded after independence. Of the total, 10.3% were still found within 100 m of the nest site in which they had hatched; otherwise, the directions in which the remainder had moved were fairly evenly scattered around the compass, although slightly higher numbers had apparently moved in easterly or southwesterly directions, but only 5% moved northwest.

Sixty-six per cent of the young birds also appeared at one time or another in the heavily grazed short grass of the Northern Plains and Home Meadow in the centre of the island which, as I showed earlier (Figure 4.5e), was also the preferred feeding area of autumn migrants.

On dispersal, some siblings moved and stayed together for varying periods, whereas other split up. In 65% of 83 broods of which two or more members were seen again, at least two siblings had moved initially in the same direction and were more or less localised within 100 m of each other. Two of brood W80 had moved 800 m when five weeks old, but the more usual distance for two siblings of this age to have moved together was between 200 and 400 m. The greatest age at which I recorded three siblings together was eight weeks, and they were localised only 200 m from the nest site (Table 16.3).

Rather fewer groups of three siblings were recorded together, but when five weeks old they were usually only 200-400 m from the nest, although one group of three had travelled 750 m at six weeks old. In the remaining 35% of the families of which two or more members were recorded later, the individuals had apparently dispersed independently. Of course, siblings which had split with other members of their family might meet them again on the Norther Plains, Home Meadow or elsewhere simply because this was a favoured area in late summer, but I have no evidence that they recognised one another.

Table 16.3, based on 118 records, shows the minimum times that young Wheatears spent at the same location, or within 100 m of it. Occasionally individuals were seen as much as 400 m away, but they had returned to their individual territories by the following day. I was able to check on one individual since it was localised in front of the Observatory buildings. The bird left the area during a gale and was found sheltering down the cliffs on the leeward side of the island 200 m from the Observatory; on the following day, after the gale had abated, it had returned to its individual territory at the Observatory.

The table shows that 54% of juveniles were recorded as spending at least two weeks in the same place before moving on to another area or being lost sight of. Once a young bird had arrived in a new area and remained there for more than a day or two, it gradually restricted its movements to one locality and, once it did this, it began to attack other Wheatears which came too close — or within its individual territory. The graph in figure 16.3 shows the

TABLE 16 3: *Minimum time juveniles localised in one area*

				Weeks					
	0-1	1-2	2-3	3-4	4-5	5-6	6-7	7-8	
No.	36	28	20	13	10	6	2	3	n = 118
%	30.5	23.7	16.9	11	8.5	5.1	1.7	2.5	

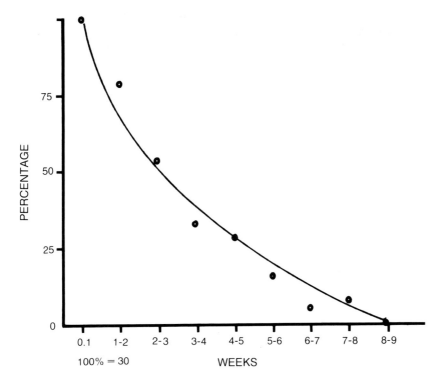

FIGURE 16.3 *Graph showing percentage of time juveniles were localised in one area*

percentage of juveniles localised in any area for increasing periods. The behaviour of young Wheatears localising themselves and establishing individual territories is described in more detail in Chapter 4.

I have very little evidence as to when the first young Wheatears crossed the sea. The youngest known was one I had ringed as a nestling on the Old Deer Park, Marloes, 4 km from the island and which crossed to Skokholm when about eight weeks old; a second, colour-ringed as a nestling on Skokholm, was identified when ten weeks old by Mrs Diana Bradley on the island of Gateholm, 4 km northwest of Skokholm. Another Wheatear in first-winter plumage was seen on Grassholm on 31 July 1948: although the origin of this individual was unknown, it must have been more than nine weeks old.

It was possible to deduce from the records of unringed young birds trapped on Skokholm that a small number of juveniles — one or two a day — were beginning to cross to the island, some perhaps from Skomer, early in July before they had begun to moult, so they were not yet six weeks old. Larger numbers of unringed birds of the year began to appear in the last ten days of July; a few were still in juvenile plumage and moulting and others were in first-winter plumage. Once they had more or less completed the moult, it was time for dispersal to change into migration. Williamson's (1957a) study of the post-nuptial moult of Wheatears trapped on Fair Isle showed that a few individuals arriving from the north had only a few secondary feathers short by a few millimetres of their full length. A record that confirms the suggestion that late July is when the change from a post-breeding dispersal to migration occurs is provided by Mead and Hudson (1985), who give details of a Wheatear ringed as a nestling on 25 May 1984 on the Isle of Mull, Strathclyde, which was found on 23 July 1984, at Bassin d'Arcachon, Gironde, France, 1350 km to the south-southeast.

MOULT

As with the study of the dispersal of juveniles, the collection of information on the moult from juvenile to first-winter plumage from ringed individuals of a known age was an incidental result of another aspect of the study of the Wheatear's life history. I did not have birds in captivity and was therefore unable to check the progress of the moult of individuals daily. I was, however, able to check changes in plumage of some individuals at various stages during the whole process, either when they were retrapped or by observations in the field, or by a combination of both.

When colour-ringed individuals were identified in the field, areas of newly-grown feathers were filled in on a printed outline of a Wheatear. These notes therefore refer almost exclusively to the moult of contour feathers, whereas Williamson was recording the moult of wing and tail feathers. When the birds were retrapped and brought into the Observatory, it was possible to record much more detail. Observations in the field were not entirely reliable, because the size of the patch of feathers, whether new-grown or juvenile, could well be affected by the way in which the neighbouring feathers lay.

Both the age at which the feather sheaths first pierced the skin and the age when the moult was judged to be virtually complete varied from individual to individual. The age of the youngest bird examined in the hand whose sheaths had pierced the skin was 31 days (14 June 1950), although another that I examined at 35 days old had feathers so advanced that they might have started earlier. The oldest juvenile that showed no sign of sheaths piercing the skin was 38 days old, whereas another of the same age had partly opened vanes which could be seen in the field on the mantle, ventral tracks and various parts of the head. At between five and six weeks old most juveniles had begun the moult, and by the end of that period new feathers were appearing through the juvenile plumage on most of the feather tracts, but the general outward appearance was still of a rather speckled juvenile with patches of a more uniform colour.

Feathers generally appeared first on the humerals (scapulars), ventral tracts (sides of breast), wing-coverts and tibia (crural tract). Feathers on various tracts on the head were often rather late in appearing, and in some individuals at least a week after the appearance of feathers on other tracts. Table 16.4 shows that in a six-week-old juvenile, as sketched in the field, 3.6% (range 0-10%) of the plumage appeared to be new.

TABLE 16.4: *Percentage of new plumage on juveniles as recorded in the field*

	Up to 6	6-7	Age in weeks 7-8	8-9	9-10	10-11
Mean %	3.6	23.6	53.4	77.7	93.5	100
S.D. = ±	3.4	12.7	22.0	21.0	8.7	

At between six and seven weeks, new sheaths were appearing on most feather tracts except the median wing-coverts and the undertail-coverts. On the mantle the length of unopened feather sheaths ranged between 0 and 11 mm, although chiefly about 1-4 mm. Seen in the field, most juveniles showed some new feathers and on average 23.6% of the plumage was recorded as new (ranging from 5%-50%). One juvenile (N4981) showed very little outward appearance of the moult when seen at 48 days old.

At seven to eight weeks old, the moult was well advanced; some feathers seemed to have attained their full length, while others had only just pierced the skin (particularly on some parts of the head and median wing-coverts). More of the sheaths had broken open and on the mantle they might measure 2 mm, while the open barbs of the feathers were already 18 mm long. On the belly some down feathers were appearing. In the hand birds tended to be very 'scurfy', which was no doubt due to the disintegrating sheaths. In the field, at seven to eight weeks, an average of 53.4% (range 20%-90%) of the plumage appeared to be new. Some individuals showed only a few small patches of juvenile plumage remaining.

Between eight and nine weeks old, the growth of the feathers on the lower back and sides of the breast were almost complete, but some juvenile feathers were still apparent on the throat. In the field, an average of 78% (range 40-100%) of the plumage was new, but, while one individual seemed outwardly to have completed the moult at 60 days, others still showed about 60% of juvenile feathers. I also noted at this time that the brown edges to the black primaries were already beginning to show signs of abrasion.

Between nine and ten weeks, the new feathers of some individuals were still not fully developed, and up to 3 mm of unbroken sheath still remained. The areas which appeared to be the last to lose their juvenile feathers were the tracts on the neck, nape, and the centre of the underparts. The oldest bird of the year which I trapped that had not completed the moult was 78 days old: some feathers, chiefly on the spinal tract, were 1-2 mm long, on the central belly some feather sheaths were 5 mm long, and a few feathers had not broken out of their sheaths; looked at in the field, new plumage covered about 93% (75-100%).

Between ten and eleven weeks, all young Wheatears examined in the field showed 100% new plumage. The latest dates on which birds examined in the hand had almost completed their moult were 28 July 1952 (for first-brood young) and 27 August 1950 (for second-brood young). The quills of one trapped on 10 September 1950, however, were still not fully developed on the head, belly and back.

The Autumn Post-nuptial Moult

The earliest dates on which I recorded both adult males and adult females showing new body feathers in the old plumage were 1 and 2 July, which suggests that moult actually began in the last week of June, as with the Fair Isle birds (Williamson 1957). In the first two-and-a-half weeks in July, I recorded the stages of moult of 40 adults (33 males and seven females), but I was able to follow the complete process in only about two individuals because the number of colour-ringed adults which had nested on the island declined rapidly from mid July onwards (presumably, they dispersed across the sea barrier after independence more quickly than did birds of the year). Such figures as I have show that, in the first seven days in July, 7.7% of the male contour feathers were new; in the second week, 16.25%; in the third week, 36.6%; and in the fourth week, 55%. After that, there are too few records to give meaningful results, but even by 18 August a few males had only 80-90% new feathers.

Kenneth Williamson (1957) studied the post-nuptial moult of flight feathers of adult Wheatears which he trapped on Fair Isle, where they are single-brooded and nest between the middle of May and early July, although a few late birds may still be attending their young at the beginning of August. He found that the moult commences during the last week in June, as on Skokholm, before the young achieve independence, and some individuals have already finished by mid August. Moult begins with the innermost primary and its coverts, and the change proceeds outwards. The tertiaries and the mid wing-coverts are the next to change, and the outermost greater coverts are a little ahead of the inner ones.

When the replacement of the primaries is well under way, Williamson states that the moult of the secondaries begins, also in mid wing and proceeding outwards towards the renewed tertiaries. The tail moult commences, together with the upper- and undertail-coverts, shortly after the innermost primaries have begun, and it lasts three to four weeks. Williamson calculated that the period of the moult was of the order of seven to eight weeks, which was the same as for the birds that I examined on Skokholm.

It would seem therefore that, in spite of the difference in timing of the breeding season between Fair Isle and Skokholm, moult began at the same period, at least so far as the males were concerned. On the other hand, although some females began at about the same period, there was some evidence that, in general, females were about 10-14 days behind the males in beginning the moult.

Autumn Plumage of Adult Male Wheatears

After the autumn moult was completed, the head pattern of an adult male usually distinguished it from females and first-winter males, even though some of the new feathers had dun-coloured tips which gave them a brownish cast even on the head. The crown looked a brownish-grey; the supercilium, often extending the whole length of the ear-coverts, was visible and in some individuals a clear white. The feathers of the lores and ear-coverts were black, but on some individuals were tipped brown.

Generally the mantle was a mid-brown, but, if these feathers were displaced by some movement of the male, the grey of the bottom half of the feathers — the summer plumage, presently hidden — showed through. The chin was white, the throat a pale sandy-buff, the breast a richer tone, the belly a pale sandy-buff, and finally the undertail-coverts were white.

The edges of the flight feathers and the ends of the tail feathers were a pale rufous to pale greyish-buff. The post-nuptial plumage of a few males lacked the brown-tipped feathers, and their plumage was almost identical to their crisp spring plumage.

First Winter Males

After its moult from juvenile into first-winter plumage, the young male is usually considered indistinguishable from females whether young or old. I concluded however, after watching them for many seasons, that it was possible to separate some — but not all — first-winter males from females. The critical points were the colour of the mantle and the supercilium. The mantle of some first-winter males appeared as dark as that of adult males, and slightly darker than the mantle of females. This shade came about because the brown tips to the feathers, which will have worn off by spring, in the autumn overlie the grey feathers of the adult male plumage, which it is sometimes possible to see when the feathers are displaced. On females, on the other hand, no matter what age, the brown tips overlay the brown adult feathers.

A second feature which, in conjunction with the darker mantle, helped to separate first-winter males from females was the pale buff supercilium, which often seemed longer and more marked than on females.

Of course, the shades of brown vary widely in different lights, and what I am describing still requires observation of the individuals, but perhaps these notes will act as a guide to clearer separation of the sexes.

Females

I found adult females which had moulted and first-winter females indistinguishable. The crown, nape and mantle were often a pale greyish-brown, rather paler than the mantle of the male, but their colours are very variable. The supercilium is a pale grey and does not usually run much further back than the eye. The lores are black, but the ear-coverts brownish to sandy-brown on some individuals. The throat and breast are sandy-buff, as on the male. The edges of the flight feathers and tips of the tail feathers are rufous to pale buff.

The underparts and head pattern can be very variable, which helps an observer in short-term identification.

Plumage Abnormalities of Wheatears

I saw only three birds showing apparent albinism when I was on Skokholm, and one on Alderney: all were on autumn migration. One appeared completely white, and the other three showed whitish patches on the nape of the neck. Examination of one individual in the hand showed that only the tips of the feathers lacked pigment; and these would be abraded away by the following spring, revealing normal spring plumage. Harrison (1985) calls this condition in which pigments are lacking 'schizochroism'.

In Northumberland, Bolam (1912) thought that varieties in Wheatear plumage were not uncommon; most of those he recorded were seen or shot during spring and autumn migration and could have been part of the normal variation that one finds in any population; one, which was a partial albino, might well have had feathers whose tips lacked pigment. Sevinga (1968) describes a white-headed Wheatear seen in September which Voous, in an editorial note, states could only have been a partial albino, and Tye and Tye (1983) show a photograph of a leucistic Wheatear.

I know of no records of a completely albino Wheatear.

FIGURE 16.4 *Relaxing*

Chapter 17

THE SOUTHWARD RETURN

As we saw in the last chapter, return migration develops from a dispersal in more or less random directions from the nest during which both adults and birds of the year have completed the moult into winter plumage. On Skokholm, they left their nesting areas and dispersed to a habitat which apparently provided better feeding; and, towards the end of July, the first movements were recorded which had a more southerly component than during their dispersal.

FROM THE NORTHWEST

Since Wheatears of the Greenland race, which inhabit northeastern Canada and its islands as far west as 95° W as well as Greenland itself, may pass through the British Isles, I shall attempt to trace their passage first. There is little detailed evidence as to when Greenland Wheatears of Baffin Island and Labrador actually leave, but Wynne-Edwards (1952) records that in 1950 Wheatears were last seen in southeast Baffin Island on 11 August, and mid August is the time that, according to Snyder (1957), they leave their Canadian breeding grounds both in the east and in the Yukon. Migrants from northern stations in Labrador apparently move further eastwards to Greenland and Iceland before beginning their southwards course to winter quarters, but those from the Gulf of Lawrence seem to fly directly across the Atlantic (Todd 1963).

Freuchen and Salomonsen (1960) tell an interesting story about Wheatear behaviour in Greenland. In September, Snow Buntings and a large number of Wheatears move south in west Greenland. They appear in the same bare, rolling, inland landscape on rocky ground with heaps of boulders and they avoid areas with dense vegetation of heath and scrub. Snow Buntings appear in small flocks and Wheatears wander about singly, but the two species migrate side by side, under the same conditions. Nevertheless, one day, all the Snow Buntings without exception change their direction to a westerly one, passing across the Davis Strait to the American mainland, while, at the same time, Wheatears shift to an easterly direction for their crossing of the Atlantic.

They leave the southern half of Greenland in September and are seldom seen there in October (Salomonsen 1927). High concentrations (up to 75 per 100 m^2) of migratory Wheatears occur in west Greenland from the last days of August (Boertmann 1979). This dispersal from breeding areas to apparently better feeding habitats is similar to the change that we saw on Skokholm, although the latter movement was very local.

With one or two exceptions, Wheatears seemed to return from their circumpolar breeding range to their winter quarters by much the same broad routes that they travelled north. The most important and spectacular exception is found in the return of the Greenland Wheatears from the northeastern Canadian Arctic and Greenland, where, instead of making the shortest possible sea crossings of the North Atlantic via Iceland, or Faroes and Scotland, some apparently head in southeasterly directions from Greenland towards the west coast of Europe.

While some Wheatears from Greenland arrive in Scandinavia, Fair Isle, western Scotland and Ireland, the majority are thought to make the direct crossing to west France and Portugal (Salomonsen 1971, 1979). Indeed, there is a growing number of birds ringed in Greenland and recovered in France and on the Iberia peninsula. Salomonsen suggests that this difference between the spring and the autumn migration is due to the utilisation of northwesterly winds by Wheatears during their Atlantic crossing southeast to Europe. In Chapter 3 I also pointed out that it was safer to make the northwards journey to Greenland in short hops because, if the Wheatear arrived in bad weather conditions, retreat would be virtually impossible.

David Snow (1953) has analysed the records of the Wheatears seen from weather ships in the Atlantic Ocean from the end of August until late September. He shows that a proportion of Wheatears fly directly across the Atlantic, from Labrador in the Canadian Arctic and Greenland, making a landfall on the west coast of Europe between the British Isles and the Iberian peninsula. He supports this statement with a map showing the position and dates when the Wheatears were recorded.

Snow comments that for the Wheatear to be met with so frequently there must be large numbers of birds at sea, a point made by Nicholson as long ago as 1928 when describing the birds seen during his Atlantic crossings. Snow emphasises that they are unlikely to be birds blown off course while making the short crossing Greenland-Iceland-Faroes-Scotland because they were recorded chiefly when the winds were westerly.

Records from Atlantic weather ships also confirm Salomonsen's (1927) statement that Wheatears start leaving Greenland in early August and that the main departures occur in September, with very few birds remaining in October (incidentally, the first migrant Wheatears are arriving in their winter quarters in Africa in September).

Spencer and Hudson (1982), describing a recovery of a Greenland Wheatear in north Wales in spring, comment that autumn recoveries are unusual: since Greenland Wheatears have a greater trans-oceanic component and northwest Europe is largely by-passed. Observations from many parts of Britain show, however, that many Greenland Wheatears still make a landfall here in autumn.

Between 30 August and the beginning of November 1961, Lee (1963) studied migration on Lewis in the Outer Hebrides by radar. He identified Wheatears from the radar record by the speed at which the 'echoes' were travelling and the direction from which they had come, and finally by checking the area surrounding the radar station for new arrivals on the following morning. The passage of Wheatears was already in full swing when study began. His observations showed that passerines arrived from three main directions: first a movement which presumably consisted of Wheatears and Meadow Pipits headed in southeast from Iceland; second, there was a movement east-southeast from Greenland which consisted mainly of Wheatears; and, third, slightly west of south, Wheatears came in from the Faroes. He also discovered that arriving migrants flew low over the sea during the second part of the night but rapidly gained height at dawn. Birds which he presumed to be Greenland Wheatears left Lewis on the same southeast bearing on which they arrived from Greenland.

As in spring, it is apparent that drift is an important factor in bringing migrants to Britain, particularly from Greenland and the Canadian Arctic. The ideal weather situation for a successful passage from Greenland or Iceland to the British Isles would appear to be an area of low pressure to the north of Britain and Ireland with high pressure building up over the North Atlantic, which conditions would produce northwesterly winds from Greenland down towards the British Isles.

On Fair Isle, between Shetland and Orkney, the local population of Wheatears was augmented by others that arrived from Greenland, Iceland and Faroes (Williamson 1965). Under strong cyclonic conditions, flights might cover as much as 2,400 km and last up to 30 hours (Cramp 1988). Following such conditions on 9, 11 and 13 September 1960, Greenland Wheatears arriving on Fair Isle, having flown for 30 hours non-stop over some 2,500 km, had lost 35-40% of their body weight. Under weaker cyclonic conditions, arrivals on Fair Isle on 12 September, after a more direct journey of little more than 1,600 km lasting about 24 hours, had lost only about 20% of their body weight (Williamson 1965). Most weight loss was due to oxidation of fat, but some water loss by respiration might have occurred, too. Further south on their journey, if they were forced down in very hot places, dehydration could be severe (Langslow 1976). It will be remembered (see Chapter 3) that birds found dead by Haas and Beck (1979) in the Sahara had exhausted their fat supplies.

On the Isle of Man on 18 October 1986, in a site relatively sheltered from the gale-force northeasterly winds, an estimated 2,000 Wheatears were going to roost in an area of 3,000-4,000 m^2 (Thorpe 1987). The author points out that this exceptionally large aggregation of Wheatears passed through the Isle of Man unusually late in the year, and suggests that the birds were of Scottish or

Scandinavian origin because they were so late in the season. Chris Mead, however, has suggested to me that, since there were large falls of other species of Scandinavian origin in other northern parts of the British Isles, it would seem more likely that they originated in Scandinavia. Such a high density of Wheatears is recorded very irregularly, and in view of this species' normal territorial behaviour on migration I assume that it was the gale-force winds that kept them penned in this sheltered area on the Isle of Man. Earlier, I mentioned that Boertmann (1979) found a similar concentration of migrating Wheatears in Greenland.

It was under cyclonic conditions similar to those at Fair Isle that Skokholm regularly recorded rushes of Greenland Wheatears, occasionally mixed with Lapland Buntings (which apparently breed in Greenland but not Iceland) and White Wagtails (which breed in Iceland but not very extensively in Greenland).

Passage migration of Wheatears of either race on Skokholm was not usually very spectacular. Increases in passage birds, however, were usually recorded from the middle of August until the end of September, when the numbers on Skokholm tended to be low, and Davis (1954) recorded stragglers on Skokholm as late as 25 October (indeed, in Britain, stragglers have been recorded in every winter month). When conditions for drift migration were favourable, large numbers — up to 240 on 9 and 10 September 1951 — were sometimes recorded on Skokholm.

Turning now to the Wheatears which breed in the British Isles, we have, first, to distinguish between a dispersal from the breeding area, which may be undertaken as soon as adults become independent of the young, either of the first or of the second brood (see Chapter 16), and the beginning of a southerly migration.

The juveniles dispersed about the island as soon as they had become independent of their parents, but apparently did not start leaving the island — and crossing the sea barrier — until they had almost completed the moult into first-winter plumage: the youngest Skokholm-ringed Wheatear recorded alive on the mainland was eight weeks old. Mead (1983) points out that this exploration period of young birds as they disperse is extremely important for learning the local star map by which they may ultimately navigate themselves back to the area of their birth.

The dispersal of the adults from the breeding sites is sometimes a drawn-out process, and they would seem to take entirely random directions. Some adults disappeared as soon as they became independent of their young, and some remained until late August. It is rather difficult to judge at what point the dispersal from nesting sites turns into the autumn migration. One way to do this is to examine the records showing the arrival of birds not bred on Skokholm.

The Skokholm Bird Observatory Reports (Conder 1950-53) show that, although some of the local breeders left the island in July, the earliest records of appreciable immigration occurred on 11 August, in 1950. Jeffery Boswall (1951) studied the departure of colour-ringed Wheatears from Skokholm. When he and his helpers began the survey on 13 August, 41 colour-ringed adults and young birds that were either local breeders or had been born on the island were still present. The numbers of island-bred young gradually decreased until the last young birds disappeared on 12 September. The 18 adults which were present on 13 August showed no sign of a decrease for a week, but then some males disappeared. Not until 30 August did the first of the nine females depart, but within three days all the females had gone, and the last male left on 6 September. Boswall's survey also showed that six first-brood young were still on the island on 13 August; one of these had left its nest on 31 May and did not leave the island until 28 August, nearly 13 weeks after leaving the nest. The peak period for the departure of local birds was between 30 August and 3 September, when the number decreased from 27 to seven. These departures occurred under anticyclonic conditions with light and variable winds and a clear sky, the ideal conditions for the start of a migratory flight.

The reports of the migration committee of the British Ornithologists' Club give a general picture of the departure of Wheatears in autumn between 1906 and 1912. These reports show that Wheatears began to leave their summer quarters in mid July, but that the passage did not get into its stride until the end of August: most Skokholm Wheatears, adults and first-winter birds, left the island in August and those recorded in September were passage migrants. The BOC reports show that by the middle of October the majority of migrants had already passed through the British Isles, but stragglers of both the typical and the Greenland races were being recorded occasionally in the middle of November. In the past 75 years, the pattern does not seem to have changed.

It is obvious from this account of the departure of a small part of the local population on Skokholm that Wheatears moved off as individuals rather than as a flock, even though several birds might leave in a night when conditions were theoretically ideal for the start of a migratory flight, with clear skies and little wind. Large falls of migrants may also occur when birds, having left on a migratory flight across the sea under ideal conditions, are suddenly caught by cyclonic conditions, such as a belt of heavy rain, and drift downwind. Large concentrations of Wheatears and other species literally fell out of the sky on to the Suffolk coast on 3 September 1965: D.J. Pearson estimated that among the 15,000 or so birds he saw on 4 km of coast near Walberswick were about 8,000 Wheatears (Davis 1966); several hundred Wheatears were present on the dunes at Winterton, Norfolk, on the morning of 4 September 1965, but almost all departed that night (Seago 1967). Davis comments that the fall of 3 September 1965 was by far the heaviest of its kind ever recorded. It is likely that most of these birds came from Scandinavia or other parts of northeastern Europe, and that originally they were heading southwest but were caught by a belt of heavy rain moving northwest across the southern North Sea.

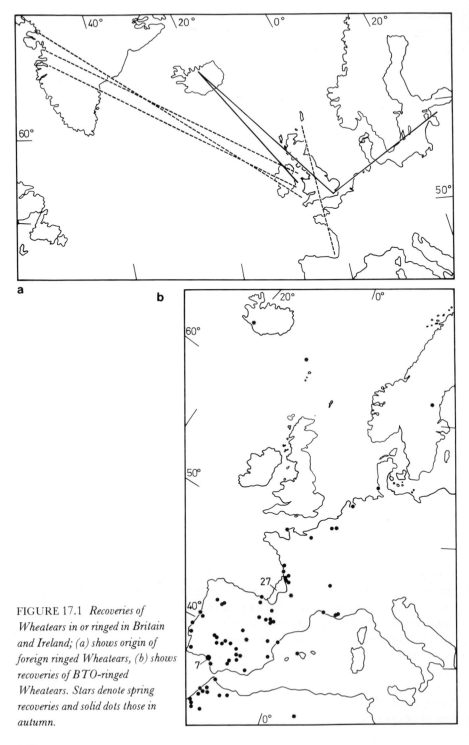

FIGURE 17.1 *Recoveries of Wheatears in or ringed in Britain and Ireland; (a) shows origin of foreign ringed Wheatears, (b) shows recoveries of BTO-ringed Wheatears. Stars denote spring recoveries and solid dots those in autumn.*

280

Once the Wheatears have left Skokholm they move in a southwesterly or south-southwesterly direction, as do populations of Wheatears from other parts of northwestern Europe. Incidentally, the Greenland Wheatears which have travelled southeast across the Atlantic to the western seaboard of Europe now change direction and, like other Wheatears, fly southwest towards their winter quarters in Senegambia. Figure 17.1 from *Ringing and Migration* shows (a) the origin of foreign-ringed Wheatears in Britain from the northwest and (b) the recoveries of BTO-ringed Wheatears. Note the numerous records in southwest France.

The BTO's ringing recoveries in western Europe and North Africa have recently been summarised in *BWP* (Cramp 1988). The authors state that, up to 1982, 111 autumn recoveries of Wheatears ringed as adults in Britain showed strong south and southwesterly directional trends, with 98% being recovered in southwest France, the western half of Spain, Portugal and Morocco. Twenty-two autumn recoveries of young ringed in Britain showed the same pattern. The southwest to south-southwest heading was also found in Wheatears ringed in Norway and in West and East Germany.

A Wheatear recovered near Redhill, Surrey, and ringed in Estonia at 24°E shows how far east some birds passing through Britain originate.

This general tendency to fly southwest takes most of the Wheatears away from the Pyrenees and to the west of the Atlas Mountains. Extreme western populations of Wheatears probably move south or even south-southeast (Zink 1973). There have been some recoveries of British-ringed birds in North Africa: from Morocco in both spring and autumn, and a few from Algeria which have, so far, all been in the spring (Mead and Hudson 1984).

Ringing recoveries from western Europe give us very little idea, with one exception, of the rate at which Wheatears travel. The exception is of a bird which we ringed on Skokholm on 16 August 1947 at 12.00 GMT. Forty-three hours later it was shot near Cap Breton, having travelled 940 km (585 miles). We do not know by what route this bird travelled nor at what time of day or night — probably at night. If it flew at about 32-40 kph (20-25 mph), it would have taken between 23 and 30 hours and it could have done it in one hop.

Wheatears pass through Malta from mid August to early November, with concentrations of up to 300 individuals being occasionally recorded (Sultana *et al.* 1975). The main passage dates for the Morocco end of the Sahara are from August to mid November, and for Algeria from August to October, and, having crossed the Sahara, the birds arrive in Senegal, Gambia, northern Nigeria and northern Cameroons and Mali and other more southern countries a little later (Tye, in press a).

The movement south from the breeding quarters to winter quarters is apparently slower than the spring movement northwards: areas north of the Sahara do not seem to be clear of migrant Wheatears until December. One still finds remarkably speedy flights: individual birds may well behave in much the same way as they did in spring, by making flights of several hundred, even a thousand, miles (1,600 km) and then resting for several days before continuing. I showed, on Alderney, that individual migrants rested for more days on average at each stop-over on their way south to winter quarters than when they were travelling north to breed. These longer intervals in autumn may well slow down the movement south. Once the Wheatear leaves northern Europe, there is perhaps very little need for a speedy withdrawal; practically all the time the bird is moving into a warmer climate.

FROM THE NORTHEAST

From the eastern and central parts of its breeding range the return of the Wheatear is not particularly well documented. Gabrielson and Lincoln (1959) show that Wheatears leave Alaska in the last half of August, even though some remain until October (latest date 21 October). More recently, Kessel and Gibson (1978) have collected more information and show that it is a fairly common migrant in western and central Alaska.

Oddly enough, we have a record of a Wheatear killed north of the Bering Strait on Captain James Cook's last voyage on the *Discovery* between August and September 1778 and presumably on its way to Africa (Stresemann 1949). In Russia, Dementiev and Gladkov (1968) give dates for the departures of Wheatears from various parts of the country; for the Anadyr Peninsula they state that Wheatears begin their return in late August and early September — indeed, that date is similar for many parts of Europe and for Skokholm itself! From northern Mongolia, Kozlova (1933) writes that the species leaves in the beginning of October.

Moreau (1972) suggests that birds migrating great circle from Yakutsk (130°E), which is on the approximate route that I would have expected Wheatears from Alaska to follow, pass east of the Trans-Caspian deserts and have a chance to feed in the central Asian oases (Tashkent) and the vegetated rims of Iran. South of this, these birds must cross the worst parts of the Arabian deserts. Further on still, the terrain over which migrating Wheatears have yet to pass remains inhospitable, but Moreau points out that they may derive some assistance from northeasterly winds and he considers that for any migrants crossing the nothern Indian Ocean towards Africa they are absolutely vital.

It would appear from the records available that the Wheatear does in fact keep very well to the north of the 40° parallel in China, avoiding, as in the spring, the high mountainous areas of Tibet and, as suggested by Moreau, perhaps coming down south through Iran and Arabia.

Ali and Ripley (1973), while saying that there are only a few sightings of Wheatears from the Indian subcontinent, suggest that the autumn movement to Africa hinges on the Himalayas. They mention a few records at Gilgit, Chitral and in Baluchistan, where, at Quetta, Meinertzhagen reported taking a female Wheatear on 18 October.

In North Africa and southern Arabia, Wheatears would seem to be most commonly reported as passage migrants in September and October; in south

Arabia on 4 September and 20 November; at Jedda frequently in September and October. There is a passage through Egypt, over open desert as well as down the Nile Valley: it appears to migrate east to west through the Sudan, perhaps to the Sahel zone of west and central Africa, and east of Lake Victoria to East Africa (Tye, in press a). The dates at which migrant Wheatears arrive in or pass through East Africa seem to compare with dates at the equivalent latitude in the west of that continent: first arrivals early in September, noticeable passage throughout October and decreasing in November.

Thus, from their almost circumpolar breeding distribution, Wheatears return to winter chiefly in Africa between 10° and 20°N in the west and south to about 30°S in the east (see Figure 3.1), although a few do winter regularly in the Euphrates and Tigris basin, Iraq (Cramp 1988). Nevertheless, some stragglers fail to arrive. Although a few individuals winter in the more southern states of America, there is, as yet, no sign of a regular wintering area for Wheatears in the New World.

FIGURE 17.2

Other individuals may be found in winter in Arabia, India, Mongolia and northern China, and rarely in eastern China and the Philippines (AOU 1983). Many of these southerly records presumably refer to single birds, such as a Wheatear in Borneo (Harrison 1951), which must have been displaced far to the south of their normal return migration.

It remains incredible to me that this small bundle of muscles, blood and fragile bones weighing about 25 g can fly the 15,000 or so miles (24,000 km) to Alaska, over harsh and inhospitable terrain; or, travelling northwest from its winter quarters, over 1,000 or so miles (1,600 km) of desert and followed by 2,000 or 3,000 miles (3,200-4,800 km) of sea. Then, after a few weeks in which it performs a whole range of activities, taking little rest, it returns the same distance with the same dangers. What for? To raise 5.5 nestlings who are condemned to the same routine as long as they survive! It is probably the longest-distance migration of any passerine. It almost defies belief.

BIBLIOGRAPHY

ALEXANDER, C.J. 1917. Notes on the zonal distribution in the mountans of Latium, Italy. *Brit. Birds* II: 71-82.
ALI, SALIM, and S. DILLON RIPLEY. 1973. *Handbook of the Birds of India and Pakistan.* Oxford University Press.
ALTUM, B. 1868. *Der Vogel and sein Leben.*
AMERICAN ORNITHOLOGISTS' UNION. 1983. *Checklist of North American Birds.*
ANDREWS, J.O. 1969. Notes on some palaearctic migrants in the Gambia. *Nig. Orn. Soc. Bull.* 6: 94.
ARMSTRONG, E.A. 1942. *Bird Display and Behaviour.* Lindsay Drummond.
—— 1955. *The Wren.* Collins.
—— 1973. *A Study of Bird Song.* Dover Publications.
ARO, M. 1968. Kaksi bigamica — Tapausta kivitarkulla. *Oenanthe oenanthe. Ornis Fennica* 45: 16-18.
ASBIRK, S. and N-E FRANZMANN. 1979. Observations on the diurnal rhythm of Greenland Wheatears *Oenanthe oe. leucorrhoa* Gm in continuous daylight. *Dansk Orn. Foren. Tidsskr.* 73: 95-102.
AXELL, H.E. 1954. The Wheatear on Dungeness. *Bird Notes.* 36: 38-41.
BAILEY, A.M. 1948. *Birds of Arctic Alaska.* Colorado Museum of Natural History.
BAIRLEIN, F. 1988. How do migratory songbirds cross the Sahara. *Tree* 8: 191-4.
BAKER, R.R. 1978. *The Evolutionary Ecology of Animal Migration.* Hodder & Stoughton.
BALDWIN, S.P. and S.C. KENDEIGH. 1932. Physiology of the temperature of birds. *Scient. Publ. Cleveland. Mus. Nat. Hist.* 3: 1-196.
BANNERMAN, D.A. 1936. *The Birds of Tropical West Africa.* London.
BELLAIRS, R. 1960. *Development in Birds*: in Marshall (1960).
BERK, K.H. 1953. Nachweis der zweiten Brut des Steinschmätzers durch beringte Tiere. *Vogelring* 22: 17 ff.
—— 1961. Bemerkungen zur Biologie des Steinschmätzers. *Vogelwelt* 82: 109-12.
BEWICK, T. 1805. *British Birds.*
BILLHAM, E.G. 1938. *The Climate of the British Isles.* Macmillan.
BIRD, C.G. and BIRD, E.G. 1941. The birds of north east Greenland. *Ibis* (14) 4: 129.
BLAIR, J. (N.D.) *Birds of the Field and Garden.* Chambers.
BOERTMANN, D. 1979. Ornithologische Observationen i Vest Grønland i somme 1972-77. *Dansk. Orn. Foren. Tidsskr.* 73: 171-6.
BOLAM, G. 1912. *Birds of Northumberland and the Eastern Borders.* Alnwick.
BORRER, W. 1891. *The Birds of Sussex.* London.
BORRETT, R.P. and H.D. JACKSON. 1970. The European Wheatear *Oenanthe oenanthe* (L) in southern Africa. *Bull. BOC.* 90: 124-9.
BOSWALL, J.H.R. 1951. Some autumn observations of the Wheatear. *Skokholm Bird*

Observatory Report 1951: 28-31.

BROOKE, M. DE L. 1979. Differences in the quality of territories held by Wheatears (*Oenanthe oenanthe*). *J. Anim. Ecol.* 48: 21-32.

—— 1981. How an adult Wheatear (*Oenanthe oenanthe*) uses its territory when feeding its nestlings. *J. Anim. Ecol.* 50: 683-96.

BROWNE, K. and E. BROWN 1956. Analysis of the weights of birds trapped on Skokholm. *Brit. Birds* 49: 241-57.

BUNDY, G. 1976. *The Birds of Libya.* British Ornithologists' Union Check list No. 1.

BUXTON, J. 1950. *The Redstart.* Collins, London.

CAMPBELL, B. 1949. *Natural History*: in North, F.J., Campbell, B., and Scott, R., *Snowdonia.* Collins, London.

—— 1953. *Finding Nests.* Collins.

—— and J. FERGUSON-LEES. 1972. *A Guide to Bird's Nests.* Constable.

—— and E. LACK. (Eds.) 1985. *A Dictionary of Birds.* T and A. D. Poyser.

CARLSON, A., L. HILLSTRÖM and J. MORENO. 1985. Mateguarding in the Wheatear *Oenanthe oenanthe. Ornis Scandinavica* 16: 113-20.

CARRUTHERS, D. 1910. The Birds of the Zarafaschen Basin in Russian Turkestan. *Ibis* 53: 461.

CASEMENT, M.B. 1979. Ocean weather ship reports on landbirds. *Sea Swallow* 29: 20.

CATCHPOLE, C.K. 1985. *Vocalization*: in Campbell and Lack (1985).

CLEMENT, P. 1987. Field identification of West Palearctic wheatears. *Brit. Birds* 80: 137-59. 187-238.

CLOUDESLEY-THOMPSON, J.L. 1975. *Terrestrial Environments.* Croom Helm.

COLE, S.J. 1980. Wheatear Movements. *Birds in Cornwall* 1980 (50):78-9.

CONDER, P.J. 1948a. The breeding biology and behaviour of the Continental Goldfinch (*Carduelis carduelis carduelis*). *Ibis* 90: 493-525.

—— 1948b. The Common Wheatear. *Skokholm Bird Observatory Report for 1948*: 6.

—— 1949. Individual distance. *Ibis* 91: 649-55.

—— 1950. Notes on the dancing display of the Wheatear. *Brit. Birds* 43: 299.

—— 1951. Wheatears on Skokholm in 1951. *Skokholm Bird Observatory Report for 1951*: 27.

—— 1952. The factors affecting the distribution of Wheatears on Skokholm. *Skokholm Bird Observatory Report for 1952*: 15.

—— 1953a. Some individual feeding habits of gulls breeding on Skokholm, *Skokholm Bird Observatory Report* 1952: 30-4.

—— 1953b. A List of the Birds of Skokholm, Pembrokeshire. *North Western Naturalist* 24: 211-19.

—— 1954. The hovering of the Wheatear. *Brit. Birds* 47: 76-9.

—— 1956. The territory of the Wheatear *Oenanthe oenanthe. Ibis* 98: 453-59.

CORNISH, A. V. 1947. *Twentieth Report of the Devon Bird Watching and Preservation Society*: 20.

CORNWALLIS, LINDON. 1975. The comparative ecology of eleven species of Wheatear (*genus Oenanthe*) in SW Iran. Thesis presented in the University of Oxford for the Degree of Doctor of Philosophy.

COTT, H.B. 1947. The edibility of birds. *Proc. Zoo. Soc. London* 161: 371-524.

COWARD, T.A. 1920. *The Birds of the British Isles.* Warne.

COX, MACHELL A.H. 1921. The breeding habits of the Wheatear. *Brit. Birds* 15: 140.

CRAMP, S. (Ed.) 1988. *The Birds of the Western Palearctic, Vol. 5.* Oxford University Press.

—— and P. J. CONDER. 1970. A visit to the oasis of Kufra. *Ibis* 112: 261-3.

—— and K. E. L. SIMMONS (Eds.). 1977. *The Birds of the Western Palearctic*, Vol. 1, Oxford University Press.

DARLING, F.F. 1947. *Natural History in the Highlands and Islands.* Collins, London.

DAVIES, N.B. 1978. In J.R. Krebs and N.B. Davies. *Behavioural Ecology.* Blackwell's Scientific Publications.

DAVIS, P.E. 1954. List of Birds. *Skokholm Bird Observatory Report for 1954.* West Wales Field Society.

——— 1966. The great immigration of early September, 1965. *Brit. Birds* 59: 353-76.

DEMENTIEV, G. P. and N. A. GLADKOV. 1966-1968. *Birds of the Soviet Union.* Israel Program for Scientific Translations.

DOWSETT, R.. 1968. Migrants at Malam' Fatori, Lake Chad, Spring 1968. *Nig. Orn. Soc. Bull.* 5: 53-6.

DRAULANS, D. and J. VAN VESSEM. 1982. Flock size and feeding behaviour of migrating Whinchats *Saxicola rubetra. Ibis* 124: 347-51.

DRENT, R. H. 1973. The Natural History of Incubation. In D. S. Farner (Ed.) *Breeding Biology of Birds.* National Academy of Sciences. Washington D.C. pp. 262-311.

DRESSER, H. E. 1871-1881. *A History of the Birds of Europe.* London.

EAGLE CLARKE, W. 1912. *Studies in Bird Migration.* Gurney and Jackson. 2 vols.

EASTWOOD, E. 1967. *Radar Ornithology.* Methuen.

EBBUT, D. P. 1967. Vocal mimicry in the Red-breasted Chat *Oenanthe heuglini. Nig. Orn. Soc. Bull.* 4: 36-7.

EDWARDS, G., E. HOSKING and S. SMITH 1950. Dancing display of the Wheatear. *Brit. Birds* 43: 9-10.

EGGEBRECHT, E. 1943. Beitrag zu Brutbiologie des Noon-Steinschmätzers (*Oenanthe p. pleschanka* Lepechin). *Orn. Monatsbericht.* 51: 127-35.

EIBL-EIBESFELDT, I. 1970. *Ethology.* New York.

ELLIOTT, H. F. L., and N. R. FUGGLES-COUCHMAN. 1948. An ecological survey of the birds of the crater highlands and rift lakes, Northern Tanganyika territory. *Ibis* 90: 394-425.

ETCHÉCOPAR R. D., and F. HÜE. 1983. *Oiseaux de Chine,* Vol 2. Paris.

FARNER, D. S. and J. R. KING. 1961. *Avian Biology. Vol. 1.* Academic Press.

FEIGE, R. 1957. Steinschmätzer brütet unter einen Weiche. *Der Falke* 4: 139.

FERNS, P. N. F. 1985. *Yawning:* in Campbell and Lack (1985).

FICKEN, M. S. 1962. Maintenance activities of the American Redstart. *Wilson Bull.* 74: 153-65.

FORBES, A. R. 1905. *Gaelic names of Beasts (Mammalia), Birds, Fishes, Insects, Reptiles, etc.* Oliver and Boyd, Edinburgh.

FORBES, J. R. 1938. Recent observations on the Greenland Wheatear. *Auk* 55: 492-5.

FORD, E. B. 1955. *Moths.* Collins, London.

FREUCHEN, P. and F. SALOMONSEN. 1960. *The Arctic Year.* Readers' Union and Jonathan Cape.

FRY, C. H. 1970. Birds in the Waza National Park, Cameroons. *Nig. Orn. Soc. Bull.* 7: 1-5.

GABRIELSON, I. N. and F. C. LINCOLN. 1959. *Birds of Alaska.* Stackpole Co. and Wildlife Management Institute, Washington, D. C.

GASTON, A. J. 1985. *Cooperative Breeding:* In Campbell and Lack (1985).

GATTER, W. 1961. Der Steinschmätzer (*Oenanthe oenanthe*) an einem Wohnhaus. *Orn. Mitt.* 13: 193-4.

GERBER, R. 1956/58. Singende Vogelweibchen. *Beiträge Vogelk.* 5: 36-45.

GIBB, J. A. 1950. The breeding biology of Great and Blue Titmice. *Ibis* 92: 507-39.

GILLHAM, M. E. 1956. Ecology of the Pembrokeshire Islands. IV. Effects of treading and burrowing by birds and mammals. *J. Anim. Ecol.* 44: 51-82.

GLADKOV, N. A. 1957. *Birds of the Mangyshlak Peninsula (Caspian). Ibis* 99: 269-74.

GLOCK, E. 1964. Grönlandsstenskvätta (*Oenanthe oenanthe leucorrhoa*) iakttagen via Abisko i slutet av augusti? *Vår Fågelvärld* 23: 348-87.

GLUTZ VON BLOTZHEIM, U. N. 1962. *Die Brutvögel der Schweiz.* Aaran, Switzerland.

GODFREY, W. E. 1986. *Birds of Canada* (Revised). National Museum of Canada. Bull. No. 203. Biological Series No 73.

GOODMAN, G. T., and M. E. GILLHAM. 1954. Ecology of the Pembrokeshire Islands. II. Skokholm, environment and vegetation. *J. Ecol.* 42: 297-327.

GOODMAN, S.M. and A.L. AMES. 1983. A contribution to the ornithology of the Siwa Oasis and the Qattara depression, Egypt. *Sandgrouse* 5: 82-96.

GORDON, SETON. 1942. Wheatear hovering. *Brit. Birds* 36: 73.

HAAS, W., and P. BECK, 1979. Zum Frühjahrzug paläarktische Vögel über die westliche Sahara. *J. Orn.* 120: 237-46.

HÅKANSSON, E., O. BRENNIKE, P. MOLAARD and P. FRYKMAN. 1981. Nordgrønlanske fugle observationer – Somrene 1976 og 1978. *Dansk Orn. Foren. Tidsskr.* 75: 51-67.

HALL, B.P., and R. MOREAU. 1970. *An Atlas of Speciation in African Birds*. British Museum (Nat. Hist.), London.

HANTGE, E. 1958. Frühjahrs Durchzug des Steinschmätzers *Oenanthe oenanthe*. *Vogelwelt* 79: 149-54.

HARRISON, C.J.O. 1975. *A Field Guide to the Nests, Eggs and Nestlings of British and European Birds*. Collins.

―――― 1985. *Plumage, abnormal*: in Campbell and Lack (1985).

HARRISSON, T. 1951. Two additions to the Borneo List, *Ibis* 93: 311-12.

HARTHAN, A.J. 1958 Thirty Years Ago. *Nature in Wales* 4: 578-80.

HARTLEY, P.H.T. 1949. The biology of the Mourning Chat in winter quarters. *Ibis* 91: 393-413.

HEDIGER, H. 1950. *Wild Animals in Captivity*. Butterworth, London.

―――― 1964. *Wild Animals in Captivity*. Dover.

HEINROTH, O., and M. HEINROTH 1928. *Die Vögel Mitteleuropas*. Berlin.

HEMPEL, C. 1957. Vom Zug des Steinschmätzers (*Oenanthe oenanthe*). *Vogelwarte* 19: 25-36.

HINDE, R.A. 1956. The biological significance of the territories of birds. *Ibis* 98: 340-69.

HOWARD, H.E. 1920. *Territory in Bird Life*. Murray, London.

HOWARD, R. and A. MOORE. 1980. *A Complete Checklist of the Birds of the World*, Oxford University Press.

HUDSON, W.H. 1900. *Nature in Downland*. London.

INGRAM, I. 1966. *In Search of Birds*. Witherby, London.

JACKSON, C.E. 1968. *British Names of Birds*, Witherby.

JACKSON, F.J. 1938. *The Birds of Kenya Colony and the Uganda Protectorate*. London.

JELLIS, R. 1977. *Bird Sounds and their Meanings*. BBC.

JENKINS, D.W. 1944. Territory as a result of despotism and social organism in geese. *Auk.* 61: 38.

JENNING, W. 1954. Polygami hos stenskvätta *Oenanthe oenanthe*. *Vår Fågelvärld* 13: 167-71.

JOST, O. 1960. Steinschmätzer *Oenanthe oenanthe* brütet in Wohnhaus. *Orn. Mitt.* 12: 10-11.

KEITH, G.S., E.K. URBAN and C.H. FRY (Eds.) In press. *The Birds of Africa. Vol 4*. Academic Press, London.

KESSEL, B. and D.G. GIBSON. 1978. *Status and Distribution of Alaska Birds*, Cooper Ornithological Society.

KING, B. 1968. Wheatears on autumn passage feeding on bumble bees. *Brit. Birds.* 61: 315.

KING, J.R., and D.S. FARNER. 1960. *Energy Metabolism, Thermoregulation and Body Temperature*: in Marshall A.J. (1960).

KISHCHINSKII, A.A. 1974. Arctic-alpine avifauna and its origin. *Zoolog. Zhurnal* 53: 1036-51. [English translation by British Library Lending Division]

KLOPFER, P. 1963. Behavioural aspects of habitat selection: the role of early experience. *Wilson Bull.* 75: 15-22.

KNEIS, P. 1983. Zum Fortragen von Eiern beim Steinschmätzer *Oenanthe oenanthe*. *Beitr. Vogelk.* 29: 118-20.

KÖNIG, C. 1965. Das 'Hohlenzeigen' des Steinschmätzers *Oenanthe oenanthe*. *J. Orn.* 106: 350-2.

KOSHELEV, N.K. 1971. [The attraction of Wheatears to artificial nests.] *Dokl. Mosk. Obsh.*

Isp'yt Priv. Zool, Bot. 1968-1969: 126-7. [In Russian]

KOZLOVA, E. V. 1933. The Birds of Southwest Transbaikalia, Northern Mongolia and Central Gobi, part v. *Ibis* (13 Series) 3: 301:32.

LACK, D. 1937. The psychological factor in bird distribution. *Brit. Birds* 31: 130-36.

—— 1954. *The Natural Regulation of Animal Numbers.* Oxford.

—— 1965. *The Life of the Robin.* Witherby, (4th Ed.)

—— and L. S. V. VENABLES. 1937. The heathland birds of the South Haven peninsula, Studland Heath, Dorset. *J. Anim. Ecol.* 6: 62-72.

LANGSLOW, D. 1976. *Weights of Blackcaps on migration.* Ringing and Migration 1: 78-9.

LEE, S. L. B. 1963. Migration in the Outer Hebrides studied by Radar. *Ibis* 105: 493-515.

LEISLER, B., G. HEINE, and K.-H. SIEBENROCK, 1983. Einnischung und interspezifische Territorialität Steinschmätzer *Oenanthe isabellina, O. oenanthe, O. pleschanka* in Kenia. *J. Orn.* 124: 393-413.

LIPPENS, L. and H. WILLE. 1972. *Atlas des Oiseaux de Belgique et d'Europe Occidentale.* Belgium.

LISTER, D. 1953. Secondary song: a tentative classification. *Brit. Birds* 46: 139-43.

LLOYD, B. 1933. *Trans. Herts. Nat. Hist. Soc.* 19: 135-9.

—— 1938. The Greenland Wheatear on spring migration at Elstree. *Trans. Herts. Nat. Hist. Soc.* 20: 335-9.

LOCKLEY, R. M. 1942. *Shearwaters.* Dent.

LÖHRL, H. 1959. Zur Frage des Zeitpunktes einer Prägung auf die Heimatregion beim Halsbandschnäpper *Ficedula albicollis. J. Orn.* 100: 132-40.

LUDWIG, E. 1965. Freibrütender Steinschmätzer *Oenanthe oenanthe. Orn. Mitt.* 17: 85.

MACGILLIVRAY, W. 1839. *A History of British Birds.* London.

MARCHANT, S. 1941. Notes on the birds of the Gulf of Suez. Part 1. *Ibis* (14) 5: 265-95.

MARSHALL, A. J. 1950. The function of vocal mimicry in birds. *Emu* 50: 5-16.

—— 1960-61. *Biology and Comparative Physiology of Birds.* 2 vols. Academic Press.

MATTHEWS, L. HARRISON. 1982. *Mammals in the British Isles,* Collins.

MAYR, E. 1935. Bernard Altum and the territory theory. *Proc. Linn. Soc. N. Y.* 45-46: 24-38.

—— and E. STRESEMANN. 1950. Polymorphism in the chat genus *Oenanthe* (Aves). *Evolution* 4: 291-300.

MEAD, C. 1983, *Bird Migration.* Country Life Books.

—— and R. HUDSON. 1984. Report on bird ringing for 1983. *Ringing and Migration* 4: 153-92.

—— and —— 1985. Report on bird ringing for 1984. *Ringing and Migration* 6: 125-72.

MEINERTZHAGEN, R. 1920. Notes on the birds of southern Palestine. *Ibis* (11) 2: 195-259.

—— 1938. On the birds of northern Afghanistan. *Ibis* (14) 2: 480-520, 671-717.

—— 1954. *Birds of Arabia.* Oliver and Boyd.

MELTOFTE, H. 1972. Ornithological observations in the Norwegian Sea, the Greenland Sea and NE Greenland. *Dansk Orn. Foren. Tidsskr.* 66: 108-12.

—— 1976. Ornithologiske observationer i Scoresbysundområdet Øst Grønland, 1974. *Dansk Orn. Foren. Tidsskr.* 70: 107-22.

MENZEL, H. 1964. *Der Steinschmätzer.* A Ziemsen Verlag, Wittenberg Lutherstadt.

MOFFAT, C. B. 1903. The spring rivalry of birds. *Irish Naturalist* 12: 152-66.

MØLLER, A. M. 1984. On the use of feathers in birds' nests: prediction and tests. *Ornis Scandinavica* 15: 38-42.

MONK, J. F. 1950. Dancing display of Wheatears. *Brit. Birds* 43: 10.

MONTAGU, G. 1802. *Ornithological Dictionary.* London.

—— 1813. *Supplement to the Ornithological Dictionary.* Exeter.

MOREAU, R. E. 1952. The place of Africa in the Palearctic migration system. *J. Anim. Ecol.* 21: 250-71.

—— 1972. *The Palearctic–African Bird Migration Systems.* Academic Press, London.

MOREL, G., and F. ROUX. 1966. Les migrateurs paléarctiques au Sénégal. *Terre et Vie* 113: 19-72, 143-76.

MORENO, J. 1984. Search strategies of Wheatears (*Oenanthe oenanthe*) and Stonechats (*Saxicola torquata*): adaptive radiation in perch, height, search time, sally distance and inter-perch move-length. *J. Anim. Ecol.* 53: 147-59.

MORSE, D. H. 1980. *Behavioural Mechanisms in Ecology*. Harvard University Press.

MUDGE, G. F., C. H. CROOKE, R. G. BOOTH and S. E. A. SMITH. 1979. *An Ecological Study of Breeding Bird Populations and Vegetation on Open Moorland Areas of Dartmoor, 1979*. RSPB.

NELSON, T. H. 1907. *The Birds of Yorkshire*. London.

NETHERSOLE THOMPSON, D. and M. NETHERSOLE THOMPSON 1979. *Greenshanks*. T. and A. D. Poyser.

NICE, M. M. 1937. *Studies in the Life History of the Song Sparrow*. 1. *Trans Linn. Soc. N. Y.* vol. 4.

—— 1943. *Studies in the Life History of the Song Sparrow*. 2. *Trans. Linn. Soc. N. Y.* vol. 6.

NICHOLSON, E. M. 1927. *How Birds Live*. Witherby, London.

—— 1928. *Bird Notes from the North Atlantic*. Brit. Birds. 22: 122-33.

—— 1930. Field notes on Greenland birds. *Ibis* (Series 12) 6: 280-314, 395-420.

—— 1977: in Cramp and Simmons (1977).

—— and L. KOCH. 1936. *Songs of Wild Birds*. Witherby, London.

NIELSON, BENT PORS. 1979. Finn Salomonsens Arbejde med ringmaerkning og fuglefredning i Grønland. *Dansk Orn. Foren. Tidsskr.* 73: 13-24.

NIETHAMMER, G. 1937. *Handbuch der Deutschen Vogelkunde*. Leipzig.

NOBLE, G. K. 1939. The role of dominance in the social life of birds. *Auk* 56: 263-73.

NORTH, M. E. W. 1950. Transcribing bird-song. *Ibis* 92: 99-114.

O'CONNOR, R. J. 1985. *Growth*: in Campbell and Lack (1985).

ORIANS, G. H. 1969. On the evolution of mating systems in birds and mammals. *American Naturalist* 103: 589-603.

PANOV, E. N. 1974. *Die Steinschmätzer des nördlichen Paläarktis. Gattung Oenanthe*. Ziemson Verlag, Wittenberg Lutherstadt.

PENNANT, T. 1766. *British Zoology*. London.

PETTIT, R. G. and D. V. BUTT. 1950. Dancing display of the Wheatear. *Brit. Birds*. 43: 298.

PEUS, F. 1952. Steppenvögel mitten in Berlin. *Vogelwelt*. 73: 1-6.

PILCHER, C. W. T., and A. TYE. (in prep.) Status and distribution of the birds of Kuwait.

PITELKA, F. A. 1959. Numbers, breeding schedule, and territoriality in Pectoral Sandpipers in Northern Alaska. *Condor* 61: 233-64.

PLESKE, T. 1928. *Birds of the Eurasian Tundra*. Boston.

PORTENKO, L. A. 1939. [Fauna of the Anadyr Region] Pts 1-11. Birds. *Tr.N.-i. inst. polyarn. zemled., zhivot.i promysl.knoz., ser. Promsyl. khoz.' lzd Glavsevmorputi*. 1939. Nos. 5 and 6.

PORTER, R. F. 1983. The autumn migration of passerines and near-passerines at the Bosphorous, Turkey. *Sandgrouse* 5: 45-74.

POULSEN, H. 1958. A study of anting behaviour in birds. *Dansk. Orn. Foren. Tidsskr.* 50: 267-98.

POUNDS, H. E. 1942. Wheatear hovering. *Brit. Birds*. 36: 94.

RAVEN, J. and S. M. WALTERS 1956. *Mountain Flowers*. Collins.

RAY, J. 1674. *A Catalogue of English Words . . . with catalogue of English birds and fishes*. London.

—— 1687. (Ed.) *The Ornithology of Francis Willoughby*. London.

RICKLEFS, R. E.1968. Patterns of growth in birds. *Ibis* 110: 419-51.

—— and F. R. HAINSWORTH. 1969. Temperature regulation in nestling Cactus Wrens: the nest environment. *Condor* 71: 32-7.

RIVIÈRE, B. B. 1930. *A History of the Birds of Norfolk*, Witherby.

—— 1933. Ornithological report for Norfolk for 1932, *Brit. Birds* 26: 318-9.

—— 1934. Ornithological report for Norfolk for 1933. *Brit. Birds.* 27: 310.

—— 1935. Ornithological report for Norfolk for 1934. *Brit. Birds* 28: 354.

ROBSON, R. W. and K. WILLIAMSON 1972. The breeding birds of a Westmorland Farm. *Bird Study* 19: 202-14.

RUSSELL, SIR JOHN 1957. *The World of the Soil.* Collins.

RUTHKE, P. 1954. Beobachtungen am Steinschmätzer. *Vogelwelt* 75: 188-91.

SALOMONSEN. F. 1927. The distribution of the Wheatear in Denmark. *Ibis* (12th Series) 3: 302-6.

—— 1950. *Grønlands Fugle.* Copenhagen.

—— 1971. Tolvete foreløbige liste over genfunde Grønlandske ringfugle. *Dansk Orn. Foren. Tidsskr.* 65: 11-19.

—— 1979. Trettende foreløbige liste over genfundne Grønlandske ringfugle. *Dansk Orn. Foren. Tidsskr.* 73: 191-206.

SALOMONSON, M. G. and R. BALDA. 1977. Winter territoriality of Townsend's Solitaires *Myadestes townsendi* in Piñon juniper Ponderosa pine ecotone. *Condor* 79: 148-61.

SAUNDERS, H. 1889. *An Illustrated Manual of British Birds.* London.

SAXBY, H. L. 1874. *The Birds of Shetland,* Edinburgh.

SCHINZ, J. 1930. Mitternachtssonne und Vogelgesang. *Beitr. Fortpfl. Vögel.* 6: 133-4.

SCHOENER, T. W. 1968. Sizes of feeding territories among birds. *Ecology* 49: 123-41.

SCHULZ, R. 1963. The winter of 1961-2 at Ottenby, Ottenby Bird Station Report. No. 38. *Vår Fågelvärld* 22: 154-60.

SCLATER, W. L. 1912. *A History of the Birds of Colorado.* Witherby.

SCZLIVKA, L. 1962. Periodično gnezdenje Kamenjara običnog. *Oenanthe oneanthe,* na Područsu Bačke Topole. *Larus* 14: 135-9.

SEAGO, M. 1967. *Birds of Norfolk.* Jarrolds, Norwich.

SELOUS, E. 1901. *Birdwatching.* Dent, London.

SEVINGA, W. 1968. Tapuit met witte Kop. *Limosa* 41: 159.

SHERIF BAHA EL DIN. 1984. Notes on the breeding of the Sooty Falcon *Falco concolor* on islands in the Red Sea and their prey. *Bulletin 12.* Orn. Soc. of the Middle East.

SIMMONS, K. E. L. 1985. *Sunning:* in Campbell and Lack (1985).

—— 1986. *The Sunning Behaviour of Birds.* Bristol Ornithologists Club.

SIMMS, E. 1978. *British Thrushes.* Collins, London.

SKUTCH, A. F. 1976. *Parent Birds and their Young,* University of Texas Press.

SNOW, D. 1953. The migration of the Greenland Wheatear. *Ibis* 95: 376-8.

—— 1958. *A Study of Blackbirds,* Allen and Unwin.

—— 1969. The moult of British thrushes and chats. *Bird Study* 6: 115-29.

SNYDER, L. L. 1957. *Arctic Birds of Canada.* Toronto.

SORAGO, D. M. 1962. Exposed nesting site of the Wheatear. *Aquila* 67/68: 257.

SPENCER, R., and R. HUDSON. 1978. Report on bird ringing for 1976. *Ringing and Migration* 1: 189-252.

—— and —— 1982. Report on bird ringing for 1981. *Ringing and Migration* 4: 65-128.

STADLER, H. 1950-51. Stimmen der Balkanvögel.v.1. Die Stimmen der mittel- und südeuropäischen Steinschmätzer. *Oenanthe oenanthe.* *Larus* 4-5: 149-84.

STANFORD, J. K. 1953. Some impressions of spring migration in Cyrenaica, March-May. *Ibis* 95: 316-28.

STEJNEGER, L. 1901. On the Wheatear (*Saxicola*) occurring in north-west America. *Proc. U. S. Nat. Mus.* 23: 473.

STEVENSON, H. 1866. *The Birds of Norfolk.* London.

STRESEMANN, E. 1949. Birds collected in the North Pacific area during Capt. James Cook's last voyage (1778 and 1779). *Ibis* 91: 244-55.

—— 1950. Interspecific competition in chats. *Ibis* 92: 148.

SULTANA, J., C. GAUCI and M. BEAMAN 1975. *A Guide to the Birds of Malta.* The Malta Ornithological Society.

SUTTER, E. 1950. Über die Flughöhe ziehender Vögel. *Orn. Beob.* 47: 174.

SUTTON, G. M. and D. F. PARMELEE 1954. Nesting of the Greenland Wheatear on Baffin Island. *Condor* 56: 295-306.

SVENSSON, L. 1975. *Identification Guide to European Passerines* (Second, revised edition). Naturhistoriska Riksmuseet, Stockholm.

SWANN, H. K. 1913. *A Dictionary of English and Folk Names of British Birds.* Witherby.

SZCZUDLOWSKI, T. 1964. Spostrzonia nad zachowansem sie bialorzytki *Oenanthe oenanthe* L. [Observations on the behaviour of the Wheatear *Oenanthe oenanthe*] *Przitglad Zoologiczny* 7: 274-6. [In Polish with English summary.]

SZLIVKA, L. 1962. *Periodische Brütendes Steinschmätzers Oenanthe oenanthe in Backa Topola.* Larus 14: 135-9.

TANSLEY, A. G. 1939. *The British Isles and their Vegetation.* Cambridge.

THIBAULT, J.-C. 1979. *Parc Naturel Régional De Corse — Les Oiseaux.* Ajaccio.

THIELCKE, G. A. 1976. *Bird Sounds.* Ann Arbour Science Library, Michigan.

THOM, V. M. 1986. *Birds in Scotland.* T. and A. D. Poyser, Calton.

THOMAS, J. F. 1922. Wheatears mobbing a weasel. *Brit. Birds* 16: 22.

—— 1925. Some results and methods of marking Wheatears. *Brit. Birds* 19: 98.

—— 1926. Marked Wheatears in three different localities. *Brit. Birds* 20: 24.

—— 1950. Dancing display of the Wheatear. *Brit. Birds* 43: 298-9.

THORPE, J. 1987. Mass migration of Wheatears. *Peregrine* 6(2): 82-3.

THORPE, W. H. 1961. *Bird-song.* Cambridge University Press.

—— 1985. *Mimicry*: in Campbell and Lack (1985).

—— and P. M. PILCHER. 1958. The nature and characteristics of sub-song. *Brit. Birds* 51: 509-14.

TINBERGEN, N. 1936. The functions of sexual fighting in birds: and the problems of the origin of 'Territory'. *Bird-Banding* 7: 1-8.

—— 1939. Field Observations of East Greenland Birds (ii). The Behaviour of the Snow Bunting in Spring. *Trans. Linn. Soc. N.Y.* 5: 1-94.

—— 1953. *Social Behaviour in Animals.* Methuen.

TISCHLER, F. 1941. *Die Vögel Ostpreussens.* Königsberg.

TODD, W. E. CLYDE. 1963. *Birds of the Labrador Peninsula and adjacent areas.* University of Toronto.

TROTT, A. C. 1947. Notes on birds seen in the Lar valley in 1943 and 1944. *Ibis* 89: 231-4.

TURNER, W. 1544. *Avium Praecipuarum.*

TYE, A. 1980. The breeding biology and population size of the Wheatear *Oenanthe oenanthe* on the Breckland of East Anglia, with implications for its conservation. *Bull. Ecol.* t.11.3: 559-69.

—— 1982. *Social Organisation and feeding in the Wheatear and Fieldfare.* PhD thesis, Cambridge University.

—— 1984. Attacks by shrikes *Lanius* spp on Wheatears *Oenanthe* spp: competition, klepto-parasitism, or predation? *Ibis* 126: 94-101.

—— 1986. Plumage stages, moults, sexual dimorphism and systematic position of the Somali Wheatear *Oenanthe phillipsi. Bull. BOC.* 106:104-11.

—— (in press a) *Oenanthe oenanthe (Linnaeus).* Northern Wheatear. Traquet Motteux, in Keith, Urban and Fry (Eds.)

—— (in press b). Vocalisations and territorial behaviour of wheatears *Oenanthe* spp. in winter quarters. *Proc. 6th Pan-African Ornithological Congress.*

—— and H. TYE. 1983. Field identification of Wheatear and Isabelline Wheatear. *Brit. Birds.* 76: 427-37.

VAN IJZENDOORN, A. J. L. 1950. *The Breeding Birds of the Netherlands.* Leiden.

VENABLES, L. S. V. 1937. Bird distribution on Surrey Greensand heaths: the avifaunal botanical correlation. *J. Anim. Ecol.* 6: 73-85.

—— 1939. Bird distribution on the South Downs, and a comparison with that of Surrey Greensand heath. *J. Anim. Ecol.* 8: 227-37.

VINCENT, A. W. 1947. Breeding habits in some African birds. *Ibis* 89: 163-204.

VINCENT, D. F. 1988. *Wheatear: Food*: in Cramp (1988).

VOOUS, K. H. 1973-1977. *List of Recent Holarctic Bird Species*. BOU and Academic Press.

WALPOLE-BOND, J. 1938. *A History of Sussex Birds*. Witherby.

WASSERMAN, F. E. 1980. Territorial behaviour in a pair of White-throated Sparrows. *Wilson Bull.* 92: 74-87.

WILLIAMSON, K. 1948. Fair Isle Bird Observatory Report for 1948. *Scot. Nat.* 61: 19-31.

—— 1949. Hovering Flight. Fair Isle Bird Observatory Report for 1948. *Scot. Nat.* 61: 26.

—— 1952. Migrational drift in Britain in autumn 1951. *Scot. Nat.* 64: 1-18.

—— 1953. Migration into Britain from the north-west, Autumn 1952. *Scot. Nat.* 65: 65-94.

—— 1957a. The annual post nuptial moult in the Wheatear *Oenanthe oenanthe*. *Bird Banding* 28: 129-35.

—— 1957b. *Fair Isle Bird Observatory Reports for 1955-56*. Fair Isle Bird Observatory.

—— 1958. Bergmann's Rule and obligatory overseas migration. *Brit. Birds* 51: 209.

—— 1965. *Fair Isle and its Birds*. Edinburgh.

—— 1968. Bird Communities in the Malham Tarn Region of the Pennines. *Field Studies* 2: 651-68.

WITHERBY, H. F., F. C. R. JOURDAIN, C. B. TICEHURST and B. W. TUCKER. 1938. *The Handbook of British Birds*. Witherby, London.

WYNNE-EDWARDS. V. C. 1952. Zoology of the Baird Expedition (1950). 1). Birds observed in central and south-east Baffin Island. *Auk* 69: 353-91.

YAPP, W. B. 1970. *The Life and Organisation of Birds*. Arnold, London.

YARRELL, W. 1871 *A History of British Birds*. Van Doorst. London. 4th edition.

YEATMAN, L. J. 1976. *Atlas des Oiseaux de France*. Ministère de la Qualité de la Vie. Environement. Paris; édité par Soc. Orn. de France, Paris.

YOC. 1978. *YOC phone-in on spring migration 1978*. RSPB, Sandy, Beds.

—— 1979. *YOC phone-in Report 1979*. RSPB, Sandy, Beds.

ZINK, G. 1973. *Der Zug Europäischer Singvögel*. Vol. 5. Vogelwarte Radolfzell, Möggingen.

BIBLIOGRAPHY

Appendix 1

SCIENTIFIC NAMES OF PLANTS MENTIONED IN THE TEXT

Bearberry	*Arctostaphyllos uva-ursi*
Bent-grass, Common	*Agrostis tenuis*
Bilberry	*Vaccinium myrtillus*
Birch	*Betula* spp.
Bracken	*Pteridium aquilinum*
Bramble	*Rubus fruticosus*
Cabbage, wild	*Brassica oleracea*
Campion, sea	*Silene maritima*
Crowberry	*Empetrum nigrum*
Dock	*Rumex* spp.
Elder	*Sambucus nigra*
Fescue, red	*Festuca rubra*
Gorse	*Ulex* spp.
Heather (ling)	*Calluna vulgaris*
Hogweed	*Heracleum sphondylium*
Juniper, common	*Juniperus communis*
Mat-grass	*Nardus stricta*
Moor-grass, purple	*Molinia purpurea*
Myrtle, bog	*Myrica gale*
Nettle, stinging	*Urtica dioica*
Pansy, wild	*Viola tricolor*
Plantain, buck's-horn	*Plantago coronopus*
Poa, annual	*Poa annua*
Radish, sea	*Rhaphanus maritimus*
Ragwort	*Senecio jacobaea*
Sage, wood	*Teucrium scorodonia*
Sedge, sand	*Carex arenaria*
Squill, spring	*Scilla verna*
Storksbill, sea	*Erodium maritinum*
Thistle	*Cirsium/Carduus* spp.
Thrift	*Armeria maritima*
Tormentil	*Potentilla erecta*
Violet, dog	*Viola* spp.
Willow	*Salix* spp.
Whortleberry, see Bilberry	
Yorkshire fog	*Holcus lanatus*

Appendix 2

SCIENTIFIC NAMES OF MAMMALS AND REPTILES MENTIONED IN THE TEXT

Cows and domestic cattle	*Bos taurus*
Fox, red	*Vulpes vulpes*
Goat	*Capra hircus*
Hedgehog	*Erinaceus europaeus*
Horse	*Equus caballus*
Marmot	*Marmota marmota*
Rabbit	*Oryctolagus cuniculus*
Rat, common	*Rattus norvegicus*
Sheep	*Ovis aries*
Snake, grass	*Natrix natrix*
Stoat	*Mustela erminea*
Souslik	*Spermophilus citellus*
Weasel	*Mustela nivalis*

*A*ppendix 3

SCIENTIFIC NAMES OF BIRDS MENTIONED IN THE TEXT

Sequence follows Voous (1977, *List of Recent Holarctic Bird Species*)

Manx Shearwater	*Puffinus puffinus*
Storm Petrel	*Hydrobates pelagicus*
Red Kite	*Milvus milvus*
Sparrowhawk	*Accipiter nisus*
Buzzard	*Buteo buteo*
Rough-legged Buzzard	*Buteo lagopus*
Golden Eagle	*Aquila chrysaetos*
Osprey	*Pandion haliaetus*
Kestrel	*Falco tinnunculus*
Red-necked Falcon	*Falco chicquera*
Merlin	*Falco columbarius*
Hobby	*Falco subbuteo*
Sooty Falcon	*Falco concolor*
Gyrfalcon	*Falco rusticolus*
Peregrine	*Falco peregrinus*
Oystercatcher	*Haematopus ostralegus*
Stone-curlew	*Burhinus oedicnemus*
Ringed Plover	*Charadrius hiaticula*
Golden Plover	*Pluvialis apricaria*
Lapwing	*Vanellus vanellus*
Pectoral Sandpiper	*Calidris melanotos*
Dunlin	*Calidris alpina*
Snipe	*Gallinago gallinago*
Black-tailed Godwit	*Limosa limosa*
Whimbrel	*Numenius phaeopus*
Curlew	*Numenius arquata*
Redshank	*Tringa totanus*
Green Sandpiper	*Tringa ochropus*
Common Sandpiper	*Actitis hypoleucos*
Turnstone	*Arenaria interpres*
Black-headed Gull	*Larus ridibundus*
Lesser Black-backed Gull	*Larus fuscus*
Herring Gull	*Larus argentatus*
Great Black-backed Gull	*Larus marinus*
Guillemot	*Uria aalge*
Razorbill	*Alca torda*

Puffin	*Fratercula arctica*
Cuckoo	*Cuculus canorus*
Little Owl	*Athene noctua*
Swift	*Apus apus*
Hoopoe	*Upupa epops*
Great Spotted Woodpecker	*Dendrocopos major*
Short-toed Lark	*Calandrella brachydactyla*
Skylark	*Alauda arvensis*
Horned Lark (Shore Lark)	*Eremophila alpestris*
Sand Martin	*Riparia riparia*
Swallow	*Hirundo rustica*
House Martin	*Delichon urbica*
Richard's Pipit	*Anthus novaeseelandiae*
Tawny Pipit	*Anthus campestris*
Meadow Pipit	*Anthus pratensis*
Rock Pipit	*Anthus petrosus*
Yellow Wagtail	*Motacilla flava*
Ashy-headed Wagtail	*Motacilla flava cinereocapilla*
White Wagtail	*Motacilla alba alba*
Pied Wagtail	*Motacilla alba yarrellii*
Cactus Wren	*Campylorhynchus brunneicapillus*
House Wren	*Troglodytes aedon*
Wren	*Troglodytes troglodytes*
Alpine Accentor	*Prunella collaris*
Robin	*Erithacus rubecula*
Nightingale	*Luscinia megarhynchos*
Red-spotted Bluethroat	*Luscinia svecica svecica*
Black Redstart	*Phoenicurus ochruros*
Redstart	*Phoenicurus phoenicurus*
Whinchat	*Saxicola rubetra*
Stonechat	*Saxicola torquata*
Isabelline Wheatear	*Oenanthe isabellina*
Northern Wheatear	*Oenanthe oenanthe*
Greenland Wheatear	*Oenanthe oenanthe leucorrhoa*
Seebohm's Wheatear	*Oenanthe oenanthe seebohmi*
Pied Wheatear	*Oenanthe pleschanka*
Black-eared Wheatear	*Oenanthe hispanica*
Desert Wheatear	*Oenanthe deserti*
Mourning Wheatear	*Oenanthe lugens*
White-crowned Black Wheatear	*Oenanthe leucopyga*
Red-breasted Wheatear	*Oenanthe bottae*
Capped Wheatear	*Oenanthe pileata*
Rock Thrush	*Monticola saxatilis*
Ring Ouzel	*Turdus torquatus*
Blackbird	*Turdus merula*
Song Thrush	*Turdus philomelos*
Whitethroat	*Sylvia communis*
Chiffchaff	*Phylloscopus collybita*
Willow Warbler	*Phylloscopus trochilus*
Goldcrest	*Regulus regulus*
Spotted Flycatcher	*Muscicapa striata*
Pied Flycatcher	*Ficedula hypoleuca*
Blue Tit	*Parus caeruleus*
Great Tit	*Parus major*

Great Grey Shrike	*Lanius excubitor*
Woodchat Shrike	*Lanius senator*
Chough	*Pyrrhocorax pyrrhocorax*
Carrion Crow	*Corvus corone*
Starling	*Sturnus vulgaris*
House Sparrow	*Passer domesticus*
Chaffinch	*Fringilla coelebs*
Greenfinch	*Carduelis chloris*
Goldfinch	*Carduelis carduelis*
Linnet	*Carduelis cannabina*
Twite	*Carduelis flavirostris*
American Redstart	*Setophaga ruticilla*
Song Sparrow	*Zonotrichia melodia*
Lapland Bunting	*Calcarius lapponicus*
Snow Bunting	*Plectrophenax nivalis*
Yellowhammer	*Emberiza citrinella*
Corn Bunting	*Miliaria calandra*

Appendix 4

Scientific names of insects and arachnids mentioned in the text

Order Collembola (springtails)

Order Dermaptera (earwigs)

Order Orthoptera (grasshoppers and crickets)
Common green grasshopper   *Omocestus viridulus*
Mole-cricket   *Gryllotalpa gryllotalpa*

Order Thysanoptera (thrips)

Order Hemiptera (bugs)
Hoppers   Homoptera

Order Lepidoptera (butterflies and moths)
Large white butterfly   *Pieris brassicae*
Small white butterfly   *Artogeia rapae*
Green-veined white butterfly   *Artogeia napae*
Small copper butterfly   *Lycaena phlaeas*
Peacock butterfly   *Inachis io*
Meadow brown butterfly   *Maniola jurtina*
Cinnabar moth   *Hypocrita jacobaeae*
White ermine moth   *Spilosoma lubricipeda*
Dark arches moth   *Xylophasia monoglypha*
Antler moth   *Charaeas graminis*

Order Diptera (two-winged flies)
House flies   *Musca* spp.
March flies   *Bibio johannis*
    Bibio marci
Craneflies   Tipulidae
Bluebottles   Calliphoridae
Gall midges   Cecidomyiidae

Order Hymenoptera (wasps, bees and ants)
Bumble bees   *Bombus terrestris* and other spp.
Ants   *Myrmica scabrinoides*
    Lasius niger
    Lasius alienus

Lasius flavus

Braconid flies	Braconidae
Ichneumon flies	Ichneumonidae
Chalcid flies	Chalcidae
Gall wasps	Cynipidae

Order Coleoptera (beetles and weevils)

Click beetles	Elateridae
Rove beetles	Staphylinidae
Leaf and flea beetles	Chrysomelidae
Weevils	Curculionidae
	Otiorrhynchus sulcatus

Order Acari (mites)

Erythraeus phalangoides

Order Araneae (spiders)

Lyniphiid spiders	Lyniphiidae
	Erigone sp.
Lycosid spiders	Lycosidae
	Lycosa sp.
Thomisid spiders	Thomisidae
	Xysticus erraticus

Appendix 5

NUMBER OF WHEATEAR PAIRS BREEDING ON SKOKHOLM

1928	10 pairs	1962	No census
1929	13 pairs	1963	No census
1930	12 pairs	1964	21 pairs
1931	9 pairs	1965	23 pairs
1932	13 pairs	1966	23 pairs
1933	13 pairs	1967	28 pairs
1934	14 pairs	1968	30 pairs
1935	12 pairs	1969	27 pairs
1936	10 pairs	1970	25 pairs
1937	9 pairs	1971	20 pairs
1938-46	No census	1972	No census
1947	14 pairs	1973	8 pairs
1948	20 pairs	1974	9 pairs
1949	24 pairs	1975	9 pairs
1950	27 pairs	1976	9 pairs
1951	38 pairs	1977	12 pairs
1952	38 pairs	1978	13 pairs
1953	28 pairs	1979	13 pairs
1954	30 pairs	1980	14 pairs
1955	35 pairs	1981	12 pairs
1956	30 pairs	1982	16 pairs
1957	30 pairs	1983	17 pairs
1958	26 pairs	1984	15 pairs
1959	36 pairs	1985	17 pairs
1960	32 pairs	1986	19 pairs
1961	12 pairs	1987	19 pairs

Appendix 6

THE STORY OF GEORGE AND MARGARET

X6875, a male nicknamed George, was hatched on Skokholm in 1947, and in 1948 and 1949 was mated with female X6716. He was always one of the earliest males to return to the island and to his territory in the northwest corner. In spring 1950, X6716 failed to return and George mated with 001448 (from nest W23²), which became known as Margaret. She had been hatched on Skokholm in 1948 but had not been recorded in 1949.

In 1951, Margaret returned to the territory she had held with George in 1950, but he had not arrived. She occupied it with a new male for four days, and the two were behaving as though they were mated. On the fifth day George returned and tried to re-establish himself, but throughout that day the unringed male was vigorously attacking and chasing George; Margaret remained close to the unringed male in the typical posture of a newly-mated female and was not observed to attack George. On the following day there had been a change, and George was attacking the unringed male and now was closely attended by Margaret. On the third day, a gale caused all three birds to leave the exposed territory. George and Margaret had moved to a sheltered but so far unoccupied area and continued as a pair, raising two broods.

In the adjoining territory a male, 10822, was mated with a first-summer female, which had lost one colour-ring, and they produced a first brood successfully; but after that, as so often happens, the male 10822 disappeared. The female remained and eventually mated polygamously with George and produced a second brood. Both females defended their own territories against each other, and George defended a 'super' territory.

In 1952, when George would have been five years old, he failed to return to the island, and so did the partly ringed female. Male 10822 returned, however, and moved into Margaret's territory and, with her, successfully raised two broods.

The story illustrates several points. First, it is desirable for a male to return early to re-occupy the territory that it has known in previous years, a familiarity which seems to be of some benefit for pairs. The return to the same territory may also ensure that they would meet the same partner, which again should be of some benefit to them. This story also shows that a female can return to and hold a territory that is known to her and which is not at that time occupied by a male. However, I do not know what would have happened if another pair had tried to take it over.

It is also interesting to note that male 10822 left a territory in which it had nested for one year and joined a female which had nested in her territory for three years. The move, however, was only a matter of 200 m into an adjoining area which he may well have hunted occasionally in the past and which might well have been accepted as an 'instinctive' territory.

Appendix 7

A LIST OF THE COMMON NAMES OF THE WHEATEAR

There were rare birds I never saw before,
The like of them I think I see no more;
Th'are called Wheat-ears, less than lark or sparrow,
Well-roasted, in mouth they taste like marrow.
Then once 'tis in the teeth it is involved
Bones, flesh, and all, lusciously dissolved.
The name of wheatears on them is ycleped
Because they come when wheat is yearly reaped
Six weeks or thereabouts they are catched there
And are well-nigh eleven months, God knows where!

John Taylor, the Water-Poet

I have compiled this list from various authors whose names are given in the normal way in the Bibliography.

Arlyng or arling (from Anglo-Saxon *aers*=rump, and *ling*, a diminutive) (Turner 1544).

Barley-bird, barley-ear (Sussex) (Jackson 1968); Bog-an-lochan (Gaelic) (Forbes 1905); Bogachan (Gaelic) (Forbes); Brù-gheal (Gaelic) (Forbes); Burrow Bird (Norf. Nat. Trust 1938); Bushchat*.

Caislincloch (Irish) (Forbes); Chack (Orcadian and Northants) (Barry 1808 and Jackson); Chacker, or Chack bird (Montagu 1813); Chackeret, chacks, chat-chock (Forbes); Chatterer*; Chattie*; Check or check-bird, checkle (Montagu and Jackson); Chetstone (Yorks) (Jackson); Chick-chack (Forbes); Chickell (Devon) (Forbes); Chicken, hedge (Montagu); Chicker (Devon) chickin, chickstane (Forbes); Chickwell (Montagu); chocharet (Jackson); Chock (Jackson); Chuck (Forbes); Clacharan or Clackaran (which in English means 'little mason') (MacGillivray 1839); Clocharet (Forbes); Clod bird, Clod-hoppers (Morton 1712); Clotbird (Turner 1544); Coney (Forbes); Coney chuck, Coney sucker (Norf. Nat. Trust 1938); Cooper (Pembrokeshire) (Forbes); Cracker (St Kilda) (Jackson); Crineachan (Gaelic) (Forbes).

Dyke-hopper (Forbes).

Ear-bird (Sussex) (Jackson); English Ortolan (Sussex) (Jackson).

Fallow-chat -finch -lunch or -smich (Forbes); Fallow Finch (Montagu); Fallow-Smich (Ray 1674); Fallow-smiter (Jackson).

Hedge-chicken or Hedge Chicker (Montagu); Horse masher or Horse smatch (Cornwall) (Jackson); Horse musher or smatch (Forbes).

Jobbler, jocktibeet, jocktie (Forbes).

Moor Warbler (Latham, quoted by Stresemann 1949).

Ortolan (Forbes).

Passage-bird (Glos.) (Jackson).

Shepherd bird (Norf.Nat.Trust, 1938); Singing skyrocket (Jackson); Smatch (Turner 1544); Snorter (Montagu); Stanechack (MacGillivray); Stanechacker (Lancs., N.Ire., Scot.) (Jackson); Stane-pecker (Forbes); Steincheck (Turner 1544); Steinkle, Stinkle (Forbes); Stinklin (Shetland) (Jackson); Stone-breaker (Cheshire) (Jackson); Stone-chat, Stonechard (MacGillivray); Stone-check (Jackson); Stone-chacker and Stone-chucker (Jackson); Stone clatter, Stone cracker (Hudson 1897).

Underground jobbler (Forbes). Utick or Bigh Utick (Cheshire) (Coward).

Whishie (Scot.) (Jackson); White arse, White ass (Blair, ND); White-ase (Forbes); one wonders if the Reverend Gentlemen could spell naughty words!); White-rump (Bewick 1805; and H. Saunders 1889, who said that the species and its congeners are known as 'white rumps' in many European languages); White-rumped Stonechat (Montagu 1839); Whitestart (Wharfedale) (Jackson); White tail (Ray 1674); Wheat-ear (John Taylor, the Water-Poet, 1654; Ray 1674); Wittol or Whittol (Forbes).

*Source lost

INDEX